BRITISH GEOLOGICAL SURVEY

British Regional Geology

The Palaeogene volcanic districts of Scotland

FOURTH EDITION

C H Emeleus
B R Bell

Contributor:

D Stephenson

NERC copyright 2005

BRITISH GEOLOGICAL SURVEY, NOTTINGHAM 2005

First published 1935
Second edition 1948
Third edition 1961
Fourth edition 2005

ISBN 978 0852725191
The grid, where it is used on the figures, is the National Grid taken from Ordnance Survey mapping. Maps and diagrams in this book use topography based on Ordnance Survey maps.

© Crown Copyright reserved Ordnance Survey licence number 100017897/2005.

Printed in the UK by Hawthornes, Nottingham

C10 7/05

Bibliographical reference

EMELEUS, C H, and BELL, B R. 2005. British regional geology: the Palaeogene volcanic districts of Scotland (Fourth edition). (British Geological Survey, Nottingham.)

Authors
C H Emeleus
Department of Earth Sciences,
University of Durham, Science Laboratories,
South Road, Durham DH1 3LE

B R Bell
Division of Earth Sciences,
University of Glasgow, Gregory Building,
Lilybank Gardens, Glasgow G12 8QQ

Contributor
D Stephenson
British Geological Survey, Murchison House,
West Mains Road, Edinburgh EH9 3LA

Front cover   Columnar jointed olivine basalt lava on the east coast of Eigg near Kildonnan, with pitchstone lava forming the ridge of An Sgùrr (393 m) on the skyline. The basalt is at the base of the Eigg Lava Formation and is one of the oldest Palaeogene igneous rocks in the Hebrides, whereas the Sgurr of Eigg Pitchstone Formation is one of the youngest (Photographer: Lorne Gill, Scottish Natural Heritage).

Contents

1 **Introduction** 1
   Summary of geology 1
2 **Pre-Mesozoic** 8
   Archaean 8
   Torridonian 9
   Moine Supergroup 10
   Dalradian Supergroup 11
   Lower Palaeozoic 12
   Caledonian igneous rocks 15
   Old Red Sandstone 15
   Carboniferous 16
   Permian 17
   Intrusions of Carboniferous and Permian age 18
3 **Mesozoic** 19
   Triassic 19
   Jurassic 22
   Cretaceous 36
4 **Pre-Palaeogene structure** 41
   Early Palaeozoic and older structures 41
   Late Palaeozoic to Mesozoic structures 42
5 **Palaeogene igneous geology: regional setting** 43
   Timing and igneous stratigraphy 44
6 **Palaeogene lava fields and associated sedimentary rocks** 49
   Petrography of the lavas 52
   Field characteristics of the lavas 52
   Lava sequences 57
7 **Dykes, dyke swarms and volcanic plugs** 78
   Dyke swarms 80
   Volcanic plugs 83
8 **Sills and sill-complexes** 85
   Little Minch Sill-complex 85
   Loch Scridain Sill-complex, Mull 87
   Holy Island and Dippin sills, Arran 89
   Microgranitic and rhyolitic sills, south Arran 90
   Composite sills of Arran and south Bute 90
   Tighvein Intrusion-complex, Arran 90
   Raasay Sill 91
   Gars-bheinn Ultrabasic Sill, Skye 91
   Other Sills 91
9 **Central complexes** 92
   St Kilda 97
   Skye 98
   Rum 110
   Ardnamurchan 120
   Mull 126
   Blackstones Bank 134
   Arran 134
   Ailsa Craig 138
   Rockall 138

10 Magmas 139
Early concepts 139
Major-element compositions 139
Contamination processes 141
Depth of magma generation 142
Ultrabasic and basic magmas of the central complexes 143
Minor intrusions 143
Silicic rocks 145
Magma mixing 146

11 Palaeogene and later structure 148
Major faults 148
Structure of the lava fields 149
Structures associated with the central complexes 149

12 Late Palaeogene and Neogene 152

13 Quaternary 153
Pre-Late Devensian glaciations 153
Dimlington Stadial 155
Windermere Interstadial 160
Loch Lomond Stadial 160
Sea level changes 163
Other Late-glacial and postglacial features 168

14 Economic geology 171

References 177

British Geological Survey publications 199

Index 201

ILLUSTRATIONS

Figures
1   Palaeogene central complexes, lava fields, sill-complexes and dyke swarms in western Scotland and north-east Ireland   2
2   North Atlantic Igneous Superprovince   3
3   Precambrian and Lower Palaeozoic rocks in the district   4
4   Cambro-Ordovician rocks of Ord on the Sleat peninsula, Skye   13
5   Mesozoic basins of the Inner Hebrides and onshore outcrops   20
6   Upper Triassic to Jurassic lithofacies of the Inner Hebrides   24
7   Upper Cretaceous of the Inner Hebrides Group at Beinn Iadain, Morvern   39
8   Time span of Palaeogene igneous activity in the Hebridean Igneous Province   46
9   Paleocene lavas in the Hebridean Igneous Province   50
10   Vertical sections of the lava fields in the Hebridean Igneous Province   61
11   Allt Geodh' a'Ghàmhna Member of the Minginish Conglomerate Formation, Skye   64
12   Zeolite zones in lavas of Mull and Morvern   72
13   Hydrothermal circulation, Mull Central Complex   73
14   Sgurr of Eigg pitchstone at Bidean Boideach   76
15   Dilation axes of the Palaeogene dyke swarms in the Hebridean area   79
16   Diagrammatic representation of the Mull Dyke Swarm   82
17   Little Minch Sill-complex, Trotternish peninsula and Shiant Isles   86
18   Dyke swarms, cone-sheets and ring-dykes   94
19   Gravity anomaly image of western Scotland   96
20   St Kilda Central Complex   98
21   Skye Central Complex   100–101
22   Marsco Hybrids at Harker's Gully, Skye   107
23   Rum Central Complex   111
24   Northern Marginal Zone, Rum Central Complex   112
25   Formation of the Central Intrusion, Rum Central Complex   119
26   Ardnamurchan Central Complex   123
27   Mull Central Complex   128
28   Mull Central Complex: Centre 1, Glen More Centre and the Early Caldera   129
29   Mull Central Complex: Centre 2, Beinn Chaisgidle Centre   131
30   Mull Central Complex: Centre 3, Loch Bà Centre and the Late caldera   133
31   Principal Paleocene igneous intrusions of Arran   135
32   North Arran Granite Pluton   136
33   Cross-section of the Central Arran Ring-complex   138
34   Normative *diopside-hypersthene-olivine-nepheline-quartz* plot for the Hebridean Palaeogene magmas   140
35   Plot of total alkalis against silica for the Hebridean Palaeogene magmas   141
36   Plot of $^{87}Sr/^{86}Sr$ against $^{143}Nd/^{144}Nd$ for basaltic rocks and crustal materials   142
37   Plot of Pb isotope data for Skye lavas and granites   143
38   Ice sheet of western Scotland, Main Late Devensian Glaciation   157
39   Direction of ice movement around Mull during the Main Late Devensian Glaciation   158
40   Limits of the Loch Lomond re-advance, central Skye and Mull   162
41   Isobases for the Main Rock Platform in western Scotland   165
42   Chromium and magnesium distribution in the surficial marine sediments off Rum   172

Tables
1   Torridon and Sleat groups on Rum and Skye   10
2   Cambrian and Ordovician strata of Skye   14

3 Permian and Triassic strata of the Isle of Arran 17
4 Ammonite zonal sequence in the Lower Jurassic and part of the Middle Jurassic of the Inner Hebrides 25
5 Great Estuarine Group and Bearreraig Sandstone Formation in the Inner Hebrides 30
6 Ammonite zonal sequence in the upper Middle Jurassic and Upper Jurassic of the Inner Hebrides 35
7 Cretaceous strata in the Inner Hebrides and the Isle of Arran 37
8 Radiometric and palynological age determinations on Palaeogene rocks from the Hebridean Igneous Province 45
9 Paleocene lava fields in the Hebridean Igneous Province 51
10 Mineralogical and petrographical characteristics of the principal lava types, Hebridean Igneous Province 53
11 Skye Lava Group: west-central Skye 63
12 Skye Lava Group: northern Skye 66
13 Strathaird Lava Formation of the Skye Lava Group 67
14 Canna Lava Formation of the Skye Lava Group: north-west Rum 70
15 Mull Lava Group 70
16 Dimensions of selected Palaeogene dyke swarms 81
17 Late-Pleistocene and Holocene events in Britain 154

Plates

Front cover Columnar basalt lava and pitchstone lava ridge, Isle of Eigg
Frontispiece Loch na Cuilce, An Garbh-choire and the Cuillin Ridge, Skye viii
1 Lewisian granodioritic gneisses, near Priomh-lochs, Rum 9
2 Torridonian strata on Mullach Mór, Rum 11
3 Permian aeolian sandstone, Corrie foreshore, Arran 16
4 Pre-Paleocene strata and Paleocene lavas at Creag a'Ghaill, Gribun, western Mull 21
5 Photomicrograph of Raasay Ironstone Formation 29
6 Honeycomb weathering in the Elgol Sandstone, Elgol, Skye 31
7 Concretions in the Valtos Sandstone Formation, Bay of Laig, Isle of Eigg 32
8 Dinosaur trackways in the Duntulm Formation, Staffin Bay, northern Skye 33
9 Portree Hyaloclastite Formation, Fiurnean, northern Skye 52
10 Basalt and hawaiite flows on Ben Scaalan, south-west Skye 55
11 Trachytic tuff on basaltic pahoehoe lava, Port Mór, Muck 56
12 Flow-folded hawaiite lava, Arnaval, south-west Skye 57
13 Columnar jointed basaltic lava, Isle of Staffa 58
14 Distorted columnar jointing, Sgurr of Eigg pitchstone 59
15 Trap features in basalt lavas, Ardmeanach, western Mull 60
16 Valley fill of tholeiitic basalt overlying hawaiite flows, Preshal Beg, south-west Skye 65
17 Fluviatile conglomerate overlain by tholeiitic andesite lava, Fionchra, Rum 69
18 MacCulloch's Tree encased in columnar basalt lava, Ardmeanach, Mull 74
19 Pitchstone lava and ash flow, Sgurr of Eigg 76
20 South Arran Dyke Swarm cutting Triassic sandstones, Kildonnan, Arran 80
21 Dolerite sills of Trotternish, northern Skye 88
22 Quartzite xenolith in basalt sill, Loch Scridain, Mull 89
23 Drumadoon Sill, Arran 91
24 Layering in the Outer Bytownite Gabbros, Cuillin Centre, Skye Central Complex 99
25 'Finger' structures in feldspathic peridotite, Cuillin Centre, Skye Central Complex 103
26 Blà Bheinn and the eastern Red Hills, Skye 106
27 Fiamme in a rhyodacite ash flow, Cnapan Breaca, Rum Central Complex 113
28 Layered ultrabasic rocks on Hallival, Rum Central Complex 115

29	Photomicrograph of anorthosite–feldspathic peridotite junction, Rum Central Complex 116
30	Slumping in bytownite troctolite, Askival, Rum Central Complex 117
31	Peridotite block dropstone, Rum Central Complex 117
32	Breccia of bytownite troctolite, Rum Central Complex 118
33	Elongate (harrisitic) olivine crystals in bytownite gabbro, Rum Central Complex 120
34	Intrusion breccia, Harris Bay, Rum Central Complex 121
35	Dolerite cone-sheets intruding Moine metasedimentary rocks, Mingary Ardnamurchan 124
36	Intrusion breccia of quartz-dolerite veined by microgranite, Ardnamurchan 125
37	Aerial view of Centre 3, Ardnamurchan Central Complex 127
38	Hills of the North Arran Granite Pluton 137
39	Hummocky morainic drift, Glen More, Mull 155
40	Glacially polished gabbro surface, Coire Lagan, Cuillin, Skye 156
41	Scouring by glacial meltwater, Loch na Keal, Mull 159
42	Raised shoreline south of Dougarie, Isle of Arran 167
43	Landslips at The Quiraing, Trotternish, Skye 168
44	Quartz sand stockpiled at the Loch Aline glass-sand mine, Morvern 174

Loch na Cuilce at the head of Loch Scavaig, An Garbh-choire and the Cuillin Ridge, Skye. The resistant gabbros and ultramafic rocks of the Cuillin Centre have been deeply eroded by glaciers to give the present rugged topography (*Photograph:* B R Bell; P521672).

Foreword to the fourth edition

Uniquely in the British Regional Geology series, the boundaries of the area covered by this volume are not easy to define, either geographically or geologically. In broad terms it covers the northern islands of the Inner Hebrides, with parts of the adjoining mainland, plus the Isle of Arran in the Firth of Clyde. This area spans several geological terranes in which most of the pre-Mesozoic rocks are already well described in other volumes of this series. It is characterised by extensive outcrops of igneous rocks of early Palaeogene age (formerly Tertiary) and by locally thick sequences of Mesozoic sedimentary rocks that, in many places, owe their preservation to a protective cover of Palaeogene lavas. Consequently, the descriptions in this volume concentrate largely upon the geological processes and products of the Mesozoic and Cainozoic eras. The term 'Tertiary' is no longer approved, having been replaced formally in both chronostratigraphical and lithostratigraphical nomenclature by the Palaeogene and Neogene systems/periods. However, it continues to be used informally by some authors.

The first edition of this book, by J E Richey, was published in 1935. It was revised and updated mainly by the original author in the second edition of 1948 and by A G MacGregor and F W Anderson in the third edition of 1961. At the time of the last edition, many of the tools and techniques now taken for granted in geological and geophysical research were in their infancy, and the concepts of global tectonics were hardly known. The purpose of this completely rewritten new edition is to provide an up-to-date, generalised account of the geology that is comprehensible and of interest to the informed amateur, undergraduate and professional geologist, planner or civil engineer. While the emphasis is on the fundamentals of the regional geology and, in particular, what can be seen in the field, this account also demonstrates how some of the major advances in our understanding of the area have been made possible through the application of new techniques and concepts, and for some of these this region has provided a crucial test bed. This is most apparent in the extensive geochemical investigations of the igneous rocks, which have contributed so much to our knowledge of magma generation and evolution, both in the Hebrides and worldwide. For this reason a chapter has been devoted to the subject of magmas, which the authors feel is necessary for a complete discussion of the processes involved in the evolution of this igneous province of worldwide importance.

Although an enormous amount of specialised research has been conducted in the area in the last 40 years by university workers from around the world, there has been no systematic resurvey by Geological Survey staff. However, following in the footsteps of Alfred Harker of Cambridge University in the early 20th century, resurveys of the Small Isles, parts of Skye and to a lesser extent Ardnamurchan and northern Mull have been conducted on behalf of the British Geological Survey by university staff. Almost all of this work has been by Henry Emeleus of Durham University and Brian Bell of Glasgow University, and we are very fortunate that they have pooled their collective experience, gained from this and other work in the area, to produce this new edition.

David Falvey, PhD
Director
British Geological Survey
Kingsley Dunham Centre
Keyworth
Nottingham

Acknowledgements

This edition of *British regional geology: the Palaeogene volcanic districts of Scotland* has been completely rewritten by C H Emeleus of the University of Durham and B R Bell of the University of Glasgow, under contract to the British Geological Survey as part of the University Collaboration Programme (Consultancy agreement number GA/94E/19). The authors would like to acknowledge assistance from J D Hudson of the University of Leicester with the Mesozoic chapter, J W Merritt and J D Peacock of the BGS with the Quaternary chapter and K Hitchen of the BGS for advice on offshore Palaeogene igneous rocks. Information from the glass-sand mine at Loch Aline was kindly provided by Tilcon (Scotland) Ltd. M A Hamilton, University of Toronto, provided the U-Pb ages quoted in Table 8.

Scientific editing and compilation on behalf of the BGS was by D Stephenson, and manuscript review by P Stone; the series editor is A A Jackson. Figures were produced by BGS Cartography, Edinburgh: page setting by A Hill.

Photographs were taken by the authors, unless otherwise stated in the captions; they have been deposited in the BGS National Archive of Geological Photographs.

one

Introduction

The varied geology of western Scotland is spectacularly displayed in the precipitous coastline of the mainland and in the islands of the Inner Hebrides. The area is noted for the widespread volcanism that occurred during the Palaeogene Period, and which has played an important role in the development of the rugged scenery (*Frontispiece*). The volcanic rocks buried and preserved thick sequences of Mesozoic sedimentary strata that now provide, where exposed by erosion, welcome tracts of fertile land in the otherwise rugged surroundings. In this book, the igneous rocks produced by the Palaeogene volcanism are described in detail, together with the Mesozoic sedimentary rocks, the Quaternary deposits and aspects of the earlier geology. A more detailed account of the Archaean Lewisian Gneiss Complex, the Proterozoic and Palaeozoic rocks and their associated structures can be found in another volume in this series, *The Northern Highlands of Scotland* (Johnstone and Mykura, 1989) and also in *The Geology of Scotland* (Trewin, 2002).

The area described in this book, and referred to here as the district, lies largely within the part of the Inner Hebrides north-west of the Great Glen Fault (Figure 1), where igneous rocks of Palaeogene age occur on the islands of Skye, Raasay, the Small Isles (Canna, Rum, Eigg and Muck) and Mull, and on Ardnamurchan and Morvern on the mainland. Also described are other islands that are dominated by Palaeogene igneous rocks, including Arran, Ailsa Craig, the Shiant Isles, Rockall and St Kilda. Since the *Third edition* of this book (Richey, 1961), the offshore search for oil and gas has resulted in a vast increase in our knowledge of the geology beneath the seas around Scotland, and the continental shelf in the Hebridean area has been described in two *BGS Offshore Geology Reports* (Fyfe et al., 1993; Stoker et al., 1993). These provide detailed summaries of the geology of the Mesozoic basins, and much information on the Palaeogene igneous rocks found offshore.

The Palaeogene igneous rocks of Great Britain, Ireland and the surrounding sea form part of the North Atlantic Igneous Superprovince that extends to East and West Greenland, Baffin Island and the Labrador Sea (Upton, 1988; Figure 2). This volcanism was a precursor to and accompanied the opening of the North Atlantic and continues to the present day on the Mid-Atlantic Ridge and in Iceland. In north-west Scotland, the igneous activity occurred during the Paleocene Epoch, from about 62 Ma to about 55 Ma, with a peak between about 61 Ma and 57 Ma (Figure 8). The remains of the central volcanoes, basaltic lava fields and innumerable minor intrusions formed at that time comprise the Hebridean Igneous Province. The intrusions extend to the Outer Hebrides and may be traced through southern Scotland into north-east England. There are also scattered occurrences in Wales, in central England, and forming Lundy Island in the Bristol Channel. Intrusions and lavas of similar age constitute an important part of the geology of north-east Ireland and include the well-known Antrim Lava Field and the central complexes of the Mourne Mountains, Slieve Gullion and Carlingford (Preston, 1982, 2001; Mitchell, 2004; Figure 1). In many publications, including the earlier editions of this book, the Palaeogene igneous rocks of the British Isles are referred to as the *British Tertiary Volcanic* (or Igneous) *Province*, but as the term 'Tertiary' is no longer used in chronostratigraphy we have chosen to replace it in the title of this book.

SUMMARY OF GEOLOGY

The Hebridean Igneous Province extends from Arran to Skye and St Kilda and thus crosses several tectonic crustal terranes (Figure 1). Prior to the Caledonian Orogeny (see below), these were all situated on the margin of the supercontinent of Laurentia, which also included the basement rocks of present-day North America and Greenland. The Archaean gneisses of the Lewisian Gneiss Complex are the oldest rocks and comprise ancient crust that has been repeatedly involved in metamorphism and deformation events (Park et al., 1994, 2002). The gneisses are widespread in the Hebridean Terrane, which constitutes the foreland to the

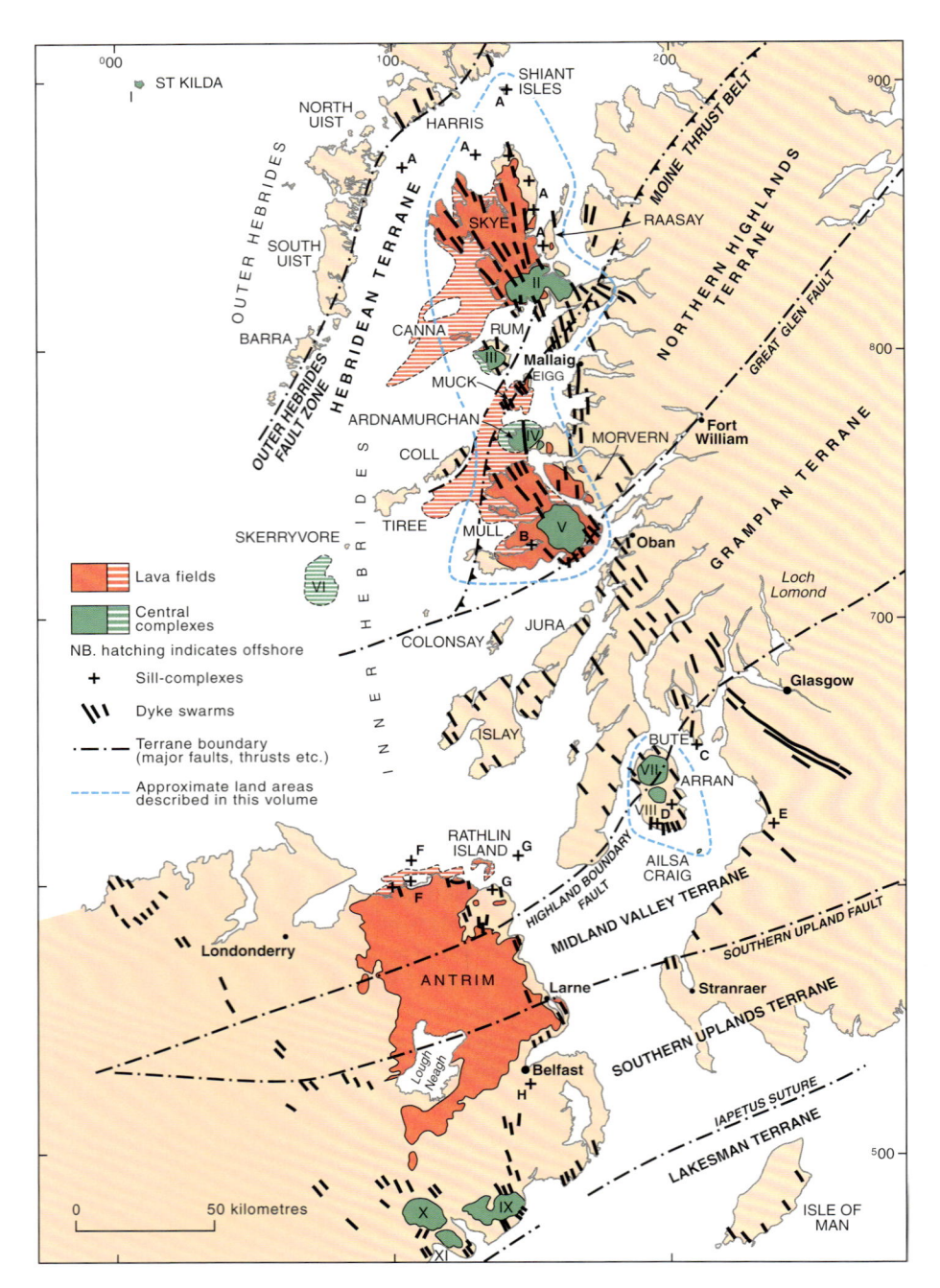

Figure 1 Sketch map showing the positions of Palaeogene central complexes, lava fields, sill-complexes and dyke swarms in western Scotland and north-east Ireland.

Central complexes:- **I** St Kilda; **II** Skye; **III** Rum; **IV** Ardnamurchan; **V** Mull; **VI** Blackstones Bank (submarine); **VII** Arran Northern Granite Pluton; **VIII** Arran Central Ring Complex; **IX** Mourne Mountains; **X** Slieve Gullion; **XI** Carlingford

Sill-complexes:- **A** Little Minch; **B** Loch Scridain; **C** Bute; **D** South Arran; **E** Prestwick–Mauchline; **F** Portrush; **G** Fair Head; **H** Scrabo

Dyke Swarms: short, thick bars show the trend of the dyke swarms, for details of the dyke distribution in the Hebridean Province, see Figure 15.

(Terrane boundaries are based on Stone and Kimbell, 1995, fig. 2.)

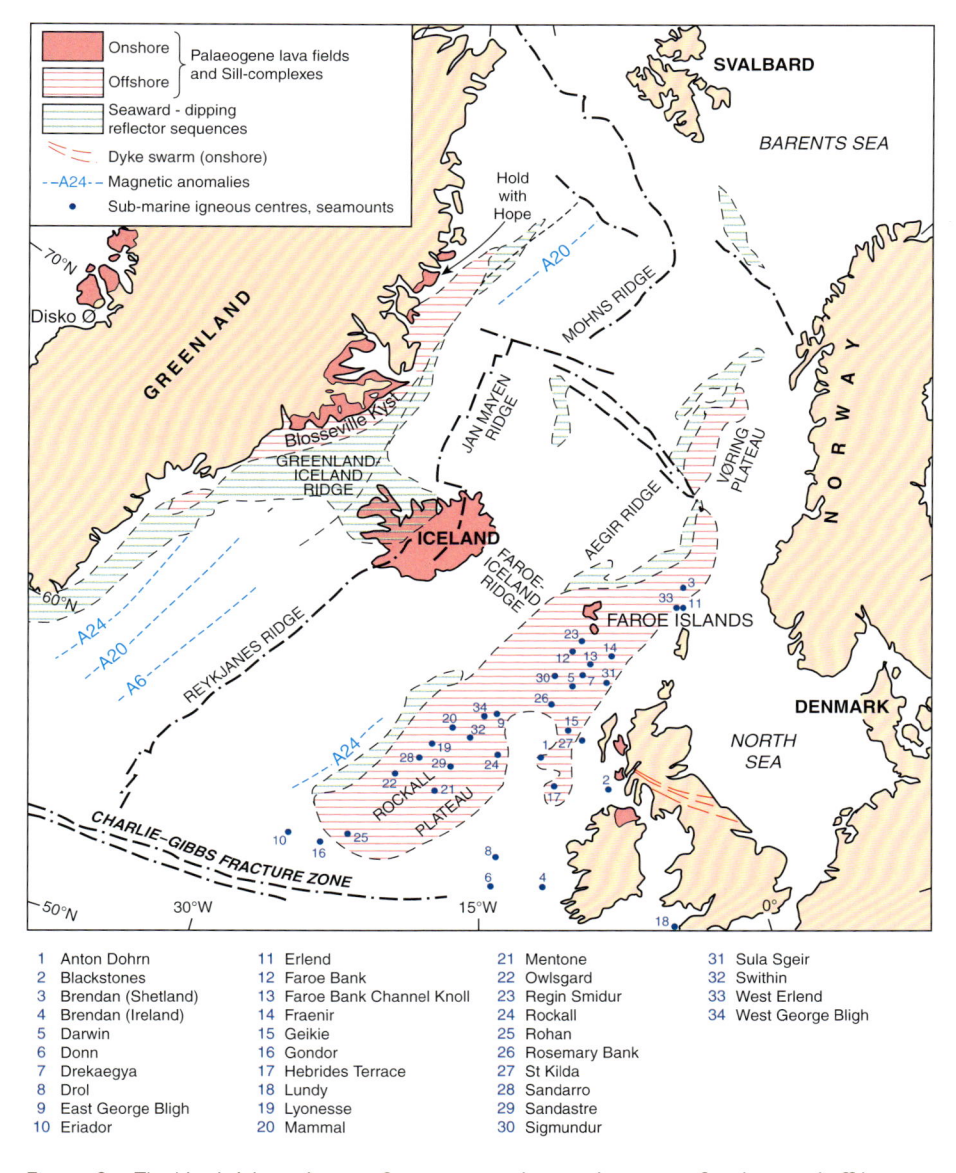

Figure 2 *The North Atlantic Igneous Superprovince showing the extent of onshore and offshore igneous activity (modified from Saunders et al., 1997, fig. 1).*

Caledonian Orogeny, and most likely underlie the younger rocks of the Northern Highlands Terrane. In the Hebridean Terrane, they are overlain unconformably by the Mesoproterozoic to Neoproterozoic Torridonian sedimentary rocks, which extend south from Cape Wrath to Skye and Rum, and eastwards to the limit of Caledonian deformation at the Moine Thrust Belt (Park et al., 1994, 2002; Stewart, 2002; Figure 3).

The Torridonian sediments were transported by rivers up to 500 km in length, which drained an area far to the north-west of the Outer Hebrides, and were deposited within a braided river system (Nicholson, 1993). Sedimentological studies and zircon age determinations show that the sediment source was not the immediately subjacent Lewisian rocks, but Precambrian gneisses of the Laurentian Shield, which range in age from 2500 to about 1100 Ma.

In Skye, and to the north, the Lewisian and Torridonian rocks are typically overlain

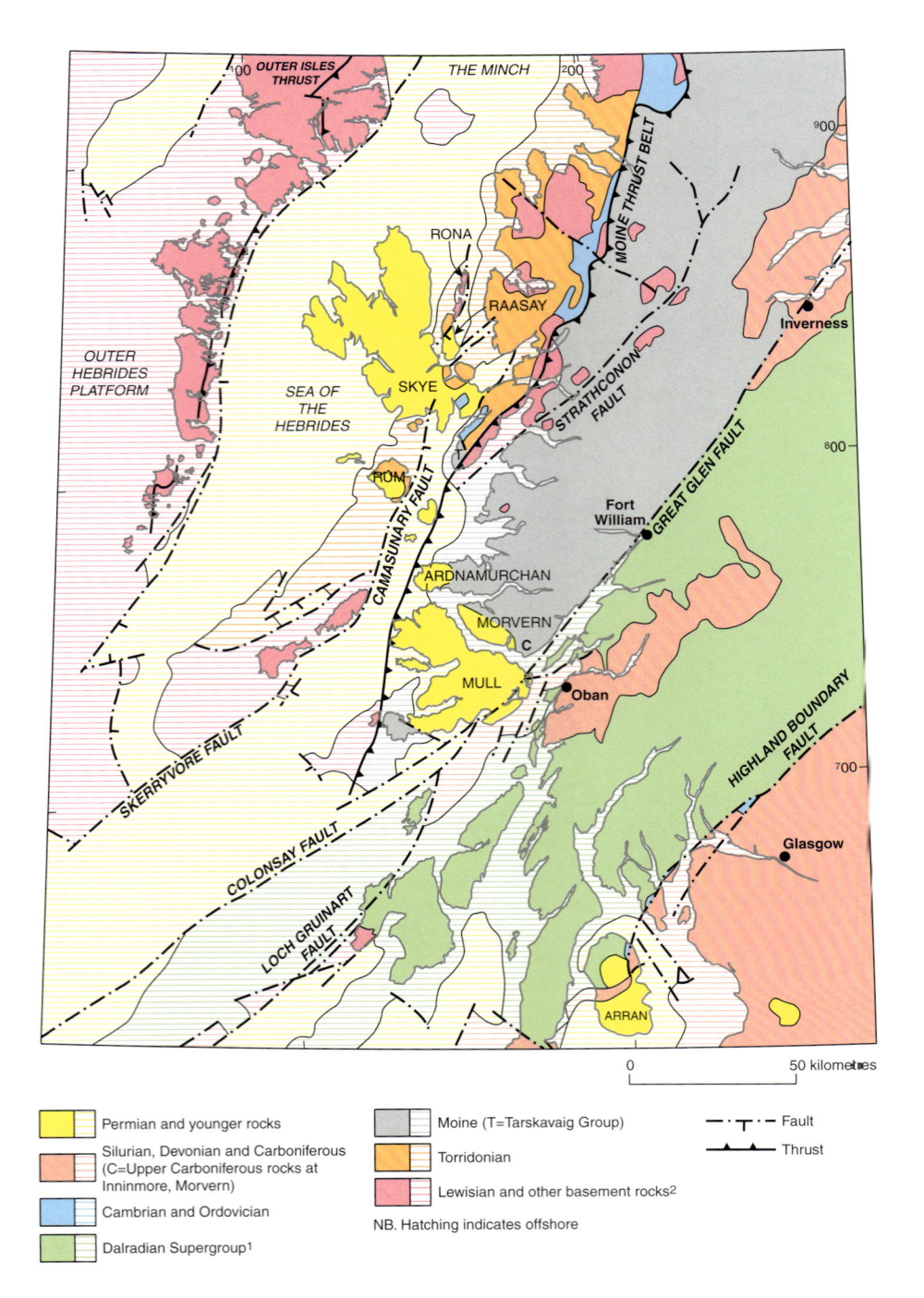

Figure 3 *Onshore and offshore occurrence of Precambrian and Lower Palaeozoic rocks in the district (based on Fyfe et al., 1993, fig.14.)*
1 includes the Bowmore and Colonsay groups on and offshore Islay; 2 includes gneisses of the Rhinns Complex south of the Great Glen Fault

unconformably by Cambro-Ordovician shelf deposits, which crop out on the foreland and within the Moine Thrust Belt. However, in south-eastern Skye the Torridonian strata have been transported tectonically over these younger rocks as part of the Kishorn Thrust Sheet.

In the Northern Highlands Terrane, there are extensive outcrops of rocks of the Mesoproterozoic Moine Supergroup. These metasedimentary rocks may be laterally equivalent to the Torridonian, from which they are separated by westerly transporting thrusts of the Moine Thrust Belt (Holdsworth et al., 1994; Strachan et al., 2002). They were deformed and metamorphosed during the Knoydartian Orogeny at around 800 Ma and were later affected by Caledonian events. On the Sleat peninsula of south-east Skye, the Tarskavaig Group, a sequence of low-grade metamorphosed psammitic rocks, may provide a link between the Torridonian and the Moine. Rocks of the Moine Supergroup also occur within the Ardnamurchan and Mull central complexes. To the south-east of the Great Glen, the Grampian Terrane is dominated by rocks belonging to the Dalradian Supergroup. However, within the area described in this book they crop out only in eastern Mull and northern Arran. They include a variety of metasedimentary and meta-igneous rocks, mainly of Neoproterozoic age, although the youngest may be earliest Cambrian. In eastern Arran, black mudstones, cherts and pillow lavas of Ordovician age belong to the Highland Border Complex (Holdsworth et al., 1994).

The Torridonian, Moine and Dalradian sedimentary sequences were all deposited in extensional basins on what was to become a passive margin of Laurentia. The Cambro-Ordovician rocks of the North West Highlands were deposited subsequently on this passive margin, while the rock of Highland Border Complex were formed within the adjacent expanding ocean. In late Neoproterozoic to Early Palaeozoic time, this Iapetus Ocean separated Laurentia from the supercontinent of Gondwana (which included the basement of present-day England, Wales, southern Ireland and western Europe) and the continent of Baltica (the basement of present-day Scandinavia and Russia). It was the complex sequence of plate-tectonic events that led to the closure of this ocean, culminating in the collision and fusion of the bounding continents to form the new supercontinent of Laurussia, that resulted in the Caledonian Orogeny. Closure probably commenced in the early Ordovician and was completed by the end of Silurian time, but post-collision uplift, with associated magmatism, continued throughout most of the Early Devonian.

The effects of the orogeny are seen mostly in the Northern Highlands and Grampian terranes (Strachan et al., 2002). There, the peak of deformation and metamorphism, known as the Grampian Event, occurred during the early to mid-Ordovician and was accompanied by intrusion of large volumes of basic magma and by crustal melting that produced a suite of granite plutons. It is possible that the Highland Border Complex was obducted onto the continental margin at this time. Later, in mid-Silurian times, large-scale thrust movements resulted in considerable crustal shortening and the whole of the Caledonian 'mobile belt' was transposed over the foreland of the Hebridean Terrane along the Moine Thrust Belt. Major movements on the Outer Hebrides Fault-zone also occurred at this time. This mid-Silurian deformation or Scandian Event was associated with the emplacement of highly alkaline magmas within and around the thrust belt. Calc-alkaline granitic plutons elsewhere in the Northern Highlands and Grampian terranes continued to be emplaced widely throughout the ensuing continental uplift phase, overlapping with calc-alkaline volcanism in late Silurian to Early Devonian times. Although by this time active subduction beneath the Laurentian margin had ceased following continent–continent collision, all of the Silurian and Devonian magmatism has subduction-related characteristics. The widespread Caledonian intrusions are represented in this district by the Ross of Mull Pluton in south-west Mull and by a wide variety of minor intrusions. The crustal terranes were juxtaposed into their present positions by large sinistral (left-lateral) movements on bounding faults, such as the Great Glen Fault, during the later stages of the orogeny. Major lateral movements on other north-east-trending faults probably occurred at the same time.

Small outcrops of late Silurian to Early Devonian sedimentary and volcanic rocks are found in eastern Mull. On Arran, the country rocks around the Palaeogene intrusions include both Lower and Upper Old Red Sandstone sandstones and conglomerates of late Silurian to Early

Carboniferous age, deposited by rivers draining from the north and north-east. The Old Red Sandstone deposits, together with the Lower Carboniferous sandstones, siltstones, limestones and basaltic lavas on Arran, all form part of the Midland Valley Terrane (Cameron and Stephenson, 1985; Figure 1). Elsewhere, Carboniferous rocks are not common. An inlier of Upper Carboniferous rocks is present on Morvern, and it has been suggested that Carboniferous rocks occur in the offshore basins (Fyfe et al., 1993). A few dykes of Late Carboniferous to Early Permian age cross the region, and quartz-dolerite plugs, for example in Morvern, may also date from this time.

Major sedimentary basins were initiated towards the end of the Palaeozoic, and continued to develop during the Mesozoic and into the Cainozoic. The basins or troughs trend approximately north–south, and include the Sea of the Hebrides–Little Minch Trough, the Inner Hebrides Trough and Blackstones Basin and the Colonsay Basin (Figure 5). During the Permian and Triassic periods, thick sequences accumulated in these troughs, consisting of both desert and fluviatile sandstones. The Jurassic Period was characterised by shallow water sedimentation, and up sequence shows a progressive change from brackish water to marine conditions. The basin margin deposits are found onshore at several places, where there is evidence that they were faulted, tilted and gently folded after Kimmeridgian times but prior to the deposition of Upper Cretaceous (Cenomanian to Middle Turonian) marine sandstones and limestones (chalk). Cretaceous rocks occur widely in the Inner Hebrides, but are typically very thin and laterally variable.

Regional uplift and ensuing erosion between the end of the Turonian and the early Paleocene (Danian) gave rise to a land surface on which the first Paleocene volcaniclastic rocks and basaltic lavas were deposited. Coarse clastic sedimentary rocks intercalated in the lower parts of the lava sequences provide indications that this land surface may have had significant relief locally. Shallow lakes, ponds and rivers were present and the basal beds of the Paleocene sequences are generally water-laid tuffaceous sandstones and siltstones, commonly with plant remains, and rare hyaloclastite deposits. These are overlain by thick, predominantly basaltic, subaerial lava successions of which the earliest appear to be those of Muck, Eigg and the basal members of the Mull succession. The eruption sites of the lavas are not known, but it is likely that they were fed from isolated vents, which developed along fissure systems now represented by the north-west-trending basaltic dyke swarms. Although evidence of dyke-related feeders is good in the Antrim Lava Field of Northern Ireland (Preston, 1982), it is less clear in western Scotland.

Several well-defined north-west-trending regional dyke swarms cross the region, intensifying in the vicinity of the central igneous complexes. The swarms reflect a regional north-east–south-west extensional stress system (England, 1988) and many of the dykes may have fed lava flows. In addition, sills of dolerite and basalt occur widely, and extensive sill-complexes intrude the Mesozoic basin deposits.

The central complexes are the deeply eroded roots of major volcanoes. With a few exceptions, they postdate the preserved lava fields and overlapped in time with the intrusion of the related dyke swarms. The central complexes represent fundamental changes in the character of the igneous activity, from the widespread feeders of the lava fields to much more localised and intense magmatism of shield volcanoes. There was a marked change too in the composition of the magmas. This is evident when the predominantly basaltic lava fields are compared and contrasted with the compositionally diverse intrusions of the central complexes, which include peridotite, olivine-gabbro, granitic rocks and minor amounts of rock of intermediate composition. Within the central complexes, granites generally form a high proportion of the surface outcrop. However, surveys of the gravity and magnetic fields over the complexes show that each centre has a substantial core of dense basic and/or ultrabasic rocks and that, despite their considerable areal extent, the granitic rocks are relatively thin and superficial bodies.

The combined evidence of radiometric ages and palaeomagnetic determinations indicates that the individual lava fields probably accumulated in little over one million years and that the active life of individual central complexes may also have been as short as one million years (Figure 8).

The siting of the central complexes has long intrigued geologists. J E Richey of the Geological Survey was responsible for much of the earlier work on this subject in Scotland, and also examined several of the central complexes in Ireland (Richey, 1932; Preston, 2001). He highlighted the possible significance of the proximity of the central complexes to pre-Paleocene faults. Additional contributory factors have been suggested, including the intersection of major faults with intra-basin ridges of Precambrian rocks, the diapiric rise of bodies of granitic magma focussing subsequent basaltic magmatism, and the intersection of dyke swarms with intra-basin ridges, possibly facilitating the initiation of granite magmatism. When a view is taken of the whole of the Palaeogene igneous activity in the British Isles, the most striking feature is the similarity of the magma composition and emplacement style across a wide range of structural settings and terranes. This suggests a region-wide stress field and a widely available heat source. This heat source is generally attributed to proto-Iceland mantle plumes (e.g. White, 1988; White and McKenzie, 1989). When a plume impinged on the area, large amounts of relatively buoyant basaltic magma were formed. As this magma rose up to and through the crust, it became ponded at traps caused by relatively abrupt density changes across major structural discontinuities. Upward movement became progressively more influenced by the local geology and structure the closer the magma approached to the Earth's surface. These factors facilitated differentiation and contamination of the magmas, and focussed their paths during ascent.

Throughout the Oligocene and Miocene epochs, significant weathering and erosion of the lava fields and central complexes occurred, with much of the detritus shed into sedimentary basins off north-west Scotland. Although Neogene sedimentary rocks are preserved offshore (Fyfe et al., 1993), the next major events recorded in the onshore geology are the Quaternary glaciations. Much of the area, including the sea bed, was covered by the extremities of the thick, Late Devensian ice cap that formed over the central and northern Highlands 29 000 to 14 700 years ago. Following the decay of this ice sheet, the higher mountains of Skye, Rum, Mull and Arran supported corrie and valley glaciers during the Loch Lomond Stadial, 12 500 to 11 500 years ago. The glaciations have left deposits of till, hummocky moraine and various glaciofluvial gravels. The accompanying changes in sea level are recorded in the raised marine beach deposits and marine notches that are common around the coastline (Plate 42). The preglacial topography was drastically modified during deep erosion caused by the local ice caps and glaciers. This contributed to the excellent exposure now found in many of the corries and cliffs of the islands, and is most dramatically expressed in the deep rock basins and jagged skyline of the Skye Cuillin (*Frontispiece*). Following the melting and retreat of the glaciers, over-steepened slopes collapsed, forming screes and some of the most extensive landslips in the British Isles (Plate 43), several of which are active to the present day.

two
Pre-Mesozoic

The general distribution of Pre-Mesozoic rocks in the western Highlands and Islands of Scotland is shown in Figure 3. A brief account of the rocks where they are closely associated with the Palaeogene igneous rocks is given here; but more detailed information can be found in Johnstone and Mykura (1989).

ARCHAEAN

Outcrops of the Lewisian Gneiss Complex are widespread in the northern part of the Hebridean Province where the principal occurrences are on Skye and the nearby islands of Raasay and Rona. The granodioritic, tonalitic and doleritic intrusive rocks from which the gneisses formed were intruded into lower crustal levels at about 3100 to 2800 Ma. Gneiss formation accompanied by granulite-facies metamorphism occurred in the Archaean prior to 2500 Ma, marking the Scourian Event (Friend and Kinny, 2001). Intrusion of a suite of basaltic dykes, the Scourie Dyke Swarm, occurred mainly about 2400 Ma. This was followed by pervasive deformation and amphibolite-facies metamorphism during the Laxfordian Event that peaked about 1700 Ma. Later retrogression, movement along shear zones and formation of crush belts occurred periodically during uplift and exhumation of the gneiss complex up to about 1100 Ma (Park, 1991; Park et al., 1994).

Skye, Rona and Raasay

High-grade, felsic, biotite-hornblende gneisses with mafic and ultramafic bands, lenses and clots formed during the Scourian Event and crop out extensively on Raasay north of the Screapadale Fault and throughout all of Rona. These gneisses were strongly reworked by the Laxfordian Event and are similar to those found in the Torridon area of the mainland. On Skye, the most extensive area of Lewisian gneisses occurs on the south-east side of the Sleat peninsula. These rocks are on the western edge of the Caledonian fold-belt and have been overthrust to the west-north-west, as part of the Moine and Tarskavaig thrust sheets. They consist of sheared and retrograde gneisses with prominent amphibolite and zoned ultramafic pods. Near Isle Ornsay and on the shores of Loch na Dal, the mafic gneisses locally contain garnet porphyroblasts, and both mafic and felsic gneisses are overprinted by needle-like crystals of actinolite up to 5 cm in length. In the west, around Tarskavaig, thin, mylonitised layers of Lewisian-like gneiss crop out close to some thrust planes. In the Eastern Red Hills Centre, gneisses unconformably overlain by Paleocene lavas crop out on Creagan Dubh (Bell and Harris, 1986), and large xenoliths of gneiss are present in the Marsco Hybrids in the Western Red Hills Centre (Wager et al., 1965; Thompson, 1969). The more felsic components of gneisses enclosed in the Marsco Hybrids have been partially melted, as have the gneiss xenoliths present in a dolerite intrusion next to Loch Sligachan.

Rum

Gneisses crop out at several localities, which lie either within the central or western parts of the central complex (Plate 1) or along the peripheral Main Ring Fault system (Figure 24). These gneisses, together with the Torridon Group that overlies them unconformably, were uplifted by as much as 2 km during the Paleocene igneous activity. The more mafic components in the gneisses are commonly altered to two-pyroxene hornfelses, whereas the leucocratic gneisses show varying degrees of partial melting. Gneiss clasts are common among the inter-lava conglomerates of the Canna Lava Formation of north-west Rum, but these have not been affected by thermal metamorphism to the same extent as those found at outcrop in the central complex.

In the central complexes of Skye and Rum, gneisses are exposed at present, either because of doming and uplift due to igneous activity or because of faulting. The outcrops provide clear proof that these areas are underlain by the Lewisian Gneiss Complex, thus corroborating geophysical (seismic) evidence.

Plate 1 Banded Lewisian granodioritic gneisses with amphibolitic layers. Near the Priomh-lochs, Rum (P580456).

TORRIDONIAN Thick sequences of undeformed, non-metamorphosed late-Proterozoic clastic sedimentary rocks that overlie Lewisian gneisses are known collectively as the Torridonian (succession). In the north-west Highlands, the lowest, Stoer Group is separated from the overlying Sleat and Torridon groups by a major unconformity that possibly represents a time-gap of over 200 Ma. Only the two younger groups crop out in the northern Inner Hebrides and their local successions are summarised in Table 1 (Stewart, 1991; Nicholson, 1993). The age of the Torridon Group is constrained by whole-rock Rb-Sr dates of 994 ± 48 Ma on phosphatic concretions in the Diabaig Formation and 977 ± 39 Ma on siltstones in the Applecross Formation (Turnbull et al., 1996). The youngest detrital zircons in the Applecross Formation have a U-Pb date of 1060 ± 18 Ma (Rainbird et al., 2001).

Skye Rocks of the Torridon Group crop out around the margins of the Eastern Red Hills Centre, for example south of Broadford, at Dunan and on the islands of Scalpay and Raasay. In addition, there are outcrops on Soay, at Camasunary, and along the north side of Soay Sound. They are generally red-brown feldspathic sandstones, commonly pebbly, and typical of the Applecross Formation. An extensive tract of Torridonian strata occurs in south-east Skye (Sleat) (Figure 3), and in isolated north-west-overthrust sheets in the vicinity of Broadford. The lowest strata belong to the Sleat Group and consist of greyish green, grey and buff gritty sandstones, commonly epidote-bearing, together with siltstones and sandy laminated siltstones. They total over 3000 m in thickness and occur in a belt up to 7 km in width stretching from the shores of Loch Alsh east of Kyleakin, south-west along the Sleat peninsula to the vicinity of Armadale. Overlying them, west of a line from about Ord to Kyleakin, are brownish red, false-bedded feldspathic sandstones of the Applecross Formation. The Torridonian strata in south-east Skye lie within the Kishorn Thrust Sheet of the Moine Thrust Belt and are much affected by faulting and thrusting, especially on the Sleat peninsula (Johnstone and Mykura, 1989).

Rum Rocks of the Torridon Group crop out in the north and east of the island where a thick succession (over 3000 m) of brown, feldspathic sandstones and pebbly sandstones belonging to the

Table 1 Torridon and Sleat groups on Rum and Skye.

	Rum Sheet 60 Rum, 1994	**Skye** Sheet 71W Broadford, 2002	
	Foreland	Foreland	Kishorn Thrust Sheet
TORRIDON GROUP	AULTBEA FORMATION Sgorr Mór Sandstone Member	AULTBEA FORMATION	
	APPLECROSS FORMATION Sresort Sandstone Member Allt Mór na h-Uamha Member	APPLECROSS FORMATION Leac nam Faoileann Member Bheinn Bhreac Member Leac-stearnan Member Sithean Glac an Ime Member	APPLECROSS FORMATION
	DIABAIG FORMATION Laimhrig Shale Member Fiachanis Gritty Sandstone Member	DIABAIG FORMATION Mullach na Carn Member Sgurr na Stri Member Bla-bheinn Member	
SLEAT GROUP			KINLOCH FORMATION
			BEINN NA SEAMRAIG FORMATION

Applecross Formation is preserved (Table 1). Within the formation, the contrasting weathering of coarse- and fine-grained beds, alternating on a scale of one to ten metres, gives a distinctive landscape in the north of the island, comprising long westward-dipping crags and intervening peaty shelves (Plate 2). Tabular sandstone bodies in the Applecross Formation are interpreted as sandbars in a braided river system, comparable with present-day examples (e.g. Nicholson, 1993). Pale grey silty sandstones and siltstones that occur towards the base of the succession are assigned to the Diabaig Formation (Table 1). Within the central complex, these fine-grained rocks are interbedded with sedimentary breccias and coarse-grained gritty sandstones, and lie unconformably on an irregular surface of Lewisian gneisses. At the top of the succession, fine-grained sandstones, commonly with heavy-mineral concentrations, are assigned to the Aultbea Formation

MOINE SUPERGROUP

A thick succession of medium- to fine-grained clastic sediments, deposited in shallow seas, has been lithified and metamorphosed to form the pelitic, semipelitic and psammitic rocks of the Moine Supergroup. In the Glenelg area, east of Skye, Moine metasedimentary rocks overlie Lewisian gneisses. The actual date of sedimentation has not been precisely determined. However, gneissose granite, dated at about 1083 Ma, cuts Moine rocks, and it has been suggested that the latter were deposited up to 200 million years earlier (Harris and Johnstone, 1991). Rocks of the Moine Supergroup crop out extensively throughout the North West Highlands (Johnstone and Mykura, 1989) and the few occurrences in the Hebridean Province are briefly described below.

Plate 2 *Westward-dipping Torridonian strata on Mullach Mór, Rum, with the Cuillin of Skye in the distance (P580458).*

Skye Psammitic, semipelitic and pelitic rocks crop out between the Kishorn and Moine thrust planes on the Sleat peninsula of Skye (Figure 3). They are intermediate in character between the Torridonian rocks in the Kishorn Thrust Sheet to the west and the rocks of the Moine Supergroup that overlie the Moine Thrust to the east, but are distinctly less metamorphosed than the latter. They have been termed the Tarskavaig Group, after the type-locality on the west side of Sleat, and are described more fully by Johnstone and Mykura (1989).

Ardnamurchan and Morvern Mildly metamorphosed pebbly sandstones, sandstones and silty sandstones of the Morar Group of the Moine Supergroup are present in and around the Ardnamurchan Central Complex. They also underlie the Palaeogene lavas and Mesozoic rocks on Ben Hiant and on the Morvern peninsula. Higher grade psammitic rocks with granite veining, belonging to the Glenfinnan Group, underlie and crop out along the eastern edge of the lava pile on Morvern.

Mull Rocks of the Glenfinnan Group are found at many localities around the margins of the Mull Central Complex, for example in the core of the Craignure Anticline and beneath the Mesozoic rocks on the coast south of Gribun. Screens of Moine rocks occur between inclined sheets and other minor intrusions in eastern Mull. Outside of the Mull Central Complex, fragments of Moine lithologies, including megablocks up to 100 m across, are common in vent infills and in other volcaniclastic rocks (Bailey et al., 1924, fig. 29). Numerous xenoliths of severely altered Moine rocks are present in the Loch Scridain Sill-complex. On the Ross of Mull, Moine rocks, possibly of both the Glenfinnan and Morar groups, crop out south of the Loch Assapol Fault where they are intruded by the late-Caledonian, Ross of Mull Pluton. From the distribution of rocks of the Moine Supergroup on Mull, it is evident that the central complex is underlain by, and intruded into, these rocks.

DALRADIAN SUPERGROUP Rocks belonging to the Dalradian Supergroup occupy much of the ground between the Great Glen Fault and the Highland Boundary Fault (Figure 3), but only on Arran are they a major part of the country rocks to Palaeogene intrusions.

Mull Grey phyllitic to slaty semipelite and black metalimestone belonging to the Appin Group (possibly the Blair Atholl Subgroup) of the Dalradian form the core of the Loch Don Anticline in eastern Mull. They are separated from Moine rocks on Mull by a continuation of the Great Glen Fault.

Arran Low-grade (chlorite zone) gritty and pebbly metasandstone, with subsidiary slaty metasiltstone and metamudstone, of the uppermost Southern Highland Group occupy much of the northern part of the island. They were displaced and folded during the emplacement of the Paleocene North Arran Granite Pluton (Figure 32) and are overlain by, or are in faulted contact with, younger rocks to the east. The original strata consisted of a sequence of turbidite deposits, and many of the original sedimentary structures are still preserved, thus aiding interpretation of Paleocene and earlier folding events.

LOWER PALAEOZOIC

Skye The lower part of the Cambro-Ordovician succession, which is comparable with that seen elsewhere in the North West Highlands, crops out in a highly complex structure around Ord on the Sleat peninsula (Figure 4; Table 2). At the base of the sequence, quartzites of the Eriboll Sandstone Formation (including the False Bedded Quartzite Member and the bioturbated Pipe Rock Member), and the An t-Sron Formation are all of early Cambrian (Comley) age. The An t-Sron Formation consists of the Fucoid Beds and Salterella Grit members. The Fucoid Beds consist mainly of dolomitic siltstone characterised by trace-fossil remains originally thought to resemble seaweed markings. The Salterella Grit Member comprises quartzite with remnants of the small gastropod *Salterella*. The overlying Ghrudaidh, Eilean Dubh and Sailmhor formations (Durness Group) range from mid Cambrian (St David's) to earliest Ordovician (Tremadoc) in age and are best exposed in the coastal sections; they consist of dolomitic limestones and dolostones that are commonly cherty.

The outcrops of Cambro-Ordovician rocks, together with associated strata of the Torridonian Kinloch and Applecross formations, occur within a structure that has been termed the Ord Window (Figure 4a). This structure is a stratigraphical outlier in that it is completely surrounded by older rocks, but it is a structural inlier in the sense that it has been interpreted as an area of the foreland that has been exposed by erosion through overlying thrust sheets of the Moine Thrust Belt. The inlier is characterised by folded, steeply dipping, fault-bounded slices, and the detailed structure is difficult to determine. A number of different models have been proposed (Figure 4b). Clough (in Peach et al. 1907) envisaged the Ord structure as a tectonic window in the core of a relatively open antiform, in which essentially right-way-up sequences are separated by two thrusts within the lower part of the Moine Thrust Belt. Bailey (1939, 1955) re-interpreted Clough's work, and suggested that the inlier consists of a pair of large recumbent folds, with inversion of strata. The two thrusts were interpreted as branches of the Kishorn Thrust on the lower limb of the recumbent Lochalsh Syncline and the lowest structures in this part of the Moine Thrust Belt. The core of the inlier could therefore be regarded as part of the foreland. Potts (1982) envisaged the rocks of the inlier as part of a recumbent fold, analogous to, but at a lower structural level than, the Lochalsh Syncline, and re-interpreted some of the thrusts as normal stratigraphical junctions. He proposed that the folded rocks were emplaced at their current level through a combination of thrusting on the western, leading edge and extensional normal faulting at the trailing, eastern edge. In this model, the inlier does not constitute a classic 'window' through a thrust sheet. As yet there is no consensus on the preferred model and there is considerable scope for further interpretation.

A thick chert-bearing dolostone sequence crops out extensively between Broadford and Loch Slapin. It is separated from the Cambro-Ordovician outcrops of Sleat by several kilometres, with only small fault-bound outcrops of the Eriboll Sandstone Formation in the intervening ground. Gastropods, bivalves and sponges have been obtained, for example near the outlet of Loch Kilchrist, and the dolostones are considered to be of early Ordovician (Arenig) age.

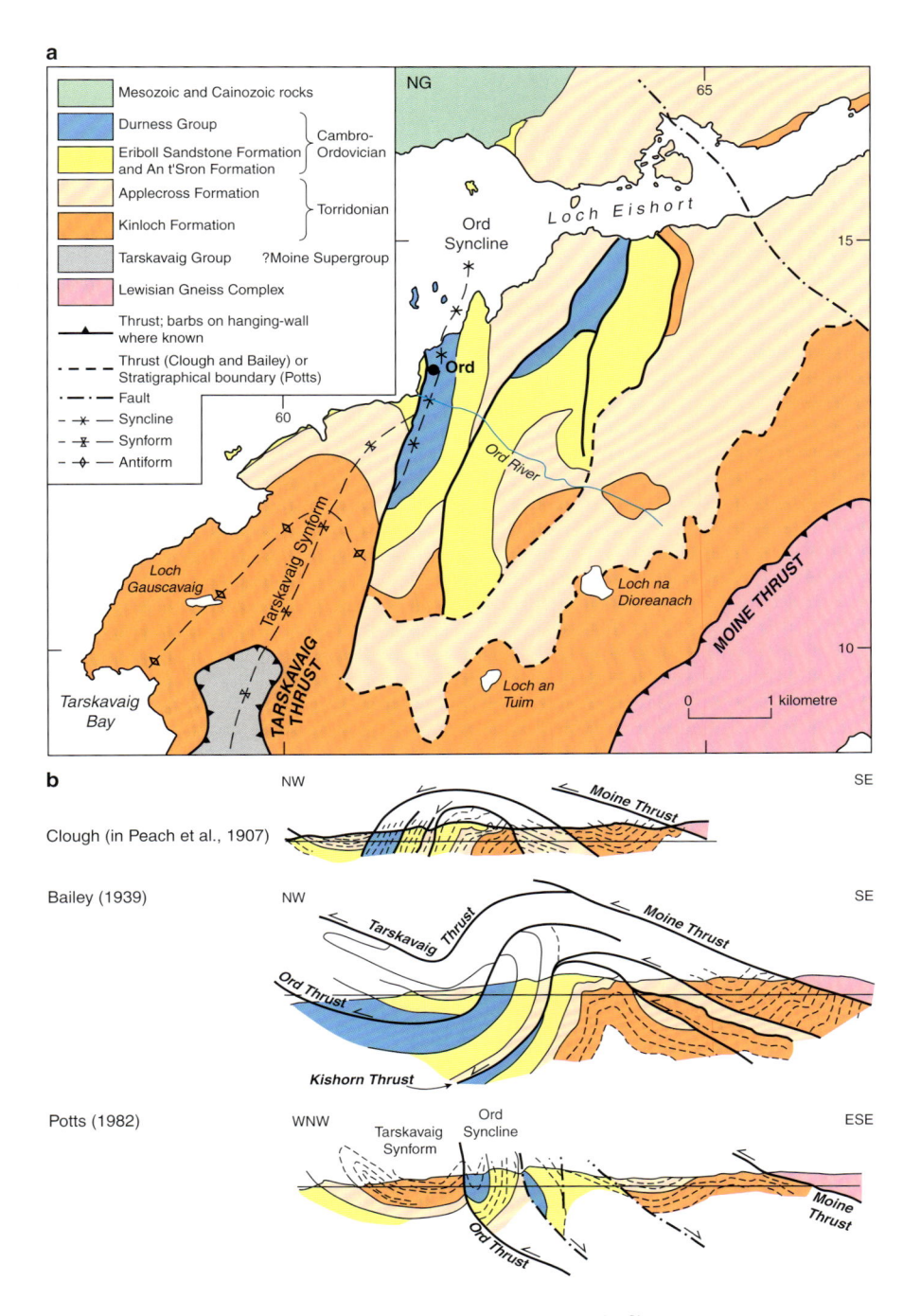

Figure 4 Cambro-Ordovician rocks of Ord on the Sleat peninsula, Skye.
a Map showing the structural inlier of Cambro-Ordovician rocks within the Kishorn Thrust Sheet at the base of the Moine Thrust Belt.
b Cross-sections illustrating alternative interpretations of the Ord structure by Clough (in Peach et al., 1907), Bailey (1939, 1955) and Potts (1982). The Clough and Bailey sections follow a common line whereas the line of Potts' section lies slightly farther to the south. The detailed structure is very complex and there is no preferred model. Several contacts, interpreted as tectonic thrusts by both Clough and Bailey, have been re-interpreted as normal stratigraphical junctions by Potts, who has also re-interpreted some other thrusts as extensional faults; the Ord Syncline was interpreted as an antiform by Bailey, but as a synform by Potts.

Table 2 *Cambrian and Ordovician strata of Skye, based on Sheet 71W Broadford, 2002.*

		Formation/member	Lithology	Thickness (m)
ORDOVICIAN	ARENIG	**Strath Suardal Formation**		
		dolostone	dark grey, massive with layers of white chert nodules	75
		Ben Suardal Member	dolostone, pale grey, massive, bioturbated with dark chert nodules	170
		Kilchrist Member	dolostone, mid to dark grey, thin chert layers, boudinage common	35
		dolostone	dark grey, massive, layers of white chert nodules	85
		Lonachan Member	dolostone, pale grey, well bedded, pale and dark smoothly rounded chert nodules	140
		dolostone	dark grey, massive, with layers of small white chert nodules	?30
		------ Junction not seen, relationship unknown ------		
	TREMADOC	**Sangomore Formation**	dolostone with beds of white chert up to 1 m thick	20
		Sailmhor Formation	dolostone, dark grey with abundant chert	80
	MERIONETH	**Eilean Dubh Formation**	dolostone, white or cream, rarely grey, chert nodules abundant near base and top	150
	ST DAVIDS	**Grudaidh Formation**	dolostone, lead grey, banded with chert layers	35
CAMBRIAN	COMLEY	**An t-Sron Formation**		
		Salterella Grit Member	sandstone, medium to coarse grained, slightly calcareous	15
		Fucoid Beds Member	calcareous sandstone and siltstone passing up into greenish fissile mudstone	17
		Eriboll Sandstone Formation		
		Pipe Rock Member	sandstone, coarse-grained and recrystallised, bioturbated with abundant pipe-like and funnel-shaped burrows perpendicular to bedding	17
		False-bedded Quartzite Member	Recrystallised sandstone, white to buff, massive, and cross-bedded with beds of quartz-pebble conglomerate particularly near the base	100
		~~ unconformity ~~~		

(Durness Group comprises Strath Suardal Formation through Grudaidh Formation)

Strath Suardal Formation crops out around the Beinn an Dubhaich Granite, in Strath, the remaining formations occur in the Ord area of Sleat

However, the succession cannot be equated directly with the upper part of the Durness Group elsewhere and hence is here termed the Strath Suardal Formation. The dolostones underlie overthrust Torridonian strata and are intruded by igneous rocks of Paleocene age including the Beinn an Dubhaich Granite, the Broadford Gabbro and numerous dykes (Figure 21b). Dolostones in contact with the granite and gabbro have been altered to a variety of calc-silicate hornfels, including the well-known yellow and green decorative 'Skye Marble', which contains diopside, serpentine and brucite. The Ordovician strata are folded into an anticlinal structure, which has the Beinn an Dubhaich Granite at its core.

Arran

A thickness of about 300 m of black mudstone and chloritised spilitic pillow lavas, together with rare beds of chert or tuff, is exposed in the North Glen Sannox River. Poorly preserved brachiopod remains obtained from the mudstone indicate an Arenig age for these deposits. The sequence rests on Dalradian metasandstones and is cut by a few sheets of altered dolerite and gabbro. It forms part of the Highland Border Complex, which crops out along the Highland Boundary Fault between Arran and Stonehaven (Stephenson and Gould, 1995).

CALEDONIAN IGNEOUS ROCKS

Mull

The Ross of Mull Pluton comprises a number of granitic bodies with a Rb-Sr age of about 414 Ma (Halliday et al., 1979); it intrudes and thermally metamorphoses the Moine metasedimentary rocks of south-west Mull. Cordierite-sillimanite hornfelses have developed from kyanite-bearing pelitic rocks adjacent to the granite and occur as xenoliths. Within the intrusion, the intricate relationships found between quartz-diorite, granite and basic enclaves suggest the co-existence and interaction of basic and silicic magmas.

Minor intrusions of Caledonian age include various felsic types and lamprophyres. The felsic dykes are most common in the vicinity of the Ross of Mull Pluton, and near the Strontian Pluton east of the Morvern lava pile (Johnstone and Mykura, 1989). The Ross of Mull granites have been extensively quarried (p.173).

OLD RED SANDSTONE

The Old Red Sandstone is mainly Devonian in age but parts may be late Silurian or Early Carboniferous.

Arran

Sandstones and conglomerates of the Lower Old Red Sandstone form a thick arcuate outcrop east and south of the North Arran Granite Pluton, and part of the succession is repeated on the south-east side of the Central Arran Ring-complex. These strata, which total about 1235 m in thickness, are faulted against Dalradian metasedimentary rocks, younger sedimentary rocks and the North Arran Granite Pluton in eastern Arran, but on the west side of Glen Rosa they lie unconformably on the Dalradian (Friend et al., 1963; McKerrow and Atkins, 1985). Clasts of schistose gritty metasandstone occur in conglomerates low in the succession, and were clearly derived from local Dalradian sources. However, the very abundant pinkish quartzite boulders and cobbles seen, for example, in conglomerates near the Glen Sannox baryte mine have no local source and match most closely Argyll Group (Dalradian) quartzites on Islay and Jura. Sparse plant remains (*Psilophyton princeps*, var. *ornatus*) have been found high in the succession in Glen Shurig (Tyrrell, 1928).

Upper Old Red Sandstone strata occur above all the sequences of Lower Old Red Sandstone but, in contrast to the mainland occurrences, no intervening unconformity has been found. In the east, the sequence is about 870 m thick and comprises sandstone with conglomerate beds containing clasts not only of local origin (vein quartz and Dalradian lithologies) but also of possible Islay and Jura quartzite (Friend et al., 1963). Similar rocks crop out south-east of the ring-complex. On the west coast, conglomerates and sandstones of the North Machrie Breccias total about 375 m in thickness, with the proportion of sandstone increasing eastwards.

The Old Red Sandstone sedimentary rocks probably represent alluvial fan deposits similar to those present on the northern edge of the Midland Valley. They were laid down by braided river systems, with the sediment load derived from a northerly source. Strongly dessicating conditions are indicated by the presence of caliche deposits (cornstones), which are especially well developed at the base of the succession at 'Hutton's Unconformity' north-east of Loch Ranza, where the carbonate pedogenesis extends for up to 50 cm into Dalradian rocks below the unconformity. There, the cornstone-bearing Upper Old Red Sandstone beds above the unconformity are assigned to the Kinnesswood Formation at the base of the Carboniferous succession.

Volcanic rocks within the Old Red Sandstone succession of Arran are limited to an (?)olivine andesite lava recorded from the Lower Old Red Sandstone south of the North Arran Granite Pluton and (altered) olivine basalt lavas interbedded with the Upper Old Red Sandstone strata north of North Glen Sannox. The olivine andesite flow has been compared with lavas found in the Lower Old Red Sandstone of the Midland Valley, whereas the olivine basalt lavas are similar to lavas found low in the Carboniferous succession of the Midland Valley (Tyrrell, 1928).

Mull   Basaltic and andesitic lavas belonging to the Lower Old Red Sandstone Lorn Plateau Volcanic Formation crop out in the core of the Loch Don Anticline in south-east Mull. Conglomerates and marly sandstones, possibly of similar age, occur on Frank Lockwood's Island, south-east of Loch Buie in southern Mull.

CARBONIFEROUS

Arran   Intensively faulted Carboniferous strata crop out on the north-east coast, and between Corrie and Glen Cloy. They range from sandstone and mudstone of the Inverclyde Group (Tournaisian) to siltstone, seatearth and sandstone of the Coal Measures (Westphalian). The sequence includes the richly fossiliferous Corrie Limestone, which is crowded with the brachiopod *Productus latissimus* and is equated with the Hurlet Limestone at the base of the Lower Limestone Formation (latest Visean) on the mainland. Olivine basalt lavas and volcaniclastic rocks are present in the lower part of the succession exposed on the Corrie foreshore and on

Plate 3   *Bedded Permian aeolian sandstone cut by silicified/carbonated veins, Corrie foreshore, Arran. (Photographer: Lorne Gill, Scottish Natural Heritage).*

the north-east coast. They overlie sandstones of the Clyde Sandstone Formation (Tournaisian) and are equated with lavas belonging to the Clyde Plateau Volcanic Formation of the Strathclyde Group. Altered basalt lavas in the Merkland Burn near Brodick Castle overlie sandstones and mudstones of the Limestone Coal Formation (Namurian).

Morvern

A small faulted inlier consisting of over 100 m of sandstones and mudstones with thin fireclays and (uneconomic) coal seams underlies Permo-Triassic beds at Inninmore (c, Figure 3). Fossil plants, including *Asterophyllites charaeformis*, *A. equisetiformis*, *Calamites cisti*, *C. schutzeiformis*, *Mariopteris muricata*, *Neuropteris gigantea* and *Samaropsis* sp., indicate an early Coal Measures (Westphalian) age for these deposits (Lee and Bailey, 1925; Johnstone and Mykura, 1989). Despite its limited extent, the occurrence is of some importance since it is one of very few outcrops of Carboniferous rocks in western Scotland: the presence of offshore occurrences has been suggested, which could have implications for hydrocarbon generation, but to date none has been proved (Fyfe et al., 1993).

PERMIAN

Arran

Permo-Triassic rocks crop out widely in the Hebrides, where they are also important constituents of the offshore basins, but it is only on Arran that the Permian and Triassic are separated in published accounts (e.g. Lovell, 1991). The approximately 500 m-thick sequence of sandstones and sedimentary breccias comprising the Corrie Sandstone and Brodick Breccia are Permian. The sandstones are of aeolian origin, with obvious dune bedding and wind polished sand grains (Frederiksen et al., 1998) (Plate 3). Fossil lightning strikes (fulgurites) have been recognised in these beds at Corrie (Harland and Hacker, 1966). The Brodick Breccia consists of coarse sedimentary breccias and conglomerates with fragments of Dalradian vein quartz and quartzose schist, and agate and basalt clasts derived from Devonian and Carboniferous lavas. They are interpreted as flash-flood deposits, formed when debris was periodically flushed out from wadis. The associated finer grained rocks are similar to the Corrie Sandstone. The overlying, diachronous and possibly interdigitating Machrie, Glen Dubh and Lamlash sandstone formations are tentatively equated with the Upper Permian to Middle Triassic Sherwood Sandstone Group (Lovell, 1991; Table 3). Flows of olivine basalt in the upper reaches of the Sliddery Water are of Early Permian age.

Table 3 Permian and Triassic strata of the Isle of Arran.

System	Formation	Thickness (m)	UK-wide group
TRIASSIC	WESTBURY*	?	PENARTH
	DERENENACH MUDSTONE*	?	MERCIA MUDSTONE
	LEVENCORROCH MUDSTONE	c. 80	
	AUCHENHEW MUDSTONE	c. 200	
	LAG A'BHEITH		
	GLEN DUBH SANDSTONE	c. 400	SHERWOOD SANDSTONE
	LAMLASH SANDSTONE		
	MACHRIE SANDSTONE		

PERMIAN	BRODICK BEDS, including:
	  BRODICK BRECCIA
	  CORRIE SANDSTONE
	  volcanic rocks near base
		c. 740
		c. 500	

* found only as blocks in the Central Arran Ring-complex

INTRUSIONS OF CARBONIFEROUS AND PERMIAN AGE

Minor intrusions of Carboniferous and Permian age are widespread through the southern Highlands and extend into the western Highlands and Islands. There are two main suites: Stephanian tholeiitic rocks and Visean to Early Permian alkali basalts and lamprophyres. On Arran, the former includes a thick quartz-dolerite dyke at Imachar. East of the Paleocene lavas on Morvern, isolated quartz-dolerite plugs and west- to west-north-west-trending quartz-dolerite dykes, which crop out between the lavas and the Strontian Pluton, are regarded as part of the Stephanian tholeiitic suite. At Gribun in western Mull, Moine rocks are intruded by a dyke of olivine nephelinite exposed at low water which contains a suite of xenoliths and megacrysts of lower crustal and possibly upper mantle origin (Upton et al., 1998). This dyke is similar to numerous other xenolith-bearing intrusions of Carboniferous and Permian age in the western Highlands and Islands and the Midland Valley. The xenoliths and xenocrysts found in these intrusions have furnished valuable information about the nature of the Lower Crust and Upper Mantle at the start of the Mesozoic (e.g. Upton et al., 1998).

three
Mesozoic

Rocks ranging in age from the Triassic to the Late Cretaceous occur throughout the Inner Hebrides and on Arran (Figure 5). They are part of the extensive basinal deposits found mainly offshore, and in many instances represent the feather edges of these successions. The basins are important structural elements in the Hebridean Igneous Province. In addition to their dominant control on Mesozoic sedimentation, the basins and associated structures appear to have influenced the emplacement of the Paleocene central complexes and possibly also the accumulation of the lava successions.

TRIASSIC Many of the New Red Sandstone occurrences mentioned below are assigned to the Triassic on grounds of lithology, unconformable relationships with older rocks, or a conformable one with overlying, fossiliferous Jurassic strata. Fossils are rare in all but the highest of these beds. In the Hebridean area, the New Red Sandstone sequences are thought to belong to the Upper Triassic and accumulated during the earliest stages of basin formation (Steel 1974a,b; Steel et al., 1975).

Arran The Lamlash Sandstone and Glen Dubh Sandstone formations are well exposed on the east side of the island. These sandstones are massive, reddish to purple, and are stratigraphically continuous with the Permian sandstones. The Permian–Triassic boundary is difficult to identify here, as it is elsewhere in Great Britain, and is arbitrarily located within the sandstones of the Sherwood Sandstone Group (Warrington et al., 1980; Table 3).

The Permian and Triassic rocks of Arran were considered virtually devoid of stratigraphically diagnostic fossils until the discovery of Triassic miospores (Warrington, 1973) in what is now termed the Lag a'Bheith Formation (Lovell, 1991). This formation comprises mudstones with fine-grained sandstones and contains carbonate concretions ('cornstones') probably resulting from evaporation under subaerial conditions.

The overlying Auchenhew Mudstone Formation covers much of southern Arran and corresponds with the areas shown as Triassic on the published maps. It extends from Largybeg Point in the east to Drumadoon in the west and there are abundant, excellent exposures on the foreshore, in cliffs and in many stream and river sections. The formation comprises a succession of fine-grained sandstones, mudstones and calcareous mudstones (marl) that has been equated with the Mercia Mudstone Group (Lovell, 1991). The presence of carbonate concretions, pseudomorphs after halite, ripple marks and dessication cracks indicate that the beds formed mainly under shallow-water conditions, but with periods of drying out and deeper flooding.

Mudstones, siltstones and fine micaceous sandstones of the Levencorroch Mudstone Formation are exposed in stream sections near Levencorroch Hill. These beds also contain carbonate concretions. The highest parts of the Triassic succession occur only as downfaulted blocks in the Central Arran Ring-complex (Warrington et al., 1980). There, they are exposed in the Allt nan Dris, where mudstones of the Levencorroch Mudstone Formation are associated with grey-green siltstones of the Derenenach Formation. The latter closely resemble the 'tea green marls' of the Collin Glen Formation at the top of the Triassic succession in Northern Ireland. Overlying black mudstones with thin limestones have yielded marine bivalves (e.g. *Rhaetavicula contorta, Chlamys valonensis, Protocardia philipiana?, Modiolus minimus?*) and these beds have been assigned to the Westbury Formation of the Rhaetian Penarth Group.

Mull Up to about 60 m of conglomerate, sandstone and cornstone are found at Gribun and on Inch Kenneth in western Mull, where the striking unconformity with the underlying Moine rocks is exposed on the foreshore and in cliff sections (Plate 4). Clasts of Moine rocks, cherty fossiliferous

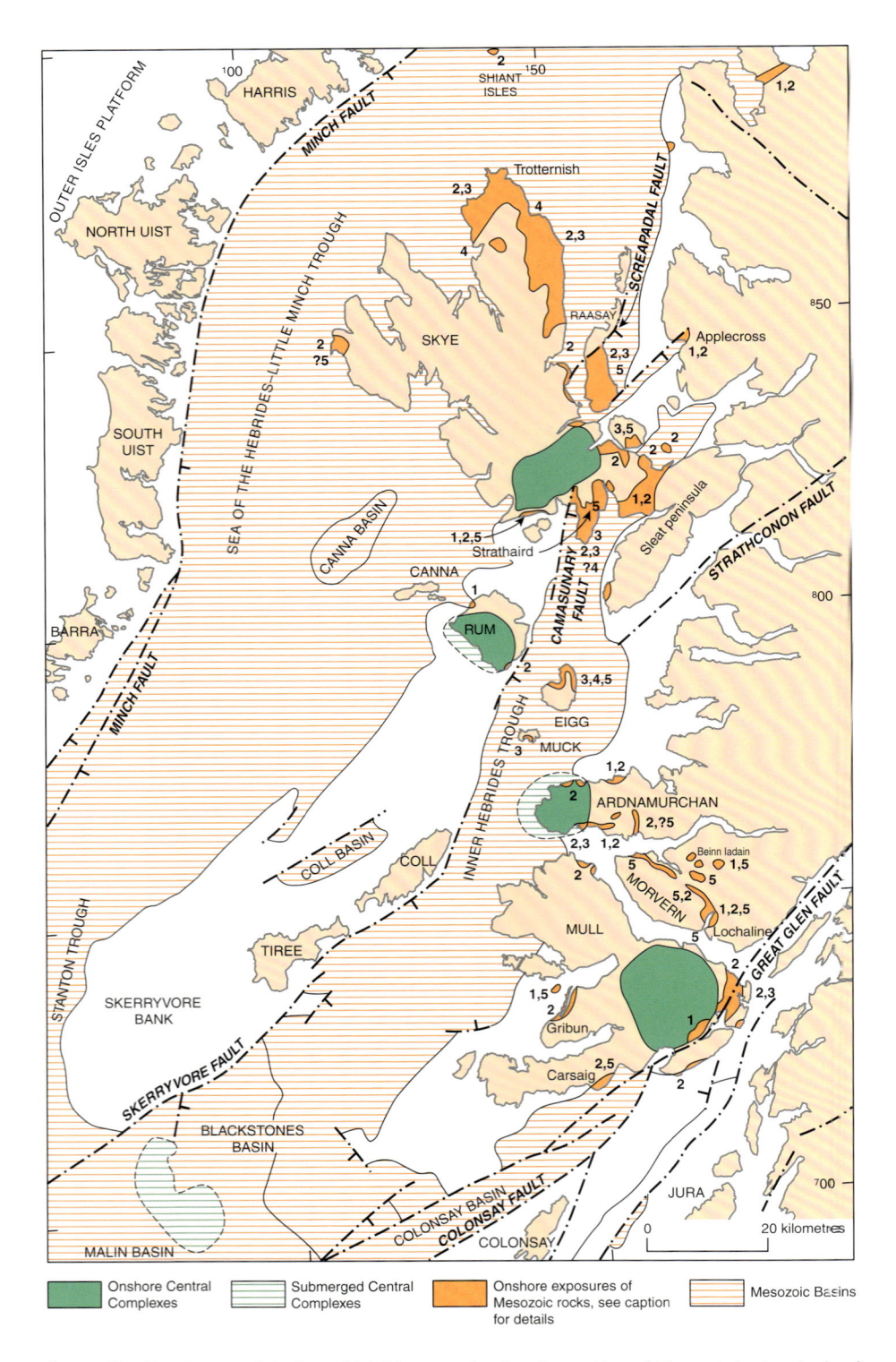

Figure 5 Sketch map of the Inner Hebrides area showing the position of Mesozoic basins, the localities where Mesozoic rocks occur onshore and the positions of the Paleocene central complexes; pre-Mesozoic rocks are uncoloured (after Fyfe et al., 1993, fig. 4). The Camasunary Fault probably links to the Skerryvore Fault.

1 Permo-Triassic; **2** Lower Jurassic; **3** Middle Jurassic; **4** Upper Jurassic; **5** Upper Cretaceous

CHAPTER THREE: MESOZOIC

Plate 4 *Pre-Paleocene strata and Paleocene lavas at Creag a'Ghaill, Gribun, western Mull.*
Platy-jointed psammites of the Moine, Upper Shiba Psammite Formation (Morar Group) are overlain unconformably by Triassic basal conglomerates and sandstones. Upper Triassic strata consist of calcareous sandstones with carbonate concretions (cornstones), and the overlying sandy limestones of the Rhaetian Penarth Group are cut by basaltic sills of Paleocene age. Upper Cretaceous silicified limestone is largely obscured by scree and Paleocene basalt lavas form the upper cliffs. The top of the upper cliff is about 260 m elevation. (Photomosaic: B G J Upton; P532641.)

limestone (Cambro-Ordovician Durness Group), red feldspathic sandstone (Torridonian), vein-quartz, quartzite, granite and rare, red andesitic porphyry (?Siluro-Devonian) indicate both local and quite distant provenance for these beds. Upper Triassic miospores have been recovered from beds near the top of the Gribun succession that also contain indistinct bivalve remains. The uppermost beds are sandy limestones of Rhaetian age, which contain fish scales and bivalves (e.g. *Cardinia* sp., *Chlamys valoniensis*, *Protocardia rhaetica*, *Rhaetavicula contorta*). In eastern Mull, Triassic conglomerates and sandstones are exposed in the cores of the anticlines that surround the central complex, from Craignure to Loch Don and Loch Spelve. In addition, there are extensive sandstone outcrops on the west shore of Loch Spelve, which extend up Glen Lussa, where the Triassic rocks occur in screens between inclined basic sheets. Clasts in the conglomerates are mainly quartzite and vein-quartz, but pebbles of Moine rocks are abundant locally where these lithologies are in situ nearby. Small outcrops of Triassic strata are also found within the central complex, where their steep dips indicate considerable disturbance by the Paleocene intrusions.

Morvern Up to 100 m of conglomerate, sandstone, calcareous mudstone, mudstone and cornstone are present beneath younger Mesozoic rocks and Paleocene lavas, and in outliers to the north of the lavas. Clasts in the conglomerates come from local Moine rocks and Caledonian felsic minor intrusions. Thickness is very variable. The Triassic beds appear to be absent to the west of Loch Teacuis, and even in the vicinity of Loch Aline and Inninmore Bay, where the thickest and most complete sections have been measured (Lee and Bailey, 1925), the beds may thin out completely. Poorly preserved bivalve remains have been found in rocks of possible Rhaetian age at Inninmore.

Ardnamurchan Sandstones, cornstones and sedimentary breccias crowded with local schist and vein-quartz fragments underlie lavas on the east side of Ben Hiant. Similar rocks, with abundant cornstone, crop out on the south coast at Mingary and underlie Lias beds on the north coast at Ockle Point and at Swordle.

Rum The Monadh Dubh Sandstone Formation of north-west Rum is 80 m thick, and consists of a fining-upwards sequence of sedimentary breccias, conglomerates, cornstones and sandstones with thin silty sandstones containing plant remains. These north-west-dipping beds unconformably overlie Torridonian sandstones, which have been permeated and replaced by cornstones. Clasts in the breccias and conglomerates were generally derived from the nearby Torridonian sandstones, although quartzite pebbles may have come from the basal Cambrian rocks of Skye. Near the middle of the succession, conglomerate-sandstone-cornstone cyclothems suggest a change from initial debris flows to fluvial deposition and soil formation (R J Steel *in* Emeleus, 1997). A general change from continental conditions to nonmarine shallow-water and estuarine deposition is indicated by the upward progression from fine-grained siltstones and calcareous sandstones with abundant plant and wood fragments, to beds containing fish scales and teeth, and conchostracan arthropods (*Euestheria minuta*). The uppermost beds may correlate with the Mercia Mudstone (Triassic) or Penarth (Rhaetic) groups of England and Wales.

Skye There is a thin, impersistent development of conglomerates, sandstones and cornstones beneath the Lower Jurassic rocks near Broadford and on either side of the shallow syncline of Jurassic rocks between Broadford and Loch Eishort. Similar rocks, predominantly conglomerate and breccia, also crop out near Tarskavaig on the Sleat peninsula and on the north side of Soay Sound. Faulted strips of Mesozoic rocks, including Triassic, are found on the northern slopes of Glamaig where they have been considerably disturbed by the granite intrusions. Pebbles and other clasts in the Triassic conglomerates are of local derivation and are predominantly of Torridonian sandstone, Cambro-Ordovician limestone, chert and, less commonly, quartzite. These New Red Sandstone strata have been assigned to the Stornoway Formation, which ranges in age from Triassic to possibly Hettangian (Morton and Hudson, 1995). However, the presence of Rhaetian strata within this formation is now disputed, and it is possible that most of the New Red Sandstone north of Mull is Jurassic in age.

Raasay There are several occurrences of the Stornoway Formation in the south of the island, near Eyre Point. The outcrops are much faulted, and are obscured towards the north-east beneath the immense slipped mass of Jurassic rocks at Beinn na' Leac, reappearing beyond the landslip at Rubha na' Leac. Marly beds and calcareous sandstones, including cornstones, are accompanied by conglomerates containing pebbles of Torridonian feldspathic sandstone, vein-quartz, and limestone (in places with chert) and quartzite of presumed Cambro-Ordovician age. The sequence is about 40 m thick at Rubha na' Leac, but less than 20 m elsewhere.

JURASSIC Thick sequences of Jurassic strata are important components of the Mesozoic offshore basins and parts of these successions extend onshore at several localities (Figure 5). Following the continental conditions that were prevalent during the Triassic Period, marine sedimentation predominated in Early Jurassic time. Shallow-water and brackish water, deltaic deposition characterised much of Mid Jurassic time, although there was a return to marine sedimentation in the Callovian, which continued into the Late Jurassic. With the exception of the Great Estuarine Group (uppermost Bajocian–Bathonian), the ammonite faunas permit correlation with successions in England. The lithological variation in time and space through the district is summarised in Figure 6.

The Jurassic rocks are commonly immature and incompletely lithified; even the earliest deposits of this age were probably buried to a depth no greater than about 1 km prior to eruption of the Paleocene lavas. A maximum of about 2 km thickness of lavas may have covered some of the Mesozoic successions but it is unlikely that the sedimentary rocks were greatly affected by heat from that source (England, 1992a). By contrast, significant thermal metamorphism has occurred in the vicinity of the central complexes. The formation of hornfels is limited to the country rocks immediately adjacent to the central complexes (less than 500 m), although hydrothermal alteration extends for several kilometres from the margins of the large complexes, for example on Skye and Mull (Taylor and Forester, 1971). Elsewhere, there is generally little more than slight induration of the sedimentary rocks, and alteration of their organic content extending from dykes and other minor intrusions for a distance of up to half their width (e.g. Hudson and Andrews, 1987; Bishop and Abbott, 1993). However, given the great abundance of intrusions in some dyke swarms and sheet complexes, the aggregated effects may have been significant for hydrocarbon maturation.

In addition to the references cited in the text, there is much detailed information about the Jurassic rocks of Skye and Raasay in Morton and Hudson (1995).

LOWER JURASSIC

In the southern part of the Hebridean area, the lower part of the Lias Group is Hettangian to lower Sinemurian (*bucklandi* Zone; Table 4), and consists mainly of limestones and fissile calcareous mudstones that are lithologically similar to beds of equivalent age in southern England. They have consequently been termed the Blue Lias Formation (Oates, 1978). In Ardnamurchan, these lithologies interdigitate with more arenaceous rocks (sandstone, sandy limestone, siltstone), which, in turn, become predominant at similar stratigraphical levels to the north on Skye and Raasay, where the beds are now termed the Breakish Formation (formerly the 'Lower Broadford Beds'). The Breakish Formation is overlain by non-calcareous sandstones, siltstones and mudstones of the Ardnish Formation (formerly the 'Upper Broadford Beds'; Morton, 1999a,b). The overlying Pabay Shales, Scalpay Sandstone, Portree Shale and Raasay Ironstone formations are recognised throughout the district.

Arran

A *remanié* mass of Lias fissile mudstone and limestone that crops out within the Central Arran Ring-complex has yielded a large bivalve fauna belonging to the *angulata* Zone (Hettangian) as well as ammonite remains (*Schlotheimia*). This is the only occurrence of Jurassic rocks on Arran. When taken in conjunction with nearby *remanié* masses of Cretaceous rocks, it provides a strong indication that Mesozoic strata were formerly much more widely distributed over western Scotland, as they are now in Antrim, a relatively short distance to the south-west (e.g. Mitchell, 2004).

Mull

With the exception of small outcrops of limestone, calcareous mudstone and sandstone on the west coast at Aird na h-Iolaire, and beds of the Blue Lias and Pabay Shale formations at and near Tobermory, Jurassic rocks are restricted to the east and south of the island. They crop out in the cores of anticlines marginal to the Mull Central Complex, from Scallastle Bay almost to Loch Spelve, in many places on the east coast from Loch Don to Loch Buie, and also at Carsaig Bay. The beds range in age from Hettangian through to Bajocian (Mid Jurassic; *garantiana* Zone) and contain a moderately abundant marine fauna. Near Port nam Marbh, there is a fairly continuous succession from the uppermost 30 m of the Pabay Shale Formation, represented by slightly calcareous siltstones with abundant *Gryphaea cymbium* and *Pecten aequivalvis*, to the Bajocian. At Torosay, the Pabay Shale Formation contains the Torosay Sandstone Member, which is over 20 m thick and is cross-bedded in places (Hesselbo et al., 1998).

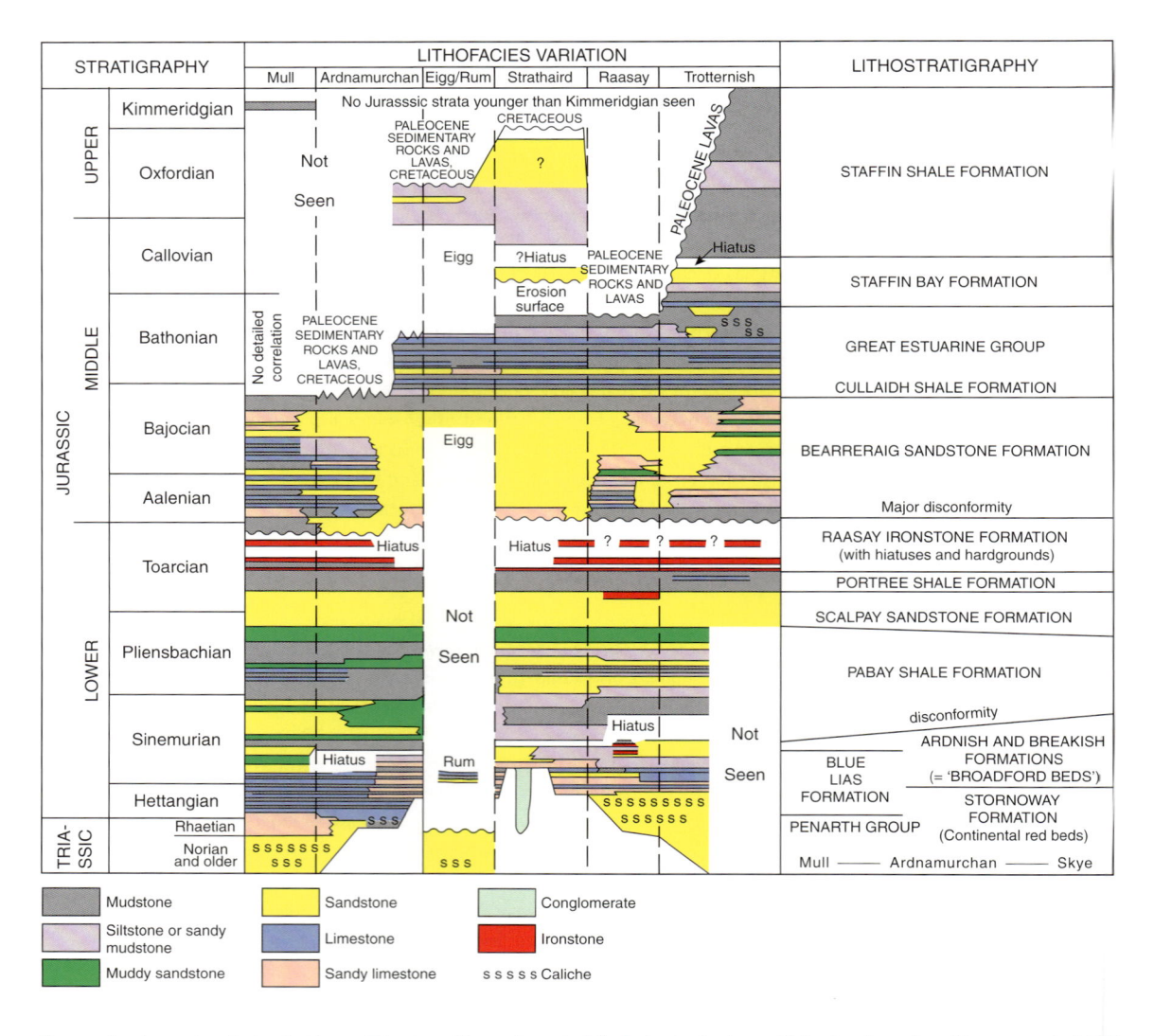

Figure 6 *Lateral variation in Upper Triassic to Upper Jurassic lithofacies in the Inner Hebrides (based on Morton et al., 1987 and Morton, 1989).*

Morvern   The most complete exposures are at Loch Aline, where the Blue Lias and Pabay Shale formations are each about 30 m thick. The sequence thins eastwards towards Inninmore; north-westwards, the Lias crops out below Cretaceous rocks on the shores of Loch Teacuis but is largely absent beneath the Cretaceous and Paleocene outliers north of the loch. An early Sinemurian hiatus is recognised in Morvern, which appears to be a local phenomenon (Hesselbo et al., 1998) and is attributed to sediment starvation. The existence of a widespread, diachronous unconformity, supposedly early Sinemurian in the south and late Sinemurian in the north, is now discounted (compare Fyfe et al., 1993).

Ardnamurchan   Despite being much disturbed by faulting and folding, and having undergone pervasive and locally severe thermal metamorphism, a fairly complete succession of Lower and Middle Jurassic rocks is preserved on the south side of the peninsula. Indurated, fossiliferous Lias mudstones, thin limestones, calcareous sandstones and thin ironstones are intruded by numerous inclined sheets; they crop out along the southern margin of the central complex

Table 4 Ammonite zonal sequence in the Lower Jurassic and part of the Middle Jurassic of the Inner Hebrides (based on Richey, 1961 and Cope, 1995). The correlation with the major lithostratigraphical divisions is based on Morton and Dietl (1989), Howarth (1992), Hesselbo et al. (1998), Morton (1999b), and Hesselbo and Coe (2000) with additional information from N Morton and J D Hudson, 2000 and 2004. Members are from the type area of the Bearreraig Sandstone Formation, in Trotternish, Skye.

Series	Stage	Zone	Locality	Lithostratigraphy	
MIDDLE JURASSIC	BAJOCIAN	Parkinsonia parkinsoni		CULLAIDH SHALE FORMATION	
		Garantiana garantiana	1, 4, 5	BEARRERAIG SANDSTONE FORMATION	GARANTIANA CLAY MEMBER
		Strenoceras subfurcatum	4, 5		
		Stephanoceras humphriesianum	1, 4, 5		RIGG SANDSTONE MEMBER
		Emileia (Otoites) sauzei	4		HOLM SANDSTONE MEMBER
		Witchellia laeviuscula	1, 4		
		Sonninia ovalis	4		UDAIRN SHALE MEMBER
		Hyperlioceras discites	4		
	AALENIAN	Graphoceras concavum	1, 3, 4, 5		OLLACH SANDSTONE MEMBER
		Brasilia bradfordensis	4		
		Ludwigia murchisonae	–		
		Tmetoceras scissum	1, 3, 4, 5		DUN CAAN SHALE MEMBER
		Leioceras opalinum	1, 4, 5		
LOWER JURASSIC	TOARCIAN	Dumortieria levesquei	3, 4, 5	HIATUS	
		Grammoceras thouarsense	–		
		Haugia variabilis	–		
		Hildoceras bifrons	1, 5		
		Harpoceras falciferum	1, 3, 4, 5	RAASAY IRONSTONE FORMATION	
				PORTREE SHALE FORMATION	
		Dactylioceras tenuicostatum	4, 5, 7		
	PLIENS-BACHIAN	Pleuroceras spinatum	1, 4, 5, 6	SCALPAY SANDSTONE FORMATION	
		Amaltheus margaritatus	1, 5, 6		2
		Prodactylioceras davoei	?1, 5		
		Tragophylloceras ibex	1, 4, 5	6	
		Uptonia jamesoni	1, 4, 5		
	SINEMURIAN	Echinoceras raricostatum	1, 4, 5	PABAY SHALE FORMATION	
		Oxynoticeras oxynotum	?3		
		Asteroceras obtusum	2, 3, 4		
		Caenisites turneri	?2, 5		
		Arnioceras semicostatum	1, 2, 3, 4, 5	'Broadford Beds'	ARDNISH FORMATION
		Arietites bucklandi	2, 5		
	HETTANGIAN	Scholoteimia angulata	1, 3, 4, 8	BREAKISH FORMATION	BLUE LIAS FORMATION *
		Alsatites liasicus	4, 5		
		Psiloceras planorbis	1, ?3, ?4		
		pre-planorbis Beds	1, 5		
				STORNOWAY FORMATION †	

Localities: 1 Mull; 2 Morvern; 3 Ardnamurchan; 4 Skye; 5 Raasay; 6 Scalpay; 7 Shiant Isles; 8 Arran
* developed only in south of district
† in Skye, Upper Triassic to Hettangian (Morton and Hudson, 1995)
– not recorded in district

from the vicinity of Port Min to Kilchoan, and form scattered outcrops eastwards as far as Loch Mudle. On the north coast, Lias beds crop out at several localities for about 4 km west of Fascadale Bay, and to the east more than 70 m of mudstone and limestone (interdigitating Blue Lias and Breakish formations) overlie Triassic rocks at Swordle. Fossils from the *angulata* and *bucklandi* zones have been obtained from this area (Oates, 1978).

A succession of beds ranging from the Pabay Shale Formation to the Middle Jurassic Bearreraig Sandstone Formation has been recognised amongst the dense cone-sheet swarms around Kilchoan and Sròn Bheag (Richey and Thomas, 1930). Rocks belonging to the Portree Shale and Scalpay Sandstone formations occur among the isolated outcrops of Jurassic strata on the north coast of the peninsula.

Rum Thin, fault-bound slivers of indurated to highly metamophosed limestone, sandy limestone, sandstone, mudstone and rare ironstone are preserved within the Main Ring Fault of the central complex on the eastern slopes of Beinn nan Stac. The limited fauna of these beds, and their lithological similarities to Lower Jurassic rocks east of Broadford, Skye, indicate that they probably correspond in the main to the Breakish and Ardnish formations, although the thin bed of altered ironstone at Dibidil may be from near the base of the Pabay Shale Formation (compare Smith, 1985).

Skye A wide outcrop of Lower Jurassic strata extends south from Broadford Bay through Strath to Loch Slapin. The type localities for the Breakish and Ardnish formations are to the east of Broadford, and for the Pabay Shale Formation on the island of Pabay. Lias rocks crop out on the south of Scalpay and on the northern slopes of Glamaig where they have been much disturbed by granites of the Western Red Hills Centre. Other outcrops occur to the west of the Camasunary Fault, where the limestones and silty beds are generally thermally altered, and on the north side of Soay Sound.

A deep, oil-exploration borehole through the Paleocene lavas and the Mesozoic rocks was drilled at the southern end of the Waternish peninsula, about 5 km north-east of Dunvegan, and provides some information about the likely Lower Jurassic succession (Hesselbo et al., 1998, fig. 23).

The Breakish Formation in its type locality at Ob Lusa consists mainly of sandy limestones, calcareous sandstones and thin fissile mudstones with an 8 m-thick, massive sandstone near the top of the succession. Bivalves are abundant in certain beds, although diagnostic ammonites are not common. The coral *Isastrea* is found in the Ob Lusa Coral Bed, near the base of the succession, and *Thecosmilia* occurs at a higher level at Ob Breakish. The lower part of the Ardnish Formation (formerly the Upper Broadford Beds) is well exposed on the Ardnish peninsula, where mudstone, siltstone and sandstone occur in coarsening-up cycles. The rocks are commonly micaceous and in places also ferruginous; they yield abundant *Gryphaea arctuata* and ammonites, especially *Coroniceras* and *Arnioceras*. An oolite containing chamosite oxidised to goethite forms the Ardnish Ironstone in the lower part of the succession (Morton and Hudson, 1995).

On the coast of Loch Slapin, south of Camas Malag, sandstones, siltstones and limestones of the Ardnish Formation rest directly on dolostones of the Ordovician Strath Suardal Formation, the Breakish Formation being absent. Here, fossils from the *semicostatum* Zone are common and the basal Jurassic layer is crowded with *Gryphaea* and other bivalves attached to the underlying Ordovician dolostone. In sea cliffs a short distance to the north-west, the overall lithofacies illustrates one of the few unequivocal exposures of a Jurassic shoreline in Britain. Clefts and hollows in a palaeokarst topography eroded in the Ordovician dolostone are infilled with a thin sequence of sedimentary breccia, conglomerate and rhythmically bedded, silty sandstone and sandstone. These rocks comprise the Camas Malag Formation (Nicholson, 1978); they are of Early Jurassic age (Amiri-Garoussi, 1982) and were laid down under marine conditions (Farris et al., 1999). On the Strathaird peninsula, part of a tibia belonging to a carnivorous dinosaur has been found in beds of probable Sinemurian age (Benton et al., 1995).

On the island of Pabay, only the upper part of the Pabay Shale Formation is exposed, from the Suishnish Sandstone Member, upwards. The beds, which are over 75 m in thickness, are of dark, micaceous fissile mudstone with red-weathering (iron-bearing) carbonate concretions, from which Murchison collected the type specimen of *Platypleuroceras (Ammonites) brevispina* in 1826. The whole formation is excellently exposed on either side of Rubha Suishnish on the shores of Loch Slapin and Loch Eishort, where a detailed succession has been established (Hesselbo et al., 1998). Soft, grey, micaceous mudstones with calcareous concretions become sandier up the succession, passing into thick grey micaceous sandstones of the Suishnish Sandstone Member. Ammonites diagnostic of the *semicostatum* Zone are present in the lower part of the succession, as well as bivalves and brachiopods, and towards the top, ammonites from the *obtusum* Zone include *Asteroceras stellare*, *Promicroceras planicosta* and *Xipheroceras* sp. (Hesselbo et al., 1998). Sinuous, lobate contacts between Paleocene basalt dykes and silty sandstones exposed in a roadstone quarry north of Suishnish (and elsewhere) indicate that the sedimentary rocks were poorly consolidated and probably wet when intruded. Sandstones of the Hallaig Sandstone Member (Hesselbo et al., 1998) are also present at Loch Eishort but these are now considered to belong to the Ardnish Formation (Morton, 1999a). The Pabay Shale Formation also crops out on south-east Scalpay and on the west side of Beinn Dearg Mhòr in Strath.

Micaceous sandstones, siltstones and calcareous sandstones of the Scalpay Sandstone Formation crop out on Scalpay and at Rubha na Sgianadin, north-west of Broadford. The formation also crops out in small areas on the west side of Loch Slapin, in shore exposures on the west side of the Sound of Raasay, north and south of Portree Harbour, and in a cliff south of Holm, in Trotternish. It belongs to the *spinatum* Zone (Upper Pliensbachian) and is some 30 m thick.

The principal occurrence of the Portree Shale and Raasay Ironstone formations is found in the cliffs and on the shore between Holm Island and Tianavaig Bay in eastern Skye, although the beds are almost everywhere obscured by landslips. North of Portree Harbour, more than 20 m of grey, micaceous, ferruginous siltstone, ooidal ironstone and dark-weathering micaceous siltstone belong to these formations. The abundant ammonites in the topmost of these beds indicate that they belong to the *falciferum* Subzone (Middle/Lower Toarcian). A thin seam of jet near Holm Island is at a similar stratigraphical level to the Jet Rock of Yorkshire. To the north, a borehole near the mouth of the Bearreraig River proved about 1.6 m of sideritic chamosite-oolite belonging to the Raasay Ironstone Formation, but it was not of economic grade. There are also two small exposures of the Portree Shale and Raasay Ironstone formations on the east coast of the Strathaird peninsula a short distance south of Faoilean on Loch Slapin.

Raasay The Breakish Formation is sparsely exposed on Raasay. The greatest development occurs at the east end of the bay at Hallaig, where about 35 m of interbedded limestone and mudstone are seen. The beds contain a bivalve-dominated fauna that includes *Modiolus*, *Cardinia* and *Liostrea*. Ammonites (*Franziceras sorlei*, *liasicus* Zone) from the lowermost beds indicate that the marine transgression affected this area earlier than the Broadford area, and was therefore diachronous (Morton, 1999b).

East of Hallaig, a 50 m-thick cliff section exposes a succession of limestones and fissile mudstones in its lower part, with calcareous siltstones and micaceous sandstones (also well exposed in the waterfall at the seaward end of Hallaig Burn), seen at higher levels. These latter beds belong to the Hallaig Sandstone Member which marks the top of the Ardnish Formation (Figure 6) (Morton and Hudson, 1995; Morton 1999a; compare Hesselbo et al., 1998). Bivalve, brachiopod and crinoid remains occur in the lower part of the member, and ammonites (*Arnioceras*, *Euagassiceras*) are present in beds about 20 m above the base, but the upper half is only sparsely fossiliferous. The Pabay Shale Formation is exposed in the Allt Fearns, south-west of

Beinn na' Leac, where it consists of dark coloured, fissile mudstones that pass upwards into micaceous, silty rocks. Bivalves (including *Gryphaea*, *Cardinia*, and *Hippopodium ponderosa*) and a variety of ammonites (including *Echioceras*, *Crucilobiceras* and *Vininodiceras simplicostata*) occur in these beds.

The Scalpay Sandstone Formation is prominent on the east coast of Raasay from Screapadal to Hallaig and south-west to Inverarish. The formation is nearly 100 m thick east of Beinn na' Leac, where micaceous siltstones and flaggy sandstones are overlain by about 30 m of massive sandstone. An abundant fauna has been obtained from black, calcareous sandstone beds with iron-rich ooids, immediately below the topmost sandstone. Bivalves are present (*Pseudopecten aequivalve*, *Gryphaea gigantea*) and several species of the ammonite *Pleuroceras*, indicating the *apyrenum* Subzone (uppermost Pliensbachian). The ooids here, and in Jurassic ironstones elsewhere in the Inner Hebrides, are generally described as chamositic, but are more likely to be composed of the mineral berthierine, which is compositionally similar but structurally different (e.g. Deer et al., 1962). A new species of glypheoid lobster (*Pseudoglyphea foersteri*) has been described from the Scalpay Sandstone. Examples were found near Hallaig and also south of Holm Island on Skye (Feldmann et al., 2002; Elliott and Feldmann, 2003).

The Portree Shale Formation comprises 2 to 3 m of dark, micaceous siltstone exposed in the Inverarish Burn and in nearby, abandoned opencast workings, where a thickness of 2.4 m of the overlying Raasay Ironstone Formation is found. The base of the Raasay Ironstone Formation consists of a thin chamosite ooid-bearing black fissile mudstone with abundant fossils (e.g. *Dactylioceras* sp.). A thin-bedded, ooidal chamosite ironstone (about 2 m thick) was the principal source of iron ore (Plate 5). The ironstone is cross-bedded in places, and was deposited in shallow water. Randomly orientated belemnites are a striking feature on bedding planes exposed on the floor of the former opencast workings. Comprehensive suites of fossils have been obtained from old iron ore dumps near Suishnish Pier, including the ammonites *Dactylioceras taxophorum*, *Hildoceras laticostata*, *Harpoceras falciferum* and *Cleviceras elegans*, all from the *falciferum* Subzone (Lower Toarcian).

Shiant Isles

About 10 m of indurated fissile mudstone is preserved between Paleocene dolerite sills. The mudstone has yielded species of *Dactylioceras* characteristic of the *falciferum* Zone. The topmost beds may be of Bajocian age.

MIDDLE JURASSIC

Onshore deposits of this age are largely confined to the northern part of the district, where the majority of the type sections occur on Skye. The marine, ammonite-bearing Bearreraig Sandstone Formation (Aalenian–Bajocian) includes the thickest sandstones of the onshore British Jurassic. The overlying deposits of the Great Estuarine Group (top Bajocian–Bathonian) (Table 5) were laid down in lagoons, lagoonal deltas and mudflats, which varied from freshwater through brackish to marine (e.g. Harris, 1992). A major hiatus and, locally, an unconformity separate the Raasay Ironstone Formation from the overlying Middle Jurassic strata in the north of the district.

Skye and Raasay

Extensive outcrops of Middle Jurassic rocks occur on the Strathaird peninsula, in Trotternish where a complete succession from the Aalenian to the Oxfordian is preserved, and in southern Raasay.

Bearreraig Sandstone Formation

The Bearreraig Sandstone Formation outcrop extends for several kilometres in near-vertical cliffs in eastern Trotternish, where it comprises massive sandstones, commonly calcareous, with rich ammonite faunas at several localities. The succession across the Aalenian–Bajocian boundary is exposed in the pipeline cutting above the hydroelectric powerhouse at the south side of Bearreraig Bay. The lithology and ammonite fauna at this locality have been described in detail by Morton (e.g., *in* Morton and Hudson, 1995) who proposed the section as the boundary stratotype for the basal

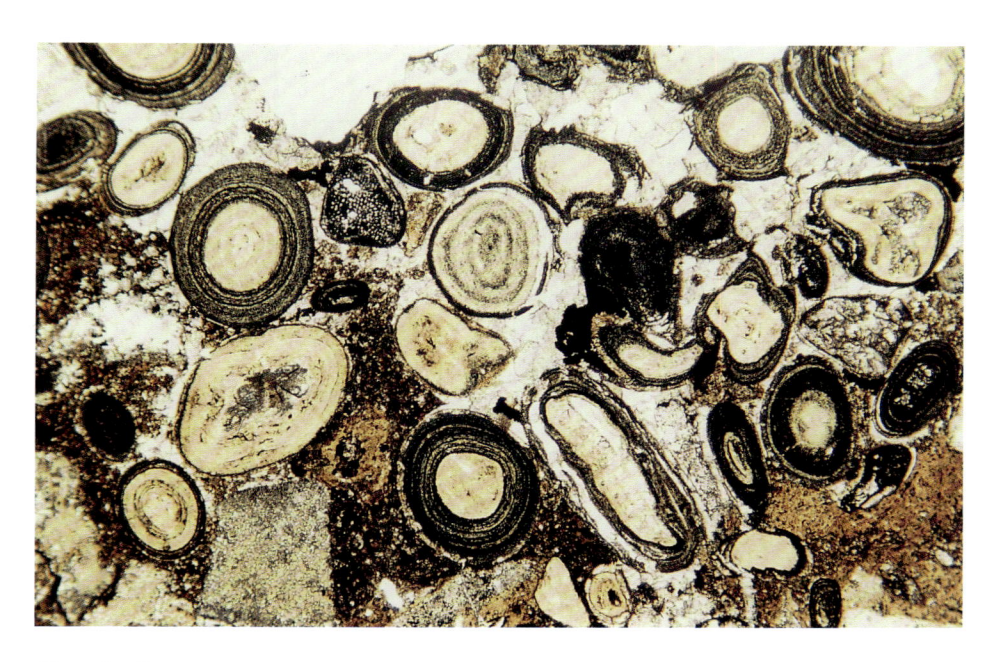

Plate 5 *Photomicrograph of ooidal chamosite ironstone with echinoderm fragments in a limestone matrix, Raasay Ironstone Formation, Raasay. Individual ooids are up to about 1 mm in diameter. Field of view is approximately 5 mm wide (plane polarised light. Sample S16931. BGS: P532649).*

boundary of the Bajocian Stage. Remains of a thyreophoran dinosaur have been recovered from the Bearreraig Sandstone at Bearreraig Bay (Clark, 2001). Below the sandstones south of Bearreraig Bay, about 21 m of micaceous mudstone of the Dun Caan Shale Member (ranging from late Toarcian, but mainly Aalenian in age), rest disconformably on the Raasay Ironstone Formation. On Raasay, the Dun Caan Shale Member crops out above the old opencast workings, but the thickest development (about 30 m) is at Gualann na' Leac, on the north-east side of Beinn na' Leac. Farther south, the Druim na Fhuarain Sandstone Member of the Bearreraig Sandstone Formation forms a broad bench on the eastern side of the Strathaird peninsula. These sandstones are predominantly cross-bedded and contain only rare ammonites. Cross-bedding indicates two flow directions, one southerly and a more common northerly one. Both may occur at the same locality and are attributed to tidal flow action redistributing the sediments. In keeping with this vigorous depositional environment, a large proportion of the fossils are preserved only as shell fragments. In southern Strathaird, up to 470 m of sandstone is present, capped by 12 m of pyritiferous fissile mudstone representing the Garantiana Shale Member. The abrupt change from sandstone to mudstone at the base of the Garantiana Shale Member marks the onset of deeper water conditions, and corresponds to a widely recognised transgression elsewhere. The sandstones thin northwards and cross-bedding appears at progressively higher horizons such that, on Raasay, the cross-bedded sandstones do not occur until well above the Aalenian–Bajocian boundary. There, the sedimentary structures indicate a north-north-east-directed water flow. One of the more spectacular outcrops of the Bearreraig Sandstone on Raasay is the deeply fissured (and possibly landslipped) mass that forms the summit of Beinn na' Leac.

Great Estuarine Group

The Cullaidh Shale Formation consists of dark bituminous fissile mudstones and black sandstones; it varies in thickness from 12 m in the south to about 2 m in the north. The formation rests, with gradational boundary, on the Garantiana Shale Member, and may include the Bajocian–Bathonian boundary. It is overlain by the Elgol Sandstone Formation (30 m thick), which comprises white, pure, non-calcareous sandstone of deltaic origin (Morton and Hudson,

Table 5 Great Estuarine Group and Bearreraig Sandstone Formation in the Inner Hebrides.

Formation	Type locality [Grid Reference]	Lithology	Fauna	Depositional conditions
SKUDIBURGH	Skudiburgh, West Trotternish [NG 379 650]	mudstone, siltstone, channel-fill sandstones	generally unfossiliferous	fluvial and alluvial
KILMALUAG	Kilmaluag, NW Trotternish [NG 427 742]	calcareous mudstone, fissile limestone desiccation cracks	gastropods, conchostracans, ostracods	freshwater lagoon
DUNTULM	Duntulm Bay, NW Trotternish [NG 410 740]	fissile mudstone, limestone, nodular limestone and shell banks	oysers and other bivalves, brachiopods	marine to brackish lagoon
VALTOS SANDSTONE	Valtos, East Trotternish [NG 514 638]	sandstone with carbonate concretions	bivalves	fluvial delta
LEALT SHALE Lonfearn Member	Lonfearn, East Trotternish [NG 520 620]	interbedded fissile mudstone and limestone, desiccation cracks	ostracods, conchostracans, bivalves	brackish to marine lagoon
Kildonnan Member	Kildonnan, Eigg [NM 490 850]	algal stromatolitic limestone at top, fissile mudstone	bivalves, vertebrates	brackish to marine lagoon
ELGOL SANDSTONE	Elgol, Strathaird [NG 520 140]	sandstone	rare plant remains	fluvial delta prograding south
CULLAIDH SHALE	Port na Cullaidh, Straithaird [NG 516 137]	dark fissile mudstone, oil-shale	fish scales, gastropods	low-salinity, anoxic sediment
BEARRERAIG SANDSTONE	Bearreraig Bay, north of Portree [NG 530 515]	sandstone, cross-bedded in south. Thin mudstone at top	ammonites (common and well preserved), belemnites, bivalves, crinoids, brachiopods	shallow marine sand sheets

1995). At the type locality, it consists of a classic coarsening-upwards deltaic cycle (Harris, 1989), and forms a prominent cliff with spectacular honeycomb weathering (Plate 6).

The Lealt Shale Formation is generally divided into two members. The top of the lower, Kildonnan Member is marked by a thin algal stromatolite band on Eigg (the type locality) and elsewhere throughout the district; it is overlain by the Lonfern Member. The mudstones and thin limestones of the formation, together with the algal band, show little variation from Eigg to northern Trotternish, indicating uniform, low-energy conditions of sedimentation. The beds are highly fossiliferous, although the diversity of gastropods, brackish water bivalves, ostracods and concostrachans is limited. Fish and other vertebrate remains are common (see under Eigg and Muck). The formation is well exposed in the coast section north of Elgol. Re-alignment of the road north of Portree has somewhat obscured the original type section at Lealt, but sections are available at the base of the steep cliff between Inver Tote and Lonfearn, one of which yielded the first dinosaur footprint discovered in Scotland (Andrews and Hudson, 1984).

The Valtos Sandstone Formation forms distinctive exposures of pale yellow sandstone that commonly contains abundant large, near-spherical, carbonate concretions or doggers over much of the area (Plate 7). In Trotternish, the formation is about 120 m thick, with two cross-bedded upward-coarsening sandstones of similar thickness, separated by about 27 m of fine-grained sandstones, silty mudstones and limestones containing *Neomiodon* shells. There is evidence of cyclic sedimentation; in places five cycles of mudstone and limestone passing up to sandstone with concretions have been recognised. The Valtos Sandstone Formation near the type locality has yielded diverse bones and fossil footprints, including a bone from a sauropod dinosaur (Clark et al., 1995). On Raasay, exposures of concretionary Valtos Sandstone occur between Dun Caan and the Screapadal Fault. The equivalent beds on Strathaird are thinner (about 24 m) and consist largely of sandy limestone, limestone and mudstone; the characteristic concretions are absent. Despite induration and alteration by the adjacent granite, concretions are still visible in the Valtos Sandstone Formation on the eastern slopes of Glas Bheinn Dearg in Strath, where it dips steeply eastwards. The Valtos Sandstone Formation is also present in north-west Skye, at Loch Bay and Waterstein. The formation has been interpreted as a system of delta lobes built out into the lagoons in which the Lealt Shale Formation had been deposited (Hudson and Harris, 1979; Harris, 1992).

The base of the Duntulm Formation is marked everywhere by the presence of limestone beds crowded with remains of the oyster *Praeexogyra hebridica*. Fossil shell banks, up to 2 m thick, were developed in marine to brackish water lagoons (Andrews and Walton, 1990). At Duntulm, a thickness of about 55 m of strata consists of interbedded mudstone and limestone with a marine to brackish water fauna. Several of the limestones contain nodular algal masses. Dinosaur tracks

Plate 6 *Honeycomb weathering in the Elgol Sandstone Formation, Elgol, Skye. Scale: hammer shaft is 30 cm (P580459).*

occur in the Duntulm Formation at Staffin Bay (Clark et al., 2004; Plate 8). Towards the top of the section there are indications of terrigenous influence and freshwater incursions as the beds are more sandy and contain the bivalves *Unio* and *Neomiodon*. There are a few poor exposures of the Duntulm Formation on Raasay. The formation is also present in the Strathaird succession, in which there is no indication of a freshwater influence or that land lay nearby. There are fossiliferous outcrops at Loch Mòr, Waternish and Waterstein Head, Duirinish.

The Kilmaluag Formation consists of calcareous mudstones with alternations of indurated layers which may be nodular; some are also dolomitic. Thin sandstones are present in the Trotternish successions, where Andrews (1985) recognised a 'clastic facies', but these are not found in the southern localities on Skye, nor on Eigg and Muck, which instead have an 'argillaceous limestone facies'. The beds are interpreted as having been deposited in short-lived lagoons and the fossil faunas indicate deposition in low-salinity environments. The gastropod *Viviparus* and the bivalve *Unio* are present, but they are not as abundant as the ostracods (e.g. *Theriosynoecum*) which gave the formation its former name, the Ostracod Limestones. The formation also occurs at Waternish and is well exposed on Strathaird, but it is not present on Raasay. A notable reptile and mammal fauna was obtained from Strathaird (Waldman and Savage, 1972; Waldman and Evans, 1994; Evans and Waldman, 1996); when discovered, this was only the second recorded occurrence of Mid Jurassic mammalian fossils in the world.

Plate 7   *Carbonate concretions in the Middle Jurassic Valtos Sandstone Formation, Bay of Laig, Isle of Eigg. The concretions are about 1 m in diameter (D1709).*

Plate 8 *Two sets of dinosaur trackways in rippled sandstone from the Middle Jurassic, Duntulm Formation, exposed near the slipway south of Staffin Bay, northern Skye. One set of tracks is pointing towards the camera and the other in the opposite direction. Individual footprints are about 50 cm in length. (Photograph: N D L Clark, Hunterian Museum, Glasgow.)*

The Skudiburgh Formation consists of red and grey-green mottled silty mudstone and dark claystone, sandstone and lenses of silty sandstone. From its variegated colours it was formerly known as the 'Mottled Clay'. At the type locality, the claystone and siltstone sequence is about 3.5 m thick and contains channel sands and small calcareous concretions, possibly caliche. The Skudiburgh Formation also marks the top of the Great Estuarine Group in the Strathaird succession. The formation is of alluvial origin; floodplain, channel and overbank deposits are present, which formed in a coastal plain, terrestrial environment during a late Bathonian regression (Andrews, 1985).

Eigg and Muck   The Middle Jurassic successions of Eigg and Muck have been described in some detail by Hudson (*in* Emeleus, 1997). The 150 m-thick succession on Eigg is the most complete, and comprises strata that range from the top of the Bearerraig Sandstone Formation up to the Kilmaluag Formation. The principal outcrops are in cliffs on the coast, from the Bay of Laig to near Eilean Thuilm in the north of the island, and discontinuous exposures also occur on the foreshore and at the base of the Paleocene lavas on the east coast. On Muck, the top of the Valtos Sandstone Formation, a full succession of highly fossiliferous Duntulm Formation strata, and the lower part of the Kilmaluag Formation are all well exposed on the foreshore at Camas Mòr.

The Bearreraig Sandstone Formation is limited to small exposures of calcareous sandstone with bivalve remains, at and below the high water mark in north-east Eigg. The overlying Lealt Shale Formation is well exposed on the north shore, although there the outcrops are greatly complicated by dolerite sheets and in places are obscured by landslips. Both the Kildonnan and the Lonfern members are present, separated by a distinctive bed of stromatolitic algal limestone, which occurs widely at this horizon in the Inner Hebrides. An abundant shelly fauna, including the bivalves *Praemytilus*, *Unio* and *Tancredia*, has been obtained from the fissile siltstones, thin limestones and sandstones of the lower, Kildonnan Member at the type section on the east coast of Eigg (e.g. Hudson and Wakefield, 1999). This section contains Hugh Miller's 'Reptile Bed' which has yielded sharks' teeth (*Hybodus*) and plesiosaur remains (*Plesiosaurus dolichoderius*) (Hudson, 1966). Vertebrate remains have also been found on the north shore.

Outcrops of the Valtos Sandstone Formation extend from the east coast near Kildonnan, around the north of Eigg to the Bay of Laig, where several cycles of sedimentation are recognised. Particularly fine examples of carbonate concretions occur at the Bay of Laig (Plate 7). The concretions were probably formed at depths of several hundred metres and at temperatures of 31° to 34°C, when meteoric pore waters dissolved carbonate from shells. The concretions started to grow about 15 Ma after burial of the host sediments and the larger, metre-sized examples may have taken several million years to form (Wilkinson, 1992).

The section in Laig Gorge, south-east of the Bay of Laig, contains exposures of both the Duntulm and Kilmaluag formations, as well as Cretaceous strata. Thick limestones crowded with the oyster *Praeexogyra hebridica*, typical of the Duntulm Formation, give way to mudstones with only thin oyster beds. The mudstones grade upwards into limy mudstones and thin limestones of the Kilmaluag Formation, which contain an ostracod and conchostracan (*Antronestheria*) fauna. There are a few small exposures of both formations north of Laig Gorge, and both must be present beneath a cover of till and landslip in Cleadale, although they are progressively cut out eastwards by the unconformable base of the Upper Cretaceous strata and the Paleocene lavas.

On the foreshore at Camas Mòr, Muck, Mesozoic strata from the upper part of the Valtos Sandstone Formation are exposed. The beds include typical concretionary sandstones, and grade up into highly fossiliferous limestones of the Duntulm Formation, in which shelly layers packed with *Praeexogyra hebrideca* are conspicuous. The lower part of the Kilmaluag Formation is seen in several small limestone outcrops totalling about 14 m in thickness. The higher beds are partly dolomitic, a feature first recognised from the mineralogy of the distinctive calc-silicate hornfelses adjacent to a large gabbro dyke (p.78). The fauna includes conchostracans of the genus *Pseudograpta* and the gastropod *Viviparus*.

Mull and Ardnamurchan

About 30 m of sandy limestone and calcareous sandstone of the Bearreraig Sandstone Formation are capped by 1 to 2 m of siltstone of the Great Estuarine Group on the east coast at Port na Marbh, south-east Mull. Similar beds crop out fairly continuously beneath the lavas on the west limb of the Loch Don Anticline, and discontinuously on the east limb as far as Duart Bay. On the Ardnamurchan peninsula, Middle Jurassic rocks are limited to massive sandstones and subordinate limestones of the Bearreraig Sandstone Formation exposed in screens amongst the basalt sheets on Maol Buidhe and at Sròn Beag. Despite the effect of thermal metamorphism, typical Aalenian fossils have been obtained from these rocks (Richey and Thomas, 1930)

UPPER JURASSIC (AND UPPER PART OF THE CALLOVIAN)

The principal developments of the Oxfordian and Kimmeridgian occur in Skye but no strata younger than the Kimmeridgian have been reported from the district. The beds were deposited in offshore marine conditions. The transgression began in the Callovian with the deposition of the fully marine Carn Mor Sandstone Member of the Staffin Bay Formation, followed by the Upper Ostrea Member, which has a marine to brackish water fauna and has yielded Callovian palynomorphs (Riding, 1992). The overlying, fully marine Staffin Shale

Table 6 Ammonite zonal sequence in the upper Middle Jurassic (Callovian) and the Upper Jurassic of the Inner Hebrides (based on Richey, 1961 and Cope, 1995). The correlation with the major lithostratigraphical divisions is based on Sykes (1975) and Sykes and Calloman (1979), with additional information from N Morton and J D Hudson (2000 and 2004).

	Stage	Zone	Locality	Lithostratigraphy (Trotternish)	
UPPER JURASSIC	KIMMERIDGIAN	Aulacostephanoides mutabilis	1 (not proved),		
		Rasenia cymodoce	1, 3		
		Pictonia baylei	3		
	OXFORDIAN	Amoeboceras rosenkrantzi	2, 3		Flodigary Shale Member
		Amoeboceras regulare	2, 3	Staffin Shale Formation	
		Amoeboceras serratum	2, 3		
		Amoeboceras glosense	2, 3		
		Cardioceras tenuiserratum	2, 3		Digg Siltstone Member
		Cardioceras densiplicatum	2, 3		Glashvin Silt Member
		Cardioceras cordatum	2, 3, 4		
		Quenstedtoceras mariae	2, 3, 4		
MIDDLE JURASSIC	CALLOVIAN	Quenstedtoceras lamberti	2, 3, 4		Dunans Clay Member
		Peltoceras athleta	2, 3		
		Erymnoceras coronatum	2, 3		Dunans Shale Member
		Kosmoceras jason	2, 3		
		Sigaloceras calloviense	2, 3		
		Proplanulites koenigi	3	Staffin Bay Formation	
		Macrocephalites herveyi	2		

1 near Craignure, Mull;   2 Strathaird, Skye;   3 Staffin Bay, Skye;   4 Laig Bay, Eigg

Formation contains an abundant fauna, including ammonites that permit zonal correlation with the Upper Jurassic strata of England (Table 6) and the Boreal Realm of Greenland and the Arctic.

Skye   The youngest Jurassic strata in the district occur in a shallow anticline beneath the Paleocene lavas of Trotternish, although there the beds are largely obscured beneath landslips (Plate 43), and in Strathaird. There are small outcrops near Uig and in the Kilmaluag River, but the most extensive occurrences are on the shoreface at the west side of Staffin Bay. These beds, which are in the toe of the Quiraing Landslip, are not metamorphosed but are much faulted, cut by minor intrusions and commonly have steep dips as a consequence of the landslipping. Lithologies present include shelly mudstone, siltstone and thin belemnite-bearing limestone of the Staffin Bay Formation (lower Callovian), and dark, silty claystone with subordinate silty sandstone and thin nodular limestone beds of the Staffin Shale Formation (Callovian to lower Kimmeridgian) (Morton and Hudson, 1995). The ammonites are mostly of Boreal provenance, but the occurrence of Tethyan forms is important in correlating the two faunal provinces of the European Jurassic. Elsewhere, Oxfordian mudstones, siltstones and sandstones are present near Strollamus, in the north and west of the Strathaird peninsula, and on southern Scalpay. These beds underlie Paleocene basaltic lavas and are commonly indurated and altered. Thin beds of altered tuff occur in the Staffin Shale Formation at Staffin Bay. They may possibly have been derived from volcanoes associated with the newly opened Rockall Trough (Knox, 1977). An isopod crustacean has been recovered from beds of Oxfordian age at Staffin Bay (Feldmann et al., 1994).

Eigg The Staffin Shale Formation is exposed among boulders on the shoreface between Clach Alisdair and the Bay of Laig. It consists of about 35 m of soft mudstone with limestone beds, and ranges from the Callovian *Quenstedtoceras lamberti* Zone to the Oxfordian *Cardioceras cordatum* Zone.

Mull A small outcrop of baked blue shale just west of Duart Bay has yielded ammonites and other fossils of Kimmeridgian age (*mutabilis* Zone).

CRETACEOUS

Cretaceous strata are present in isolated small outcrops, which generally owe their preservation to a protective covering of Paleocene lavas (Table 7). They were deposited unconformably over rocks varying in age from late Precambrian to Jurassic. It may therefore be inferred that a period of folding, uplift and erosion occurred between Late Jurassic and Early Cretaceous times in this district, as happened elsewhere in the north and west of the British Isles. These events produced a gently undulating surface with river valleys. This landscape was eventually submerged and low-energy shallow-water marine sediments were deposited. These consist of fine- to medium-grained sandstones, calcareous sandstones and limestones, and less commonly conglomerates and sedimentary breccias. There are marked differences in lithologies and in the successions from nearby areas, and on Eigg and on Mull there is clear evidence that some deposits have been reworked. Tectonic activity may have occurred throughout much of Late Cretaceous time. In Skye, some of the movement on the Camasunary Fault postdates the Cretaceous deposits, and on Mull, late movements on the Great Glen Fault probably affected the Cretaceous rocks at Torosay (Mortimore et al., 2001).

Thermal alteration of the Cretaceous rocks by the overlying Paleocene basaltic lavas makes precise dating of the deposits difficult, since microfossils and spores have commonly been destroyed. The Cenomanian age of glauconitic sandstones at the base of the successions in Mull and Morvern is firmly established (Lowden et al., 1992), but the age of overlying beds is less certain. At the top of the succession, the Beinn Iadain Mudstone Formation and part of the Clach Alasdair Conglomerate Member, may well be partly of Paleocene age (Mortimore et al., 2001), although Hancock (2000) suggested a Maastrichtian age for these beds on Mull, which he termed the Gribun Conglomerate Formation (see below). The brown mudstones have generally been interpreted as decomposed ash, deposited immediately prior to the main basaltic lava outpourings (e.g. Richey, 1961).

The presence of rounded 'millet seed' sand grains in many of the Hebridean Cretaceous rocks, together with the silica-cementation of some beds (interpreted as 'silcretes') and the presence of silicified chalk led to the suggestion that desert conditions prevailed around the chalk seas (Bailey, 1924), although others have disagreed with this interpretation. A study of the silicification within the Loch Aline White Sandstone Formation at the Lochaline Mine, Morvern showed that the silica-cementation, which has resulted in frosted and faceted sand grains of all sizes, is restricted to certain well-defined beds, suggesting that it was caused by groundwater circulation, concentrated along highly permeable layers (Lowden et al., 1992). The presence of lignite and lateritic mudstone overlying the chalk were taken by Mortimore et al. (2001) to suggest hot, wet conditions towards the end of Cretaceous time, which would have required a dramatic and sudden change in climate if this had been preceded by desert conditions.

The most complete successions of Upper Cretaceous rocks in the district are found at Gribun, Mull (over 15 m thick) (Plate 4), and on Ben Iadain, Morvern (over 21 m thick) (Figure 7) (Mortimore et al., 2001 and references therein). Offshore deposits of Cretaceous age are very limited (Fyfe et al., 1993).

Table 7 *Cretaceous strata in the Inner Hebrides and the Isle of Arran.*

Locality	Litho-stratigraphy	Thickness (m)	Lithologies	Underlying strata
Arran: *Central Arran Ring-complex*	?	3 (max)	limestone, white sandstone	*remanié* masses in ring-complex
Mull: *Gribun*	Gribun Chalk Fm Coire Riabhach Phosphatic Fm Loch Aline White Sandstone Fm Morvern Greensand Fm	15	chalk, white sandstone, glauconitic sandstone	New Red Sandstone to Blue Lias Triassic to Hettangian
Mull: *Torosay*	?Gribun Chalk Fm	3	chalk	Pabay Shale Fm Sinemurian
Mull: *Auchnacraig*	Gribun Chalk Fm Loch Aline White Sandstone Fm Morvern Greensand Fm	5	glauconitic sandstone	Lias Group to Bearreraig Sandstone Formation Hettangian to Bajocian
Mull: *Carsaig*	Loch Aline White Sandstone Fm Morvern Greensand Fm	13	glauconitic sandstone, ?chalk	Scalpay Sandstone Fm Pliensbachian to Toarcian
Morvern: *Beinn Iadain*	See Figure 7 for details	22	chalk, white sandstone, glauconitic sandstone	Moine psammite and pelite. Blue Lias to south Sinemurian
Morvern: *Loch Aline*	Loch Aline White Sandstone Fm Morvern Greensand Fm	20	white sandstone, glauconitic sandstone	Blue Lias to Pabay Shale Fm Hettangian to Sinemurian
Ardnamurchan: *East of Bheinn Buidhe*	-	5	*remanié* chalk flint and sand grains	In Paleocene mudstone on Moine gneisses
Ardnamurchan: *East of Ben Haint*	?Loch Aline White Sandstone Fm	1	silicified pure quartz sand	Moine psammite and pelite
Eigg: *Laig Gorge*	Strathaird Limestone Fm (Laig Gorge Sandstone Mbr)	5	chalk, sandstone and conglomerate	Kilmaluag Formation Bathonian
Eigg: *Clach Alasdair*	Strathaird Limestone Fm (Clach Alasdair Conglomerate Mbr)	<1	silicified sandstone, glauconitic sandstone	Staffin Shale Fm Oxfordian
Eigg: *Allt Ceann a' Gharaidh*	?	1.5	glauconitic sandstone, flint conglomerate	Valtos Sandstone Fm Bathonian
Skye: *Strathaird*	Strathaird Limestone Fm	3	chalk, gritty sandstone	Staffin Shale Fm Oxfordian
Skye: *Strollamus*	Strathaird Limestone Fm	3	chalk	Staffin Shale Fm Oxfordian
Skye: *Soay Sound*	Morvern Greensand Fm	5	red sandstone, glauconitic sandstone, cherty sandstone	Broadford Beds, Hettangian to Sinemurian
Skye: *Waterstein Head*	?	?	glauconitic sandstone	?Duntulm Fm Bathonian
Scalpay: *Allt Stapaig*	Strathaird Limestone Fm in part	5	sandstone, gritty calcareous sandstone	Staffin Shale Fm Oxfordian

Skye Three small outcrops of distinctive limestone, sandstone and conglomerate occur immediately below the Paleocene lavas around Beinn Leacach on the Strathaird peninsula (e.g. Hudson, 1983). Two of these outcrops include up to 2 m of blue, sandy limestone containing fragments of the bivalves *Inoceramus* and *Pecten*. The limestone, together with the underlying coarse-grained calcareous pebbly and gritty sandstones (about 1 m thick), have been assigned to the Upper Cretaceous (Turonian) Strathaird Limestone Formation. The formation also crops out south of Strollamus, where a 3 m-thick bed of white limestone overlies Oxfordian strata. On the north side of Soay Sound, up to 16 m of red and green glauconitic sandstone, capped by hard sandstone with cherty concretions, have yielded microfossils of probable Cretaceous age, and have been assigned to the Cenomanian Morvern Greensand Formation. Chaotic assemblages of clasts and slabs of Cretaceous sandstone and chert are found hereabouts; they formed in pre-Paleocene times, when Cretaceous strata collapsed into a cavity formed by solution-weathering of the underlying Lower Jurassic limestones. Glauconitic sandstones recovered from landslips below the Paleocene lava cliffs of Waterstein Head in north-west Skye are also probably Cretaceous in age.

Scalpay and Raasay A Cretaceous sequence, about 5 m thick, underlies basaltic lavas in the Allt Stapaig valley in the south of Scalpay, close to the Strollamus outcrops on Skye. The upper half of the succession consists of sandstone with a bed of chert nodules at the base. Underlying the sandstone is limestone with cherty masses and, at the base, about 0.5 m of hard calcareous gritty sandstone containing the bivalves *Exogyra* and *Pecten asper*. The lower beds may be equated with the Strathaird Limestone Formation. A thin bed of glauconitic sandstone on the western side of Dun Caan, Raasay is probably of Cenomanian age (Lee, 1920).

Eigg At Laig Gorge, about 2.5 m of conglomerate fines upwards into coarse- and fine-grained sandstones, forming the Laig Gorge Sandstone Member. Rounded pebbles in the conglomerate include quartzite, vein-quartz and phosphorite (which may contain ooidal chamosite). Hudson (1960) has recorded a derived Oxfordian ammonite, *Cardioceras* (*Scarburgiceras*) from these beds. The overlying 2.5 m-thick beds consist of sandy limestones overlain by flint-free micritic limestones. The limestones contain a planktonic fauna including the foraminifera *Hedbergella praehelvetica*, which indicates an early to mid-Turonian age. Rare fragments of *Inoceramus* bivalves also occur. At Clach Alasdair, west of the Bay of Laig, glauconitic sandstones and silicified flint conglomerates form a layer less than 1 m thick between calcareous siltstones of the Staffin Shale Formation and overlying Paleocene lavas. These beds form the Clach Alasdair Conglomerate Member and, together with the Laig Gorge succession, they are assigned to the Strathaird Limestone Formation. Their Cretaceous age has been determined from a poorly preserved microfauna (Harker, 1908). In northern Eigg, exposures in the Allt Ceann a' Gharaidh comprise about 50 cm of coarse conglomerate with pink to grey, possibly baked, flint fragments up to 10 cm in diameter, overlain by white sandstone containing rare glauconite; wedges of chalk conglomerate are present and lignite occurs towards the top of the section (Mortimore et al., 2001). These deposits overlie sandstone of the Valtos Sandstone Formation, and are capped by Paleocene lavas.

Ardnamurchan No in-situ fossiliferous Cretaceous deposits have been described from this area but remanié flint pebbles, rounded quartz grains and a shark's tooth (*Oxyrhina* (?)*mantelli*) have been found in mudstone beneath Paleocene lavas on the south-east side of Bheinn Buidhe in eastern Ardnamurchan (Richey and Thomas, 1930). A bed of highly indurated pure quartz sandstone, about 1 m thick, overlies Moine psammite and underlies red mudstone (possibly of the Beinn Iadain Mudstone Formation) and lavas, about 2 km east of Ben Hiant. This may be the equivalent of the indurated sandstone of Cretaceous age that is seen on the north side of Morvern and in the upper part of the succession at Gribun.

AGE	GROUP, FORMATION OR MEMBER		LITHOLOGY

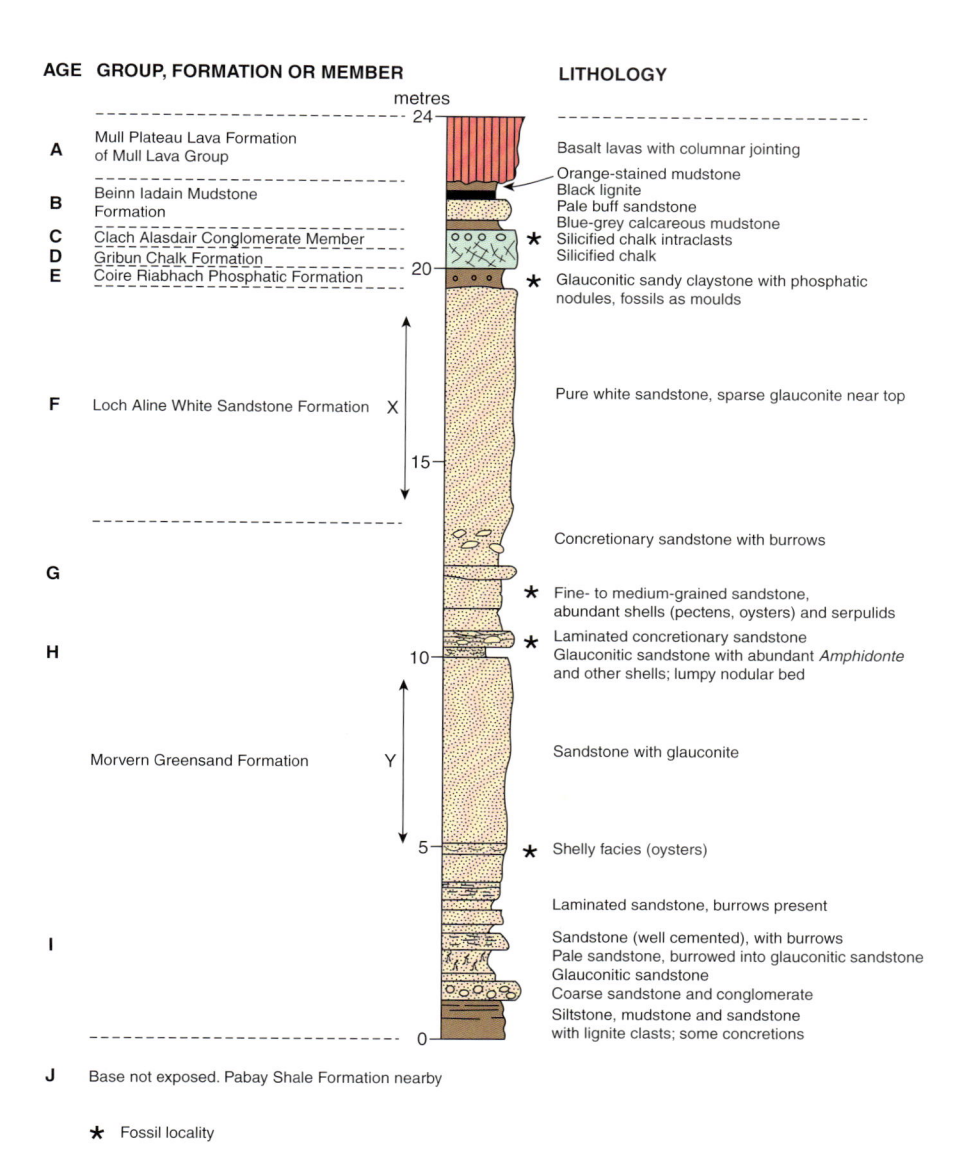

Figure 7 *Upper Cretaceous of the Inner Hebrides Group, exposed at the south-east corner of Beinn Iadain, Morvern (after Mortimore et al., 2001, fig.6.29). The thick sandstones of the Loch Aline White Sandstone and Morvern Greensand formations are obscured beneath steep grassy slopes at X and Y, respectively.*

Age of the deposits: **A** Lavas of Paleocene age; **B** Beinn Iadain Formation, generally assumed to be of Paleocene age; **D** Judd (1878) recorded Belemnitella (Campanian) from the Chalk, but the exact horizon is uncertain; **E** Sponges dated as Santonian; **F** undated, generally considered to be Turonian (but may be Cenomanian; Braley, 1990); **G** probably upper Cenomanian; **H** probably middle Cenomanian (compare with similar, better dated, site in Northern Ireland); **I** uncertain, possibly lower Cenomanian; **J** Lias and Permo-Triassic rocks seen in track section leading to Beinn Iadain

Mull   At Gribun, Cenomanian glauconitic sandstone (6 m thick) is overlain by 3 m of white sandstone and 3 m of silicified chalk with flints (possibly Santonian). The chalk is overlain by breccia containing angular pebbles and cobbles of silicified chalk, sandstone with chert nodules, and 2 m of weakly bedded, red mudstone. The beds above the chalk have been formally designated the Gribun Conglomerate Formation by Hancock (2000), but have been equated with the Clach

Alasdair Conglomerate Member and Beinn Iadain Mudstone Formation by Mortimore et al. (2001). The succession is overlain by Paleocene lavas (Plate 4). A rich bivalve fauna is known from the glauconitic sandstone, including *Rhynchonella*, *Exogyra conica*, and *Pecten asper*. *Ostrea* and *Pecten* have been found in the chalk. Glauconitic sandstone occurs at Auchnacraig, south of Loch Don, and also at Carsaig where it reaches about 13 m, its greatest thickness in the district.

Morvern Siliceous and glauconitic sandstones, characterised by species of *Exogyra* and *Pecten* bivalves, underlie the Paleocene lavas of Morvern. At the Lochaline mine there are excellent exposures of a white sandstone (about 15 m thick), formed from pure quartz sand that is generally poorly cemented, except in hard, silica-cemented bands (see above). The locality is an important source of high-grade glass-sand (p.174). The thickest succession is on Beinn Iadain, where the sandstones are overlain sequentially by about 1 m of silicified chalk, the Clach Alasdair Conglomerate with intraclasts of silicified chalk, and clastic sedimentary rocks with lignite of the Beinn Iadain Mudstone Formation (Figure 7).

Arran Rocks of Cretaceous age on Arran are restricted to chalk with flints and white sandstone contained in several *remanié* masses within the Central Arran Ring-complex. Microfossils and fragments of *Inoceramus* bivalves recovered from these rocks indicate a Late Cretaceous age (Tyrrell, 1928). Most of the limestone is thermally altered but some pieces are merely indurated and resemble the chalk of Antrim.

four
Pre-Palaeogene structure

The islands and seaboard of western Scotland record a long and complicated geological history prior to the Palaeogene and the structures developed during this period appear to have influenced Palaeogene tectonic and igneous events to a considerable degree. Two early structural elements dominate the area: firstly, the thrusts and faults that formed during the Early Palaeozoic Caledonian Orogeny and which define the main terrane boundaries (Figure 1); secondly, the generally north-south-trending sedimentary basins (troughs) filled with Late Palaeozoic and Mesozoic sedimentary rocks together with the intervening ridges that are largely made of Precambrian rocks (Figures 3; 5). It is therefore convenient to consider the pre-Palaeogene structure of the district in two parts: events prior to the Late Palaeozoic, and from the Late Palaeozoic to the end of the Mesozoic.

EARLY PALAEOZOIC AND OLDER STRUCTURES

The intense deformations associated with the Caledonian Orogeny in Scotland were responsible for a number of notable features in the district. In eastern and southern Skye, the Moine Thrust Belt separates rocks of the foreland, Hebridean Terrane (Lewisian gneiss, Torridonian and Cambro-Orodovician sedimentary rocks) from rocks of the Moine Supergroup to the east (Figure 3). South of Broadford, movements on the Ben Suardal Thrust Plane carried Torridonian sandstones over Ordovician limestones and dolostones at Beinn Suardal. The Moine Thrust has been proved as far south as the Point of Sleat, Skye, but farther south it is either concealed beneath Mesozoic rocks of the Inner Hebrides Basin or has been removed by erosion. It has been suggested that the Moine Thrust Belt may occur beneath the Ross of Mull, with the Lewisian Gneiss Complex and Iona Group rocks of Iona being separated from the Ross of Mull Pluton by a steep normal fault with a downthrow to the east-south-east (Potts et al., 1995).

Widespread uplift and deep erosion of the Lewisian basement occurred prior to and during deposition of the Torridonian strata, the basal members of which blanket irregular palaeosurfaces in northern Raasay and within the Rum Central Complex. The evidence for uplift, tilting and erosion of the Torridonian strata before the deposition of the Cambro-Ordovician rocks, so clearly demonstrated in the North West Highlands, is largely obscured on Skye by tectonic contacts within the Moine Thrust Belt. To the south, steeply dipping, north-striking Moine metasedimentary rocks, which were intensely deformed during the Caledonian Orogeny, are invaded by intrusions of the Ardnamurchan Central Complex. Still farther south, the Great Glen Fault and the Highland Boundary Fault, which form two of the major terrane boundaries in Scotland, were active during the Palaeozoic; the Great Glen Fault was subsequently subjected to further significant movement (p.148).

On Mull, the Great Glen Fault is readily defined between Craignure and north-west of Loch Spelve, where Moine rocks crop out, and Loch Don, where phyllite and limestone of the Dalradian Appin Group occur. The volcanic breccias of Centre 1 in the Mull Central Complex provide further proof of this fault in eastern Mull. Moine clasts are common in the breccias, whereas there are none of Dalradian lithologies. South of Loch Don, the fault departs somewhat from its north-east-trending course, deviating to the south-east along Loch Spelve and Loch Buie, where it has been deflected by the Mull Central Complex.

The Highland Boundary Fault and the associated Highland Border Complex cross Scotland from Stonehaven in the east to Bute in the west. In northern Arran, where Dalradian rocks crop out extensively, the pillow lavas, cherts and mudstones in North Glen Sannox are assigned to the Highland Border Complex. Various faults in north-east Arran have been proposed as continuations of the Highland Boundary Fault, for example the Corloch Fault and the fault at the eastern edge of the North Arran Granite Pluton, exposed in the White Water above Corrie. On the south side of the granite, the contact between Dalradian rocks and Devonian

(Old Red Sandstone) sandstones and conglomerates was also once regarded as the continuation of the Highland Boundary Fault, deflected during forcible emplacement of the granite pluton (Tyrrell, 1928). Subsequently, however, the Devonian strata have been shown to lie unconformably on the Dalradian rocks south of the granite (Friend et al., 1963) and so movement on these faults appears to have been at least in part contemporaneous with granite emplacement (England, 1992b). Remnants of the Highland Boundary Fault are present in the North Glen Sannox area, but the course of the fault to the south-west is unclear. There is no evidence of significant movement on this fault on Arran during the Palaeogene, or subsequently. Folding and associated cleavage development in metasedimentary rocks of the Dalradian Supergroup predates folds imposed during emplacement of the Paleocene granites of North Arran (p.151; Figure 32).

LATE PALAEOZOIC TO MESOZOIC STRUCTURES

The Sea of the Hebrides and Inner Hebrides basins, or troughs, are amongst the most striking pre-Palaeogene structures in the district (Figure 5). On Skye, there is unequivocal evidence that the main movement on the Camasunary–Skerryvore Fault, which forms part of the western boundary of the Inner Hebrides Basin, occurred prior to the eruption of the Paleocene lavas. At Camasunary, Middle Jurassic rocks are downfaulted by as much as 200 m against Lower Jurassic and Torridonian strata to the west, but there is only minor displacement of the Paleocene lavas that overlie both. To the south, the thin Lower Jurassic succession on Rum is separated from the thick Middle Jurasic rocks of Eigg by the southern continuation of the Camasunary–Skerryvore Fault. Farther south, for example east of Coll, the proven movements on this fault postdate the Paleocene lavas. On Skye, the northern continuation of the Camasunary–Skerryvore Fault is disturbed and largely obscured by the central complex, which may have caused a westerly offset of the fault. Splays of the fault may continue on Raasay as the Screapadal Fault (Figure 5) and as a major fault in south-east Scalpay.

On Rum, the general moderate west-north-west to north-west dip of the Torridonian and Triassic strata predates the emplacement of the Paleocene central complex. A major fault displacing Torridonian strata on Bloodstone Hill predates Paleocene lavas and intrusions, and a monoclinal structure in the Triassic rocks of north-west Rum was probably synsedimentary in origin (R J Steel *in* Emeleus, 1997). It is likely that the major north-trending Long Loch Fault on Rum had considerable pre-Paleocene movement, and on Eigg movement on the normal Laig Gorge Fault affects the Mesozoic rocks in the opposite sense to that found in the overlying Paleocene basaltic lavas. Throughout the district there is evidence of a period of uplift prior to deposition of the Upper Cretaceous strata. Small, scattered outcrops of these rocks overlie a wide variety of older rocks that range in age from Proterozoic to Jurassic (Table 7). Further, there is widespread evidence that uplift and denudation at the close of the Cretaceous, or very early in the Paleocene, gave rise to an undulating land surface that may have had significant local relief. Hence, the earliest Paleocene lavas and the associated sedimentary strata rest on a variety of older rocks and some of the Paleocene clastic sediments were clearly derived by vigorous contemporaneous erosion.

five
Palaeogene igneous geology: regional setting

The Palaeogene rocks that comprise the lava fields, dyke swarms, sill-complexes and central complexes of the Hebridean Igneous Province are part of the much larger, North Atlantic Igneous Superprovince (Figure 2). The magmatism was the consequence of a large mantle plume, which first impacted on the base of the continental lithosphere towards the end of the Cretaceous Period. Magmatism continues today on Iceland and along the Mid-Atlantic Ridge but represents a much smaller, less productive stage in the evolution of the plume.

Seismic reflection and refraction studies have revealed that the north Atlantic Ocean between north-west Europe and Greenland comprises thick sequences (up to 6000 m) of seaward-dipping reflectors. These are attributed to extrusive igneous rocks, erupted at the time of rifting and the initiation of ocean floor spreading (White, 1988). Much of the volcanic activity occurred above sea level, because of the dynamic support provided by the plume (White and Mackenzie, 1989). This is demonstrated by the presence of inter-lava palaeosols recovered from deep-sea boreholes (e.g. Viereck et al., 1988). In addition, large amounts of igneous rock, up to 1500 m thick, were underplated during the initial stages of rifting. The landward equivalents of the offshore lavas include the thick and laterally continuous flood basalt lavas of west and east Greenland, and the less extensive lava fields of the Faroe Islands, north-west Scotland and north-east Ireland.

The vast majority of the lava sequences, both offshore and onshore, were erupted subaerially, predominantly from fissure-type feeders, now represented by laterally continuous linear dyke swarms. The north-west-trending dyke swarms that are associated with the Skye and Mull lava fields of western Scotland developed almost at right angles to the rift margin and indicate contemporaneous north-east–south-west extension of the north-west European margin.

The central complexes represent sites where magma emplacement into the upper continental crust along the rift margin reached anomalously large proportions. Along the north-west European continental margin, the central complexes occur within a relatively narrow, north-trending zone about 40 km in width (Figure 1). The zone can be traced from Lundy, in the Bristol Channel, north through the central complexes in north-east Ireland to those that occur along the western coast of Scotland. The St Kilda archipelago is formed by a central complex that lies to the west of this zone. It is most appropriately grouped with the many submerged central complexes now recognised to the west of Shetland, which occur on the south-east side of the Rockall Trough and within the Faroe–Shetland Basin (Ritchie and Hitchen, 1996; Figure 2). In broad terms, the continental crust was thinned (stretched) during the initial stages of rifting and associated volcanic activity (compare with Meissner et al., 1986) in a north–south zone running through western Britain and north-east Ireland, and also farther to the west. Crustal extension permitted access to the surface for mantle-derived magmas; some were erupted directly to the surface while others ascended in stages, ponding at the mantle-crust boundary or in reservoirs at various levels within the crust (Morrison et al., 1985).

The central complexes of western Scotland appear to be sited where older Palaeozoic (Caledonian) or Mesozoic structures cross the north–south zone. For example, the North Arran Granite Pluton occurs on the trace of the Highland Boundary Fault, the Mull Central Complex occurs on the trace of the Great Glen Fault, and the Skye and Rum central complexes are situated on, or close to, the Camasunary–Skerryvore Fault. Typically, the central complexes occur within areas of basement rocks exposed in the elevated footwalls of Mesozoic faults (Roberts and Holdsworth, 1999; Figure 3).

The lava fields of western Scotland are thickest within fault-bound basins containing syn-rift terrigenous Triassic sedimentary rocks (up to 500 m in thickness) overlain by marine to brackish Jurassic sandstones, limestones and mudstones (Binns et al., 1974; Walker, 1979). The

Skye Lava Field, comprising the thick sequences of flows preserved on Skye and Canna, is contained within the Sea of the Hebrides Basin, bounded to the north-west by the Minch Fault. The sedimentary rocks of the basin have a feather-edge against Lewisian and Torridonian rocks of the adjoining Skye–Rum Horst. To the south, the Mull Lava Field is contained within the Inner Hebrides Basin, bounded to the north-west by the Camasunary–Skerryvore Fault (Figure 3), with Precambrian rocks of the Skye–Rum Horst forming the footwall. The fault displaces lavas of the Eigg Lava Formation and Mull Lava Group but not those of the younger Skye Lava Group. The south-east limit of the Inner Hebrides Basin was not fault-controlled, and there lavas of the Mull Lava Field overstep Jurassic strata and thin remnants of overlying Cretaceous rocks to lie directly upon basement rocks of the Moine Supergroup north and south of Loch Sunart.

The Mull Lava Field is considered to extend north to include the thin sequence of lavas preserved in the Ben Hiant area on the south side of the Ardnamurchan peninsula. The lava sequences of Muck and Eigg also appear to form a northern extension to the Mull lavas (Fyfe et al., 1993, fig. 40) and may be contemporaneous with the earlier part of that succession. The basaltic lavas on Eigg are capped by a thick rhyodacitic pitchstone lava, which was erupted into a palaeovalley eroded into the earlier flows. The rhyodacite lava is considered to be the youngest flow preserved onshore in the Hebridean Province (Figure 8).

The timing of emplacement of the linear regional dyke swarms relative to the lava fields is difficult to deduce directly from field relationships. There are few recorded examples of dykes feeding flows, and on Skye, Mull, Muck, Eigg and in south-east Rum intense dyke swarms cut the lavas. Furthermore, on Skye, Rum and Mull, earlier members of the central complexes are commonly intruded by numerous north-west-aligned dykes. However, from detailed mapping within the Hebridean Igneous Province (e.g. Williamson and Bell, 1994) and throughout the North Atlantic Igneous Superprovince (Upton, 1988), it is likely that the vast majority of the lavas were erupted from fissure systems, now represented by dykes, and from plugs developed on the dykes (Kerr, 1997). The magmas were also intruded to form sill-complexes, which were preferentially developed in the sedimentary rocks of the Mesozoic basins. Only towards the ends of the individual pulses of igneous activity were lavas possibly erupted from more-localised volcanic edifices overlying the central complexes.

The age of the lavas in relation to the central complexes can be deduced reasonably easily in some places. Clear intrusive contacts between members of the central complexes and earlier lavas are quite common, and are accompanied by pronounced high-grade thermal and hydrothermal aureoles. Conversely, where a lava field developed subsequent to the emplacement of a central complex, the interflow sedimentary deposits may contain material derived from the complex, if it was unroofed and eroded during the construction of the lava pile.

TIMING AND IGNEOUS STRATIGRAPHY

The age relationships of the rocks within this province have not always been readily resolved by radiometric and other age data (compare with Mussett et al., 1988). However, the relative ages are now becoming clearer through a combination of more precise radiometric dating techniques, palynological dating of interflow sedimentary rocks, and palaeomagnetism (e.g. Pearson et al., 1996; Bell and Jolley, 1997; Hamilton et al., 1998; Chambers and Pringle, 2001; Chambers et al., 2005). Published precise radiometric age determinations and biostratigraphical ages based upon palynofloral assemblages are summarised, together with other data, in Table 8 and Figure 8. The lavas of the province span an interval of at least three million years, from about 61 Ma to at least 58 Ma. The interval corresponds to Chron 26R of the magnetic time scale (Berggren et al., 1995). All of the Hebridean onshore central complexes appear to have been initiated during this interval, and it is apparently only in Skye, and possibly Mull, that significant activity occurred at later dates.

There is only one precise radiometric age determination for the Skye Lava Field, although there is a considerable body of palynological data. The Portree Hyaloclastite Formation

Table 8 Radiometric and palynological age determinations on Palaeogene rocks from the Hebridean Igneous Province. At the time of compilation, the radiometric and palynological data were not fully reconciled (Jolley et al., 2002). Ages are in million years (Ma).

Mull Central Complex			
Loch Ba Felsite, Centre 3	58.5 ± 0.1	U-Pb	*
	58.48 ± 0.18	Ar-Ar	6
Corra Bheinn Gabbro, Centre 2	58.3	U-Pb	*
Mull Lava Group			
Staffa Lava Formation	55–54.5	Palynology	3
	60.56 ± 0.29	Ar-Ar	6
Mull Plateau Lava Formation	58.35 ± 0.19	Ar-Ar	6
Trachytic tuff, Calgary Bay	60.57 ± 0.24	Ar-Ar	#
	60.5 ± 0.3	U-Pb	*
Skye Central Complex			
Eastern Red Hills Centre			
Beinn an Dubhaich Granite	55.89 ± 0.15	U-Pb	*
Pitchstone dyke cutting Beinn na Caillich Granite	55.7 ± 0.1	U-Pb	*
Western Red Hills Centre			
Marsco Granite	58.4 ± 2.1	U-Pb	*
Southern Porphyritic Granite	57 ± 0.5	U-Pb	*
Loch Ainort Granite	58.58 ± 0.13	Ar-Ar	6
Srath na Creitheach Centre			
Ruadh Stac Granite	57.5	Ar-Ar	5
Cuillin Centre			
Coire Uaigneich Granite	59.3 ± 0.7	Rb-Sr	4
Pegmatitic facies of Outer Gabbro, A1.	58.91 ± 0.08	U-Pb	2
Skye Lava Group			
Conglomerate underlying Talisker Formation, and other sedimentary rocks within Skye Lava Group	58.0–58.25	Palynology	3
Sleadale Trachytic Tuff, Skye Lava Group, below Preshal Beg Conglomerate Formation	58.91 ± 0.1	Ar-Ar	8
Canna Lava Formation			
Hawaiites	60.00 ± 0.23	Ar-Ar	7
Conglomerates within lavas, Rum and Canna	58.0–58.25	Palynology	3
Arran			
North Arran Granite Pluton	58.5, 58.4	Ar-Ar	5
Ardnamurchan Central Complex			
Great Eucrite, pegmatitic gabbro, Centre 3	59.05	Ar-Ar	5
Tonalite, Centre 3	58.6 ± 0.2	U-Pb	*
Rum Central Complex			
Alkaline segregation in Central Intrusion	60.53 ± 0.08	U-Pb	2
Biotite, pegmatitic facies of Western Layered Intrusion	60.1 ± 1.0	Ar-Ar	2
Western Granite	60.01 ± 0.45	Ar-Ar	7
Eigg Lava Formation			
Sanidine in tuff near base of formation on Muck	60.65 ± 0.07	Ar-Ar	7,1
Zircon in same tuff	61.15 ± 0.25	U-Pb	7
Sgurr of Eigg Pitchstone Formation			
Pitchstone lava	58.72 ± 0.07	Ar-Ar	7

All radiometric age determinations quoted at the 2 sigma level, except for references 5 and 6 which are at the 1 sigma level.
References: 1 Pearson et al., 1996; 2 Hamilton et al., 1998; 3 Jolley, 1997; 4 Dickin, 1981; 5 Chambers, 2000; 6 Chambers and Pringle, 2001; 7 Chambers et al., 2005; 8 Bell and Williamson, 2002
* unpublished U-Pb analyses by M A Hamilton, Jack Satterly Geochonology Laboratory, Department of Geology, University of Toronto
# unpublished Ar-Ar analysis by S P Kelley (Open University)

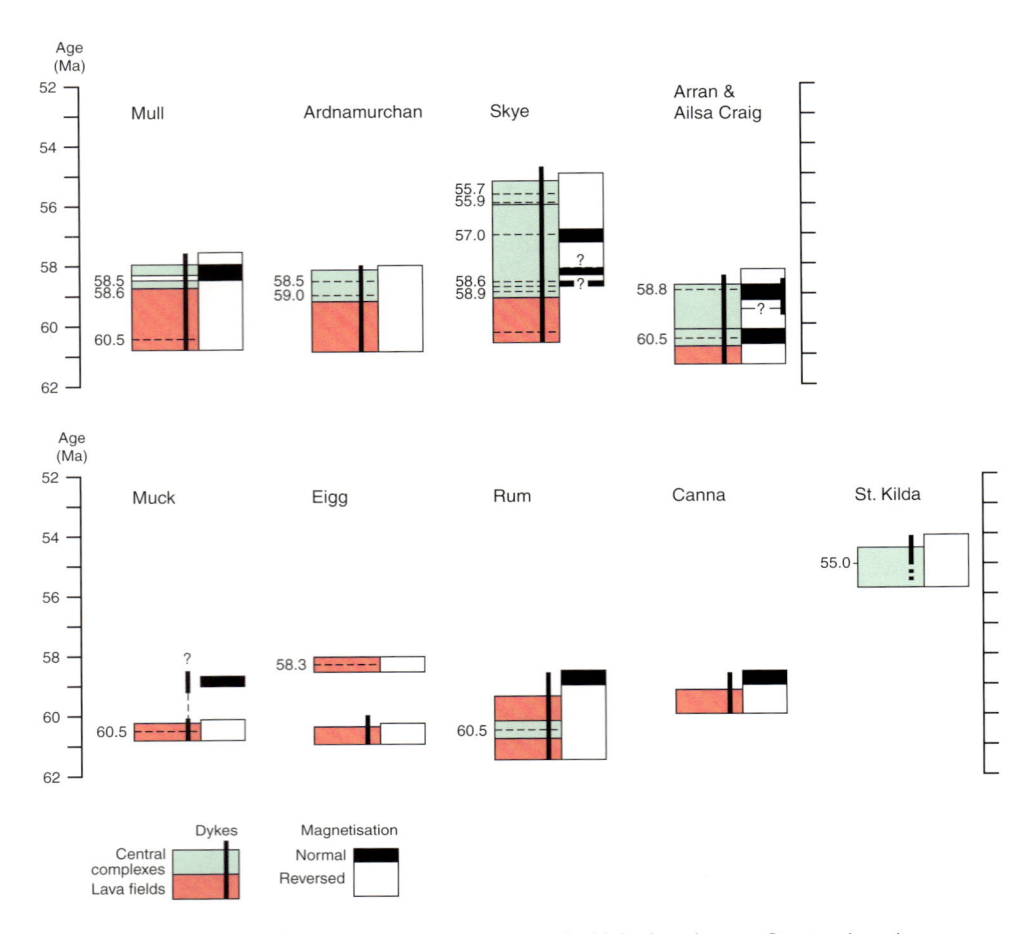

Figure 8 Time span of Palaeogene igneous activity in the Hebridean Igneous Province based on Mussett et al., 1988, with additional radiometric age determinations by the U-Pb and Ar-Ar techniques (see Table 8). Reversed and normal polarities are assigned to 26N and 26R except for the Skye Central Complex, which may extend to 24R.

(Palagonite Tuffs of Anderson and Dunham, 1966) at the base of the Skye Lava Group and the Preshal Beg Conglomerate Formation near the top of the group contain pollen assemblages that indicate an age range of 58.23 to 58.0 Ma. (Jolley, 1997), and an Ar-Ar age of 58.91 ± 0.18 Ma has been obtained from the trachytic Sleadale Tuff beneath the Preshal Beg Conglomerate (Bell and Williamson, 2002). Petrologically distinct lavas of the Talisker Formation, which directly overlie the Preshal Beg conglomerates, are the youngest lavas within the Skye Lava Field (p.65). Although the Talisker lavas and the Preshal Beag conglomerates are separated from the underlying lavas by a clear erosional break, which coincides with a significant change in magma-type, the time interval represented was probably short (Hamilton et al., 1998).

The lavas of the Canna Lava Formation preserved in north-west Rum, which belong to the Skye Lava Group (Williamson and Bell, 1994; Emeleus, 1997), are demonstrably younger than the Western Granite, one of the earliest members of the Rum Central Complex. The lavas rest upon a weathered surface of the granite, and interstratified conglomerates on Rum, Canna, Sanday and south-west Skye contain clasts derived from the granite and from other, younger, members of the Rum Central Complex. Thus, the lavas were erupted in the interval of less than 1.6 million years, between the emplacement and unroofing of the Rum Central Complex (60.5–60.1 Ma) and the emplacement of the gabbros of the Cuillin Centre on Skye (58.9 Ma), which intrude the Skye Lava Group.

In the Small Isles, the Sgurr of Eigg Pitchstone Formation is one of the latest eruptive events in the Hebridean Province (58.7 Ma), whereas the tuffs at the base of the Eigg Lava Formation on Muck have given the oldest ages (60.6 Ma). Neither formation has yet yielded useful material for palynological dating, although plant remains are well known from conglomerates beneath the Sgurr of Eigg pitchstone (p.77).

The Mull Lava Field is invaded by the Mull Central Complex. Radiometric age determinations and geochemical signatures (Chambers and Pringle, 2001) indicate that construction of the Mull and Skye lava fields may have been more or less contemporaneous. This is, however, difficult to reconcile with the palynological evidence. The Mull flora constitutes a widely recognised, well-defined and short-lived assemblage that flourished in warm, subtropical conditions, the late Paleocene thermal maximum, and is completely different from the older flora obtained from Skye and the Small Isles (Jolley, 1997). The late Paleocene thermal maximum is dated at 54.98 Ma on the geological time scale of Berggren et al. (1995), but on Mull rocks containing the diagnostic pollen flora are overlain by lavas that have yielded isotopic U-Pb and Ar-Ar dates between 57.5 and 60.54 Ma (e.g. Chambers and Pringle, 2001). Furthermore, within the post-lava Mull Central Complex, the Corra-bheinn Gabbro and Loch Bà Felsite intrusions are dated by the U-Pb method at 58.3 and 58.5 Ma, respectively (Hamilton et al., 1998).

The Skye Central Complex contains a large number of intrusive units, the oldest of which are various layered gabbroic and ultrabasic rocks, together with a number of confluent cone-sheets that comprise the Cuillin Centre. A U-Pb age of 58.9 Ma was obtained from a gabbroic pegmatite formed early in the development of this centre (Table 8). Amongst the cone-sheets is a relatively late-stage suite of dolerites and basalts, which were emplaced at very shallow depths, and are compositionally identical to flows of the Talisker Formation (Walker, 1993b; Bell et al., 1994). It may therefore be inferred that the later stages in the formation of the Cuillin Centre, which represents the eroded roots of a basaltic central volcano, were synchronous with and relate to the outpouring of the distinctive flows of the Talisker Formation. The Talisker lavas flowed onto a land surface formed from older and compositionally different formations of the Skye Lava Group. There is a close correspondence between the radiometric age for early activity in the Cuillin Centre and that for the eruption of the trachytic tuff underlying the Preshal Beg Conglomerate Formation, and the biostratigraphical age of the Preshal Beg Conglomerate Formation. Consequently, the Cuillin Centre was developing contemporaneously with at least part of the Skye Lava Field. The younger granite centres of the Skye Central Complex yield ages that range from about 58.5 to 57 Ma for intrusions of the Srath na Creitheach and Western Red Hills centres, to about 56 Ma for the Beinn an Dubhaich Granite and a pitchstone dyke that intrudes the nearby Beinn na Caillich Granite, both belonging to the Eastern Red Hills Centre (Figure 8). These ages suggest that the growth of the Skye Central Complex may have spanned up to three million years, compatible with the complex palaeomagnetic record (Mussett et al., 1988).

The Ardnamurchan Central Complex was emplaced into basaltic lavas. No reliable age data are available for these hydrothermally altered flows, although those around Ben Hiant, along the eastern side of the central complex, are considered to be an outlier of the Mull–Morvern Lava Field. Radiometric age determinations on the Great Eucrite and tonalite intrusions of Centre 3 give ages of about 59 Ma.

The radiometric age of 60.6 Ma obtained from the oldest lavas in the Eigg Lava Formation is similar to those from rocks near the base of the Mull Lava Field (Table 8). Investigations of the offshore geology between western Mull, Ardnamurchan and the Small Isles show that the lavas of Mull are linked to those of Muck and Eigg (Fyfe et al., 1993, fig. 40).

The episodes of lava eruption in the Hebridean Igneous Province are summarised in Figure 8 and appear to constitute a few discrete pulses within a period of less than three million years. However, the age and duration of activity within the central complexes are currently less well constrained. The Rum Central Complex postdates the Eigg Lava Formation but predates the Skye Lava Group. The majority of the Skye Central Complex most likely postdates the Skye

Lava Group, but the youngest Talisker Lava Formation may have been erupted from a central volcano now represented by parts of the Cuillin Centre. The Mull Central Complex postdates the Mull Lava Group. The St Kilda Central Complex, the Blackstones Central Complex, the North Arran Granite Pluton, the Central Arran Ring-complex and the Ailsa Craig Granite have all been dated by radiometric techniques, but some have significant errors. They also have the disadvantage of not being associated with lavas to offer a precise comparison with pollen and spore age data.

Evidence that the plume-driven igneous activity in the Hebridean Igneous Province occurred sporadically rather than continuously may be provided by the temporal spacing of detritus fans in offshore Palaeogene sedimentary rocks (White and Lovell, 1997). The detritus fans resulted from increased uplift and erosion in the sedimentary source areas, which may, in turn, have been caused by pulses of magmatic underplating from a mantle plume. The underplating may also have triggered the igneous activity, as the correlation between major periods of fan deposition and maxima in igneous activity in the Hebridean Igneous Province is quite striking. The very clear break in magmatism between emplacement of the Rum Central Complex and accumulation of the Canna Lava Formation could be one result of pulsed activity.

six

Palaeogene lava fields and associated sedimentary rocks

Two main lava fields are preserved in western Scotland: the Mull Lava Field (Bailey et al., 1924; Kerr, 1995b) and the Skye Lava Field (Anderson and Dunham, 1966; Williamson and Bell, 1994). The lavas of Canna, Sanday and north-west Rum, comprising the Canna Lava Formation, are considered to be outliers of the Skye Lava Field (Emeleus, 1997). A sequence of somewhat older flows, termed the Eigg Lava Formation, is preserved on the islands of Eigg and Muck and on the south-east part of Rum (Emeleus, 1997). These older lavas may be outliers of the Mull Lava Field (see p.69); in the same area, a much younger glassy porphyritic rhyodacite flow comprises the Sgurr of Eigg Pitchstone Formation (Emeleus, 1997). The distribution of the lavas is shown in Figure 9 and the various successions are summarised in Table 9.

Eruption of the lavas and various pyroclastic materials was predominantly subaerial, although during the initial stages of the volcanic activity, the contemporaneous land surface had local lakes and river systems, leading to rare eruptions of magma into water to produce pillow lavas and hyaloclastite breccias. The land surface was essentially a peneplain, but with some relief. Locally, higher ground was associated with the Moine Supergroup in the south-west of Mull and eastern Morvern, and with the Torridonian strata in central Skye and around the Rum Central Complex (Emeleus, 1985; 1997; Williamson and Bell, 1994). On Rum, the relief probably increased as the central complex developed. Sedimentary beds within the early, Eigg Lava Formation are virtually free of clasts above granule grade, whereas cobbles and boulders derived from the Rum Central Complex and its surroundings are abundant in the conglomerates interbedded with lavas of the younger Canna Lava Formation. It is likely that a similar situation pertained in Skye as the Skye Central Complex developed.

The climate during the Palaeogene, as indicated by the palynoflora, was generally warm temperate, and marked a period of significant cooling following the tropical conditions of the Late Cretaceous (e.g. Jolley, 1997). Where Upper Cretaceous (Maastrichtian) chalk is still preserved below the lavas, for example in west and south-west Mull and on the Morvern peninsula, residual deposits of 'clay with flints' are found together with a karstic surface with a relief of up to several metres. Elsewhere, further evidence of pre-lava relief is provided where Palaeogene, coal-bearing strata and channel-fill conglomerates typical of fluviatile environments are preserved locally within small pockets between the lavas.

The initial phases of volcanic activity in both the Skye and Mull lava fields were marked by the eruption of basaltic scoria from fissure-controlled cones, and by the extrusion of lavas into shallow water with the formation of hyaloclastite deposits (Plate 9). Magma was also intruded as shallow sills into unconsolidated, water-saturated sediments (mainly silt and mud) of Palaeogene age. On Mull, the presence of basaltic pillow lavas has been used to infer the presence of a water-filled caldera during the development of a later central volcano (Bailey et al., 1924).

Pauses in the predominantly subaerial volcanic activity are marked in the lavas by the development of weathered tops that pass up into relatively thick and laterally continuous red (oxidised) palaeosols (boles). During these periods, lacustrine sediments (clay, silt and peaty mire) accumulated in shallow depressions on the land surface, together with fluviatile channel-fills that formed condensed sequences of conglomerate, sandstone, siltstone and coal. These sedimentary sequences have been used to subdivide the lavas of Skye and Rum into a number of coherent formations (Anderson and Dunham, 1966; Emeleus, 1985, 1997; Williamson and Bell, 1994; Tables 11; 12; 14). On Mull and Morvern, the scarcity of sedimentary beds or other

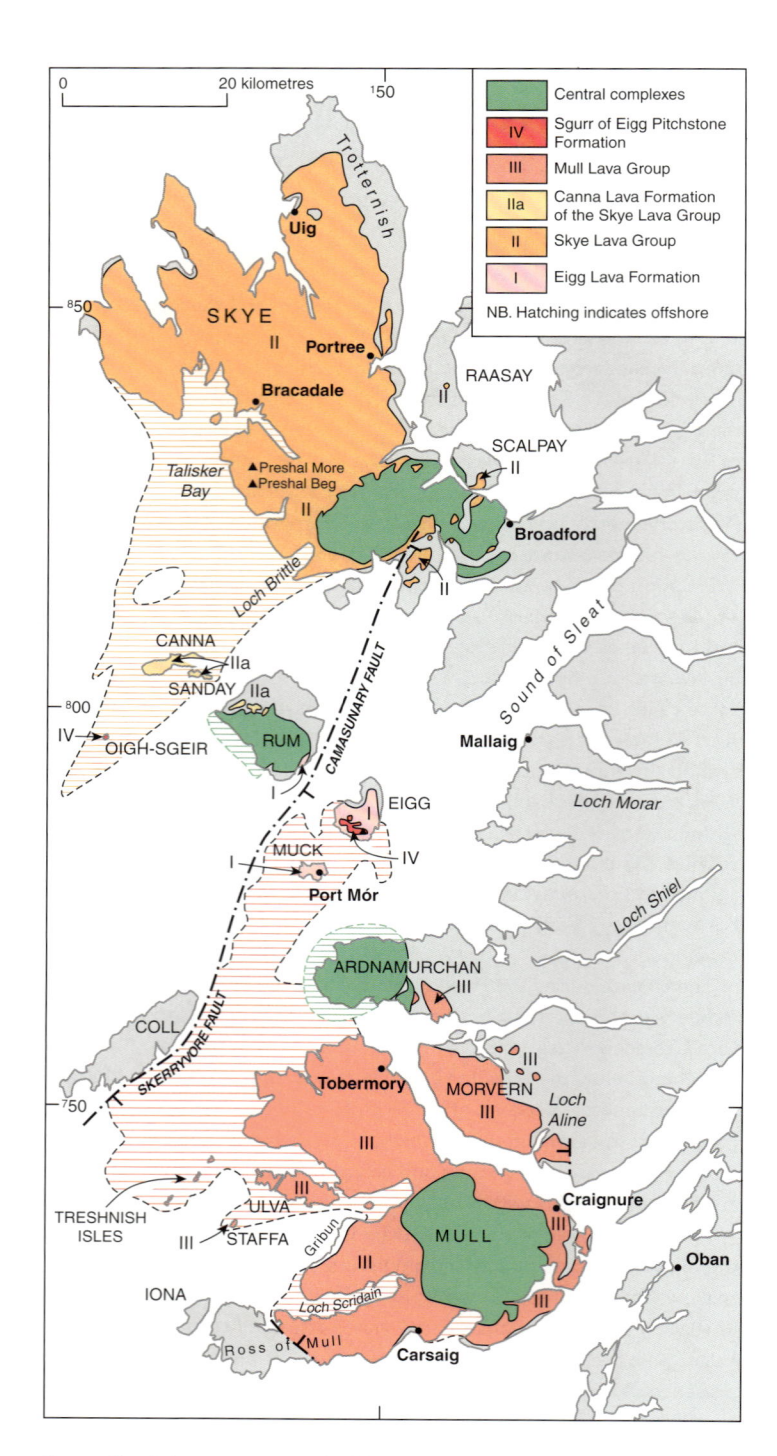

Figure 9 *Paleocene lavas in the Hebridean Igneous Province.*
Not shown are masses of lavas (mainly basaltic) enclosed within the central complexes. In addition, extensive lavas occur offshore west of Lewis and north-west of St Kilda, and small masses are present within the Central Arran Ring-complex.

Table 9 Paleocene lava fields in the Hebridean Igneous Province.

Lava field	Sector	Formations	Members and other notable subdivisions
Skye	North [1]	Osdale Bracadale Beinn Totaig Ramasaig Beinn Edra Portree Hyaloclastite	Glen Osdale conglomerates and siltstones Trotternish Shale
	West-central [2,3]	Talisker Preshal Beg Conglomerate Gleann Oraid Loch Dubh Eynort Mudstone Fiskavaig Glen Caladale Cruachan Minginish Conglomerate Bualintur Rubh' an Dùnain	 Sleadale, Cnoc Scarall, Arnaval Loch an Sguir Mhoir, Biod Mór, Ben Scaalan Rubha nan Clach, McFarlane's Rock Skridan, Sgurr Buidhe, Stac a'Mheadais, Tusdale Glen Brittle, An Cròcan Culnamean, Allt Geodh' a'Ghàmhna, Allt Mór Creag Mhór, Meacnaish, An Leac
	North-central [3]	Fionn Choire Roineval Coire nan Circe Sligachan	 Meadale, Carbost Conglomerate, Harport Am Mam, Coire Daraich, Sron a' Ghrobain, Allt Dearg Meall Odhar, Cnoc an t-Sithean
	South-central [4]	Strathaird	Upper Lava, Picrite Basalt, Ophimottled Basalt, Basal Lava
	Canna/ Sanday [5]	Canna Lava	Eilean a'Bhàird, Beul Làma Sgorr, Carn a'Ghaill, Cnoc Bhrostan, Tighard
	NW Rum [5]	Canna Lava	Orval, Guirdil, Upper Fionchra, Lower Fionchra
Eigg	Eigg [5] Eigg [5]	Sgurr of Eigg Pitchstone Eigg Lava	 Cora-bheinn, Cnoc Creagach, Brutach Dearg, Glacan Dorchadois, Gleann Charàdail, Laig
	SE Rum [5]	Eigg Lava	undivided
	Muck [5]	Eigg Lava	Beinn Airein–An Stac, Port an t-Seilich, Basal
Mull	Mull [6]	Mull Central Lava Mull Plateau Lava Staffa Lava	Undivided Ben More Pale, Ben More Main Quinish Lava, Geochemical subdivisions [7] Ardtun Conglomerate and equivalents MacCulloch's Tree/Fingal's Cave/Carraig Mhór Flow Undivided sequence in Ross of Mull Malcolm's Point Conglomerate Gribun Mudstone and equivalents
	Ardnamurchan and Morvern [8]	Mull Plateau Lava	undivided

1 Anderson and Dunham (1966); 2 Williamson and Bell (1994); 3 British Geological Survey, 2000, Minginish, Scotland. Sheet 70, Solid and Drift Geology 1:50 000 Series; 4 Almond (1964);
5 Emeleus (1997); 6 Bailey et al. (1924); 7 Kerr (1995a,b); 8 Richey and Thomas (1930)

Plate 9 Hyaloclastite breccia in the Portree Hyaloclastite Formation, Skye Lava Group. Fiurnean, northern Skye. Scale: hammer shaft is 55 cm (P580460).

distinctive stratigraphical markers has so far limited subdivision of the lavas (Table 15). The conglomerates within the lava sequences also provide provenance information for the periods during which active erosion occurred. Locally, thick accumulations of debris formed during construction of the lava fields, for example at Maclean's Nose, south of Ben Hiant on the Ardnamurchan peninsula (Richey and Thomas, 1930), and at Compass Hill on the east side of Canna (Harker, 1908; Emeleus, 1997).

The original limits of the lava fields are not easily estimated. However, Preston (1982) pointed out that the coastal cliffs, up to 300 m high, which mark the onshore truncation of the lava piles on, for example Skye and Mull, taken together with the horizontal attitude and thickness of the lava piles, suggests that they extended at least 50 km beyond their present-day limits. Extensive areas of basaltic lavas are known to exist beneath the sea south-west of Skye and west and north-west of Mull (Figure 9).

PETROGRAPHY OF THE LAVAS

The lava fields are made up predominatly of basaltic flows with a lesser quantity of flows showing more evolved compositions, ranging from hawaiite and tholeiitic basaltic andesite to trachyte and rhyodacite. The general petrographical and mineralogical characteristics of lavas of the Skye and Mull fields, which have been summarised by Williamson and Bell (1994), Bell and Williamson (1994) and Kerr (1995a), and those of the Small Isles by Emeleus (1985, 1997), are outlined in Table 10. The geochemistry of the lavas is discussed in Chapter 10.

FIELD CHARACTERISTICS OF THE LAVAS

A variety of field and hand specimen characteristics have been identified for each of the main lava rock types (e.g. Williamson and Bell, 1994) and these generally permit field identification without immediate recourse to microscopic examination.

Table 10 *Mineralogical and petrographical characteristics of the principal lava types of the Hebridean Igneous Province.*

Lava type	Mineralogy and petrography
Alkali basalt and transitional olivine basalt (both commonly mapped as olivine basalt)	Phenocrysts and microphenocrysts of olivine (Fo_{87-70}), plagioclase (An_{70-55}) and rare clinopyroxene. Groundmass of plagioclase laths, commonly ophitic diopsidic augite/titanaugite ($En_{43}Fs_{14}Wo_{43}$) and minor titanomagnetite and apatite. Chrome-spinel as inclusions in olivine and rarely as microphenocrysts
Picrobasalt	Abundant phenocrysts of olivine (Fo_{89-85}) enclose chrome-spinel. Matrix of olivine, plagioclase laths, ophitic diopsidic augite and titanomagnetite. Groundmass minerals of similar compositions to alkali basalt
Hawaiite	Phenocrysts of compositionally zoned plagioclase (An_{51-45}), olivine (Fo_{75-50}) and rarer augite ($En_{45}Fs_{11}Wo_{45}$). Abundant microphenocrysts of magnetite. Groundmass of plagioclase ($An_{60}Ab_{39}Or_1$–$An_{15}Ab_{70}Or_5$), ophitic to subophitic clinopyroxene ($En_{38}Fs_{20}Wo_{42}$), minor biotite, amphibole, apatite and interstitial alkali-feldspar ($An_2Ab_{60}Or_{38}$). Plagioclase phenocrysts and xenocrysts to 25 mm long in 'Big Feldspar' varieties
Mugearite	Phenocrysts of zoned olivine (Fo_{40-18}), plagioclase (An_{60-30}) and iron-rich clinopyroxene. Groundmass of flow-aligned plagioclase laths ($An_{22}Ab_{68}Or_{10}$), equigranular clinopyroxene ($En_{38}Fe_{16}Wo_{46}$), apatite needles and granular titanomagnetite, with minor iron-rich olivine, biotite, amphibole and interstitial alkali feldspar. 'Big Feldspar' mugearites occur
Benmoreite	Abundant microphenocrysts of flow-aligned, zoned plagioclase (andesine–oligoclase) and anorthoclase. Fine-grained groundmass of flow-aligned alkali-feldspar and sodic plagioclase laths, granular clinopyroxene and magnetite, and apatite needles. Rare olivine, generally serpentinised
Trachyte	Phenocrysts of anorthoclase or sodic sanidine ($An_1Ab_{48}Or_{51}$), iron-rich clinopyroxene ($En_{21}Fs_{33}Ca_{46}$) and rare olivine (Fo_{20}) and microphenocrysts of apatite. Abundant small, flow-aligned sodic alkali feldspar laths, and interstitial sodic pyroxene and amphibole
Tholeiitic olivine basalt	Sparse microphenocrysts of zoned olivine (Fo_{91-55}) and plagioclase (An_{90-66}). Groundmass of plagioclase laths (An_{82}), clinopyroxene ($En_{43}Fs_{12}Wo_{45}$), olivine and homogenous titanomagnetite. Minor apatite and amphibole. Calcium-poor pyroxene absent
Tholeiitic basalt	Rare phenocrysts of plagioclase (An_{87-76}) and altered olivine. Fine-grained plagioclase laths (An_{72}), granular aggregates of augite ($En_{47}Fs_{12}Wo_{41}$), rare pigeonite and opaque oxides. Intersertal glass commonly altered to chlorite
Tholeiitic basaltic andesite	Phenocrysts of plagioclase ($An_{57}Ab_{41}Or_2; An_{30}Ab_{63}Or_7$), augite ($En_{43}Fs_{17}Wo_{40}$), and rare orthopyroxene ($En_{65}Fs_{31}Wo_4$). Fine-grained groundmass of plagioclase laths ($An_{40}Ab_{56}Or_4$), granular clinopyroxene ($En_{40}Fs_{23}Wo_{37}$), magnetite and intersertal glass of sodic granite composition
Tholeiitic andesite 'Icelandite'	Phenocrysts of plagioclase ($An_{60}Ab_{39}Or_1; An_{38}Ab_{58}Or_4$), clinopyroxene ($En_{42}Fs_{18}Wo_{40}$), orthopyroxene ($En_{66}Fe_{30}Wo_4$), titanomagnetite and apatite. Very fine-grained groundmass of plagioclase laths, magnetite grains and granular pyroxene
Rhyodacite pitchstone	Phenocrysts of plagioclase ($An_{32}Ab_{60}Or_8$), grid-twinned anorthoclase ($An_6Ab_{51}Or_{43}$), clinopyroxene ($En_{44}Fs_{18}Wo_{38}$), orthopyroxene ($En_{68}Fs_{29}Wo_3$), magnetite, ilmenite, apatite and rare quartz. Phenocrysts (especially feldspar) commonly embayed. Groundmass of flow-banded glass, commonly devitrified to dusty, felsitic mass

Mineral compositional data are from Bell and Williamson (1994), Emeleus (1985, 1997), Esson et al. (1975), Kerr (1998), Sutherland (1982) and Tilley and Muir (1964)

Flow thickness and extent	Typically, the most compositionally primitive flows are the thinnest. Alkali olivine basalts and transitional olivine basalts, with or without olivine phenocrysts, are usually less than 5 m thick, and commonly less than 2 m. Where seen in cliff sections, they obviously interdigitate and are not laterally extensive. Palaeosols or laterites are thin, poorly developed or absent, testifying to the relatively rapid and near-continuous nature of lava eruption. Plagioclase-phyric olivine basalts are typically more massive and thicker as, commonly, are tholeiitic basalt and tholeiitic olivine basalt flows, although flows of any composition may appear unusually thick where ponded in palaeovalleys. Unusually thick, magnesium-rich lavas occur in north-west Mull. For example, at Port Haunr, a lava, 16 to 30 m thick, contains flat-lying zones rich in amygdales, and has alternating olivine-rich and olivine-poor bands, all suggestive of lava pulses during continuous eruption (Kent et al., 1998).

Individual flows of more evolved composition, such as hawaiite and mugearite, typically have a greater volume and are commonly 8 to 12 m thick with some greater than 20 m (Plate 10). They have remarkable lateral persistence, despite being derived from more viscous magmas. For example, the outcrops of two mugearites in northern and eastern Eigg are about 12 km in length. The unusual thickness (relative to area covered) of these far-travelled lava flows suggests high effusion rates. The relatively rare benmoreite and trachyte flows are typically thick. In west-central Skye, the benmoreite of Cnoc Dubh Heilla is at least 30 m thick, and the trachyte of Cnoc Scarall in Glen Eynort is in excess of 100 m. It is, however, possible that the thickness of the latter has been accentuated through ponding. Evidence of flow termination is not common; rare examples include the flow fronts observed on Muck, near Torr nam Fitheach and Port an t-Seilich, and on Canna at Cùil a' Bhainne.

Amygdaloid structures	Amygdales of secondary hydrothermal minerals are ubiquitous but are best developed within the thinner, more vesiculated, unevolved lava types, such as the alkali olivine basalts. Amygdale minerals include calcite, quartz, analcime, and the zeolites chabazite, stilbite, mesolite, thomsonite, gyrolite and apophyllite. These flows generally have well-developed, basal flow structures and less common pipe amygdales; the latter are especially concentrated where the flow has over-ridden thin palaeosols and mudstones. Within a lava field there may be a zonal distribution of the amygdale minerals related to depth of burial or the effect of later central complexes, as was demonstrated in the classic study of the Mull Lava Field by Walker (1971) (see p.71; Figure 12).

Hawaiites and mugearites are generally less vesicular, indicative of a lower dissolved volatile content upon eruption, although some trachytes are distinctly vesicular. The tholeiitic andesites of north-west Rum are noted for their amygdales of bloodstone (green-stained chalcedonic silica flecked with red, oxidised pyrite) and banded agate, both of which occur in the beach deposits at Guirdil.

Palaeosols	The development of palaeosols (or boles) to many of the lavas is a particularly useful feature that enables individual flows (both simple and compound) to be distinguished. Such material is indicative of subaerial weathering during the Palaeogene, in a warm temperate climate, with moderate to high rainfall on free-draining ground. Typically, the palaeosols are weathered and obscured inland, but are well exposed in the sea cliffs. They range in colour from dark, chocolate-brown through dull reddish brown to bright ocherous red, and from grey to mauve. They are generally not preserved in the vicinity of the central complexes, where pervasive hydrothermal alteration has commonly reduced lavas and palaeosols alike to dull, grey rocks.

The thickness of the palaeosols varies considerably. Flow tops may be merely stained a reddish brown with no obvious palaeosol between successive flows but, more typically, flows are separated by a few millimetres to several centimetres of apparently structureless, massive material. In a few instances, the thicker beds preserve crude rhizolith-like structures where mineral matter has replaced plant roots. Such features are taken to indicate that the palaeosols formed in situ. Some

Plate 10 *Thin basalt flows (dark) and thicker, impersistent hawaiite flows (pale) on Ben Scaalan, south-west Skye. Scale: the summit flow is about 12 m thick (P580461).*

of the thickest and best developed palaeosols occur in intimate association with thin mudrocks. The latter are generally preserved as lensoid bodies on the undulating, eroded tops to flows and palaeosols, and point to minor reworking of soil and volcanic ash by running water, most likely rain water run-off rather that organised fluvial systems. Commonly these thicker palaeosols (with or without associated mudstone) are immediately overlain by thick, differentiated flows of hawaiite or mugearite, suggesting prolonged periods between eruptions. Particularly good examples are seen at Talisker Bay, Beinn nan Dubh Loch, Fiskavaig and Biod Mór in west-central Skye.

Palaeosol-like layers with indistinct bedding and rudimentary pisoidal structure, commonly containing pristine minerals such as sanidine, biotite and pyroxene, are interpreted as the product of contemporaneous weathering of crystal and vitric tuffs and not a consequence of deep-weathering of in-situ lava. Examples are quite common and include those found close to Lochan nan Dunan in northern Trotternish, Skye, and in south-east Muck (see p.60) (Bell et al., 1996; Emeleus et al., 1996a; Plate 11).

Flow structures

These structures are best, but not exclusively, developed within the more evolved lavas, and the flow banding is generally parallel to the dip of the flow. The commonest type is due to the alignment of groundmass and microphenocryst feldspars, which imparts an obvious fissility and banded appearance to the rock, and is a useful aid to field identification (e.g Harker, 1908); flow-alignment of plagioclase macrophenocrysts is not generally noted. Where the base of the flow is irregular, flow folding of the foliation is common (Plate 12). Large-scale examples include the mugearite at McFarlane's Rock, west-central Skye, whereas smaller-scale folds are present in the tholeiitic andesite on Fionchra, Rum. Systematic and abrupt changes in the inclination of flow-banding and jointing are seen at several localities. Usually, an otherwise flat to low-angle pattern is replaced by a near-vertical one. This could indicate either a flow dipping in an intrusive manner under its own surface, or reflect the possibility of the lava having flowed over uneven topography. There are many examples, including those seen at Loch Dubh, Cnoc Scarall and Coir' an Rathaid, in west-central Skye.

Plate 11 Bedded, oxidised trachytic tuff overlying a broken pahoehoe surface of basalt lava. Port Mór, Muck. Sanidine and zircon from this deposit have been dated at 60.65 Ma and 61.15 Ma respectively (Table 8) (Hammer shaft: 30 cm in length) (P580462).

Flow direction may be established on a local scale where the base of a flow has over-ridden unconsolidated material such as sediment. There are good examples in flows overlying the Minginish Conglomerate Formation in the Allt Mór south of Loch Eynort, west-central Skye. There, the flows have both 'nosed' into soft sediment, incorporating strings of sandy material, and also have inclined 'flame-like' masses of sediment along their bases. Flow direction may also be indicated by inclined or deflected pipe-amygdales and by discoid amygdales.

Some basalts, hawaiites and mugearites have brecciated tops, typical of a'a-type lava flows, and this autoclastic material can easily be misidentified as monolithological pyroclastic breccia. Typically, such a'a-type lavas are indicative of cooler, volatile-depleted magma at some distance from its point of eruption. A mugearite at Rubha Cruinn on the north side of Talisker Bay, west-central Skye, illustrates this very well and also shows a basal breccia carpet.

Jointing The extent to which columnar jointing is developed in lava flows appears to depend, amongst other things, on the composition of the magma. The best examples of columnar jointing are usually found in tholeiitic basalt or tholeiitic olivine basalt lava flows, and the most celebrated example of a columnar jointed flow in the Hebridean Igneous Province is the tholeiitic basalt at Fingal's Cave on Staffa (Plate 13). Details of the columnar structure are well exhibited on Preshal More and Preshal Beg in west-central Skye, where tholeiitic olivine basalt flows are over 120 m thick. Each of these flat-lying flows has a lower 'colonnade' in which vertical columns are typically six-sided and between 0.3 and 1 m across. The overlying 'entablature' shows a complex, highly irregular jointing pattern, and a thin zone characterised by a subhorizontal flow-foliation separates the colonnade from the entablature. Curved ball and socket joints commonly divide the columns of the colonnades along their lengths. Inclined columns at several localities indicate the proximity of the sidewalls of the valley in which the magma was ponded, but which have been subsequently removed by erosion. Spectacular jointing also occurs elsewhere in lavas of the Staffa Lava Formation in south-west Mull, at Ardtun and on the Ardmeanach peninsula, where there are clusters of radiating columns in lavas overlying hyaloclastite deposits on the foreshore south of the MacCulloch's Tree locality.

True columnar jointing is less common within the unevolved alkali olivine basalt flows, which are generally characterised by a more irregular 'blocky' or prismatic jointing. Columnar jointing is commonly present in hawaiite, mugearite and tholeiitic andesite flows, and in flows of more evolved composition such as trachyte and silicic pitchstone. The Sgurr of Eigg rhyodacitic pitchstone flow contains superb columns, smaller in diameter than those of the basalts and highly irregular in places (Plate 14).

Weathering

Weathering style is also a useful field characteristic of lava type. The minerals in basalt lavas rich in olivine tend to alter more readily than those in the more evolved types, such as hawaiite and mugearite. Coupled to the more obviously amygdaloidal and usually coarser grained nature of the olivine basalts, this results in the degradation of rock faces and a tendency for the basalt lavas to form hollows or lower ground ('slacks') or terraced slopes between more resistant flows. Within individual flows, the amygdaloidal top and base weather more readily than the massive centre, resulting in typical 'trap-featuring' (Plate 15).

LAVA SEQUENCES

The principal lava sequences are summarised in Figure 10 and Table 9. Details are given in Tables 11–15. They are:

- the Eigg Lava Formation on Eigg, Muck and south-east Rum
- the Skye Lava Group on Skye, and including the Canna Lava Formation on Canna, Sanday and north-west Rum
- the Mull Lava Group on Mull, Morvern and eastern Ardnamurchan

No lavas occur onshore in the St Kilda archipelago and none are seen on Arran apart from the foundered masses within the Central Arran Ring-complex (p.137).

Plate 12 *Flow-folded hawaiite lava, Arnaval, south-west Skye. Scale: hammer shaft is 30 cm (P580463).*

Plate 13 Tholeiitic basalt lava resting on bedded volcaniclastic deposits, Isle of Staffa. A classic example of a lava divided into a lower, columnar jointed 'colonnade' and an upper irregular 'entablature' (see text). Scale: the cliff is about 35 m in height (P580464).

Plate 14 *Distorted columnar jointing in the Sgurr of Eigg pitchstone lava. Columns are nearly vertical in the lower part of the body but vary from horizontal to inclined in the upper part. A prominent sheet of pale coloured felsite cuts the inclined columns (GN79).*

EIGG LAVA FORMATION

The Eigg Lava Formation comprises the sequence of lavas that crops out on Eigg and Muck and, as down-faulted slivers, in south-east Rum (Allwright, 1980; Emeleus, 1997; Table 9). The lavas are intruded by a north-west-trending dyke swarm (Chapter 7), which in turn is cut by the Rum layered intrusions (Chapter 9). Consequently, it may be deduced that the eruption of the Eigg Lava Formation predates at least part of the Rum Central Complex. Precise radiometric ages of 60.6 Ma both for trachytic tuffs interbedded with the basal lavas on Muck (Chambers, 2000; Chambers et al., 2005) and for the Rum Layered Suite (Hamilton et al., 1998) indicate that all of this happened in a very short time (Figure 8).

Eigg

The lavas on Eigg lie unconformably on the Upper Cretaceous Strathaird Limestone Formation and the Middle Jurassic Great Estuarine Group. The volcanic succession is about 400 m thick and dips gently towards the south-west. The lowest units, exposed in north and north-east Eigg, are tuffs, typically only a few centimetres thick. The majority of the flows are of alkali olivine basalt and hawaiite. Mugearites occur especially about 50 m above the base of the formation (Figure 10), and near the top there is a distinctive group of feldspar-phyric basaltic hawaiite flows. Interspersed with the lavas are thin, massive, red to orange mudstones and siltstones, some of which are of volcaniclastic origin. Fluviatile deposits are rare, but a thin, laterally impersistent pebbly conglomerate occurs on the foreshore west of the old pier in the south-east of Eigg.

Muck

The volcanic succession on Muck comprises about 140 m of near-horizontal or gently north-west-dipping lavas. They are intruded, as on Eigg, by many members of the north-north-west-trending Muck Dyke Swarm (Chapter 7). The lowest units comprise an assemblage of laterally impersistent olivine basalt, feldspar-phyric basaltic hawaiite, hawaiite and mugearite lavas and sedimentary deposits, which overlie the Middle Jurassic Kilmaluag Formation at Camas Mór. The higher flows are mainly of olivine basalt but with a flow of feldsparphyric basalt near the summit of Beinn Airein (Figure 10). Bright red to orange mudstones are commonly interleaved with the lavas, especially

Plate 15 Trap featuring in the basalt lava succession of Ardmeanach, western Mull (P580465).

near the base of the succession. Some represent transported material, whereas others grade downwards into reddened flow tops and most likely developed by in-situ lateritic weathering. An example of the former crops out south-east of Port Mór (Plate 11), where 20 to 30 cm of red interflow siltstone overlies the reddened, brecciated pahoehoe top of an olivine basalt lava. The crystal component of the siltstone (sanidine, clinopyroxene, sphene) was derived from an explosive eruption of peralkaline trachytic magma, during the early stages of development of the lava sequence (Emeleus et al., 1996a). Radiometric age determinations (see also Pearson et al., 1996; Chambers, 2000; Chambers et al., 2005; Figure 8), obtained from the sanidine crystals are considered to accurately date the start of igneous activity in the Hebridean Igneous Province at 60.6 Ma.

A coarse deposit, termed the Camas Mór Breccia, crops out beneath shoreface boulders at the east end of Camas Mór. It is dominated by clasts, generally up to 30 cm across, of grey and brown limestone (Duntulm and Kilmaluag formations, respectively), in a matrix consisting of much smaller clasts of black carbonaceous siltstone and grey and brown impure limestone. Locally, the breccia contains prominent, thin interbeds rich in quartz granules. The bedding is deformed beneath particularly large limestone clasts, 0.5 to more than 2 m in diameter. The breccia is generally lacking in igneous clasts and predates the lavas (Emeleus, 1997).

Rum Metamorphosed and fractured basalt, intruded by north-west-trending basaltic dykes, occurs as a fault-bound mass, about 1 km in length, within the Main Ring Fault on the south-east slopes of Beinn nan Stac, and on the north-east side of Dibidil (Emeleus, 1997). The basalt unconformably overlies Lower Jurassic Broadford Beds (Smith, 1985) and is considered to be a remnant of the Eigg Lava Formation.

SKYE LAVA GROUP The Skye Lava Group crops out in north and west-central Skye, borders much of the Skye Central Complex, and extends south to the islands of Canna, Sanday and north-west Rum (Figure 9). The position of the Skye Lava Group in the igneous stratigraphy of the Hebridean Igneous Province is considered in Chapter 5. The various parts of the group are not easily correlated, but detailed successions for the lavas of west-central and northern Skye, Strathaird (Skye), and for north-west Rum and Canna are given in Figure 10 and Tables 11, 12, 13 and 14.

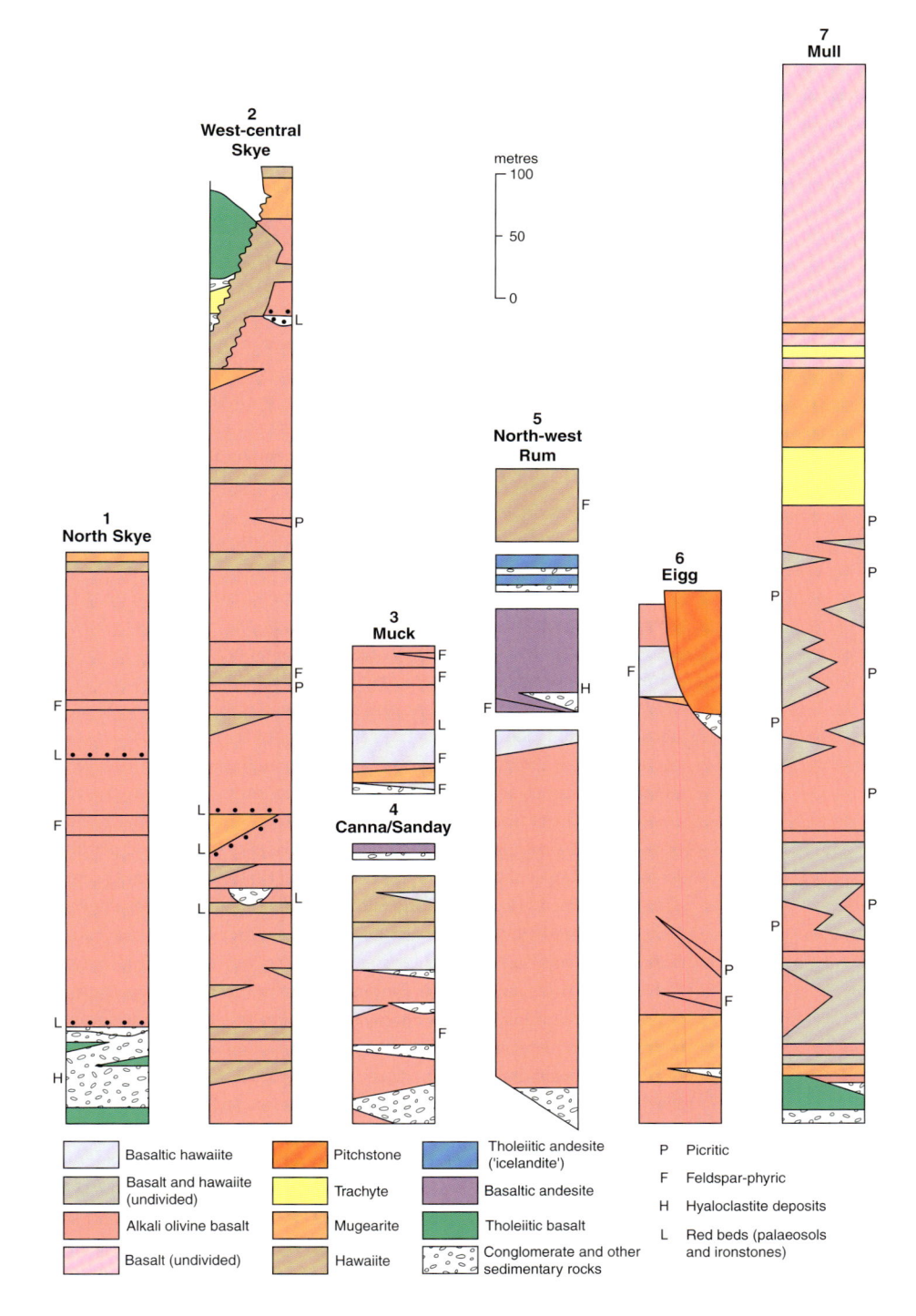

Figure 10 Generalised and composite vertical sections of the lava fields in the Hebridean Igneous Province.

For further details of individual lava fields, refer to: **1** Anderson and Dunham, 1966; **2** Williamson and Bell, 1994 and BGS Minginish, Scotland Sheet 70 (2002 edition); **3, 4, 5, 6** Emeleus, 1997; **7** Kerr, 1995b. See also Table 9.

CHAPTER SIX: PALAEOGENE LAVA FIELDS AND ASSOCIATED SEDIMENTARY ROCKS

West-central Skye

The lavas of west-central Skye are divided into a number of formations and members on the basis of distinct lithological associations, originally termed groups and formations by Williamson and Bell (1994). Interlava sedimentary units have been used to separate the stratigraphical divisions (Figure 10; Table 11). Numerous faults dissect the predominantly flat-lying flows. The faults, many with only small displacements, are typically orientated parallel, normal and slightly oblique to the north-west-trending Skye Dyke Swarm (see Chapter 7).

The stratigraphically lowest sequence in the group forms the Rubh' an Dùnain Formation south-east of Loch Brittle. On the north shore of Soay Sound its basal deposits, the An Leac Member, rest upon an irregular pre-lava land surface cut across Upper Cretaceous, Lower Jurassic, Triassic and Torridonian sedimentary rocks. The member is about 10 m thick, and comprises laterally impersistent units of volcanic breccia, lapilli-tuff and hyaloclastite. Large clasts and rafts of pre-lava lithologies occur within the volcaniclastic deposits. This initial hydromagmatic phase of activity also occurred in north Skye (see below), suggesting that the early Paleocene land surface was covered with shallow, water-filled depressions into which the magmas were erupted. The first lavas are intercalated with lacustrine mudstones and siltstones in the overlying Meacnaish Member. The succeeding Bualintur Formation, which crops out on the west side of Loch Brittle, may possibly interdigitate with the upper part of the Rubh' an Dùnain Formation. Interbedded with the lavas of this formation are deposits of heterogeneous, poorly sorted stratiform volcaniclastic breccia, conglomerate and sandstone, for example in the inaccessible sea cliffs around Sgurr an Duine.

Overlying the Rubh' an Dùnain and Bualintur formations, the Minginish Conglomerate Formation comprises three distinctive, geographically separated members: Culnamean, Allt Geodh' a' Ghamhna and Allt Mòr (Figure 11). All comprise volcaniclastic conglomerate–sandstone–siltstone sequences together with subsidiary mudstones and coals; the conglomerates contain 'exotic' clasts. Plant remains are widespread and, in places, build up into leaf-rich beds. These sedimentary sequences are typical of fluvial environments, and in particular those recorded from braided river systems. Sandy, matrix-supported conglomerates and associated pebbly sandstones at the base of the formation were most likely the products of mass-flow or debris-flow processes, initially triggered by volcanotectonic activity, and modified by high water-flow regimes, possibly during flash-flooding. Clasts of granite and porphyritic felsite were derived from unroofing of the uplifted massif overlying the Rum Central Complex that lay to the south. Quartzite, green gritty sandstone and feldspathic sandstone were derived from the east, sourced from an uplifted block to the east of the Camasunary–Skerryvore Fault composed of various Torridonian and Cambrian lithologies associated with the Moine Thrust Belt.

Higher in the sequence, the Eynort Mudstone Formation was deposited during another hiatus in the volcanic activity. Following a period of intense weathering of the lavas, mud and volcaniclastic material transported by the combined effects of rain-wash and wind action were deposited in shallow pools and lakes. Four separate outcrops, all from the same stratigraphical level within the lava pile, have been afforded member status.

In the upper part of the sequence, the Gleann Oraid Formation is dominated by evolved flows, including trachytes and a benmoreite. The initial flows of the lowermost Arnaval Member vary in their nature and thickness and may have filled in topographical depressions produced prior to their eruption. The Sleadale Member contains at least one thick trachyte flow of limited lateral extent, possibly indicating an original dome-like form. The trachytic tuff at the base of the member contains exceptionally fresh anorthoclase and biotite crystals from which an Ar-Ar age of 58.91 Ma has been obtained (Bell and Williamson, 2002; Table 8). The thick trachyte flow forming the Cnoc Scardall Member occupies a broad palaeovalley excavated out of the subjacent flows of the Skridan and Arnaval members. Close to the presumed base of the Gleann Oraid Formation at Dùn Ard an t-Sabhail is a composite hawaiite flow with a nearly aphyric base and an upper plagioclase-macrophyric unit. Similar composite flows have been identified from the Beinn Totaig Formation of north Skye (see below; Anderson and Dunham, 1966) and it is possible that flows of the two formations (Gleann Oraid and Beinn Totaig) may be interleaved

Table 11 The Skye Lava Group: west-central Skye (after Williamson and Bell, 1994).

Formation*	Thickness	Details
Talisker	At least 120 m	Two tholeiitic olivine-basalt flows of Preshal More basalt type, akin to Mid-Ocean Ridge Basalt (MORB)
Preshal Beg Conglomerate	20 m to (locally) 40 m	Volcaniclastic conglomerate, breccia, gritty and tuffaceous sandstone
Loch Dubh	70–80 m	Plagioclase- and olivine-microphyric hawaiites and thin alkali basalt flows
Gleann Oraid	100 m+	Cnoc Scarall Member* Single, compound porphyritic trachyte flow.
	About 125 m	Sleadale Member Flow(s) of vesicular, porphyritic trachyte overlying a 3 m-thick trachytic tuff
	230 m	Arnaval Member Flow-banded mugearites and hawaiites, and interbedded alkali basalt flows, a single benmoreite flow and a 10 m-thick bed of stratiform volcaniclastic breccia and conglomerate
Eynort Mudstone	About 15 m	Loch an Sgùirr Mhóir Member Red-brown clay and massive structureless mudstone
		Gleann Oraid Member Minor occurrences of thin siltstones and tuffaceous sandstones, below Gleann Oraid Formation lavas
		Biod Mór Member Lignites and thin beds of fissile mudstone
		Ben Scaalan Member Interbedded mudstones and thin volcaniclastic siltstones, with thin intercalated basalt flows. Overlain by orange-brown-weathering mudstones, ironstones and impersistent carbonaceous fissile mudstones. Capped by an impure lignite
Fiskavaig	About 130 m	Rubha nan Clach Member Alkali basalt flows and a single, laterally extensive hawaiite flow
	Up to 55 m	McFarlane's Rock Member Evolved flows (mugearite, hawaiite) alternating with thin olivine basalts. Includes a 20 m-thick flow-folded mugearite at base
Glen Caladale	150 m	Skridan Member Thin-bedded, (compound) and massive alkali basalt flows
	Up to about 65 m	Sgurr Buidhe Member Alkali basalt flows capped by one or more flows of hawaiite interbedded with thin basalt flows
	Up to 25 m	Stac a' Mheadais Member A single, laterally extensive, mugearite flow with a strongly lateritised top
	Over 160 m	Tusdale Member Alternating olivine + plagioclase-phyric alkali basalt flows intercalated with hawaiites and a single flow of picrobasalt
Cruachan	About 300 m	Glen Brittle Member Interlayered hawaiite and alkali basalt flows
	About 80 m	An Cròcan Member Thin amygdaloidal alkali basalt flows overlain by flows of hawaiite and mugearite. Single flow of picrobasalt near top
Minginish Conglomerate	10–30 m	Three geographically separated members are recognised Culnamean, Allt Geodh' a' Ghàmhna and Allt Mór. All comprise volcaniclastic conglomerate–sandstone sequences (see text)
Bualintur	About 250 m	Alternating flows of alkali basalt and laterally extensive hawaiite. Deposits of heterogeneous, poorly sorted stratiform volcaniclastic breccia, conglomerate and sandstone are interbedded with the lavas
Rubh' an Dùnain	At least 230 m	Creag Mhór Member Alkali basalt flows (+ olivine and/or plagioclase phenocrysts). Plagioclase-phyric hawaiite flows near top of sequence
	About 75 m	Meacnaish Member Mugearite and alkali basalt flows, with intercalated lacustrine fissile mudstones and siltstones
	About 20 m	An Leac Member Laterally impersistent units of volcaniclastic breccia, lapilli tuff and hyaloclastite deposits

* The formations and members were termed 'groups' and 'formations' in the original account

LITHOLOGY	DESCRIPTION	ENVIRONMENT FACIES
	Thin alkali olivine basalt lavas with scoriaceous tops (c. 7 m) overlying a massive basaltic lava with pillow structure towards base (c. 5 m)	Lava of the Cruachan Formation
	Thin white tuff (3 cm)	
	Coal (5 cm)	
CONGLOMERATE 3	Sandstone and siltstone with ? plant remains (diffuse carbonaceous streaks) and rootlets (20 cm)	Overbank and quiescence ponds
	Coal (1-5 cm), thinly interbedded with siltstone. Draped over underlying pebbles (1-5 cm), very sharp basal contact	Sheet deposit, high-energy flow regime, ? torrential
	Massive conglomerate with well-packed, rounded pebbles and cobbles (up to 15 cm) of granite, porphyritic felsite, quartzite and red feldspathic sandstone, in a pale sandy matrix (3.2 m)	Mass-flow deposit
	Sandstone with micaceous partings (20 cm)	Overbank flood
	Coal (2 cm) (irregular to localised development only)	Swamp pool
CONGLOMERATE 2	Sandstone with plant remains (1.8 m)	Low-profile scour channel grading to overbank
	Conglomerate with a sandy matrix, small amount of silicic igneous rock clasts, abundant clasts of arenaceous sedimentary rock, rare clasts of amygdaloidal and plagioclase-macrophyric basalt, clasts up to 30 cm but more commonly 10-15 cm; thin lenses of white sandstone up to 5 cm thick in lower part (2.3 m)	Debris flow
		Channel fill and scour, mixed dilute debris flows
	Lenticular bed of sandstone with fine-grained, sharply defined, laminated base (1.1 m)	Within-channel dune or gradational overbank
CONGLOMERATE 1	Fining-up conglomerates with densely packed and crudely imbricated clasts of feldspathic sandstone up to 30 cm; green siltstones: sandstone wedge (scour fill) thickening towards the north (2.75 m)	Sheets and channel-fill deposits. Intermittent flow, high to low energy
	Massive conglomerate, clasts as above. Irregular silty sandstone wedges	Debris flow with local reworking of fine-grained material
		- erosional base -
	Highly amygdaloidal basaltic lava forming the top of the cliff at c. 125 m OD (10 m)	Lava of the Bualintur Formation

(MINGINISH CONGLOMERATE FORMATION)

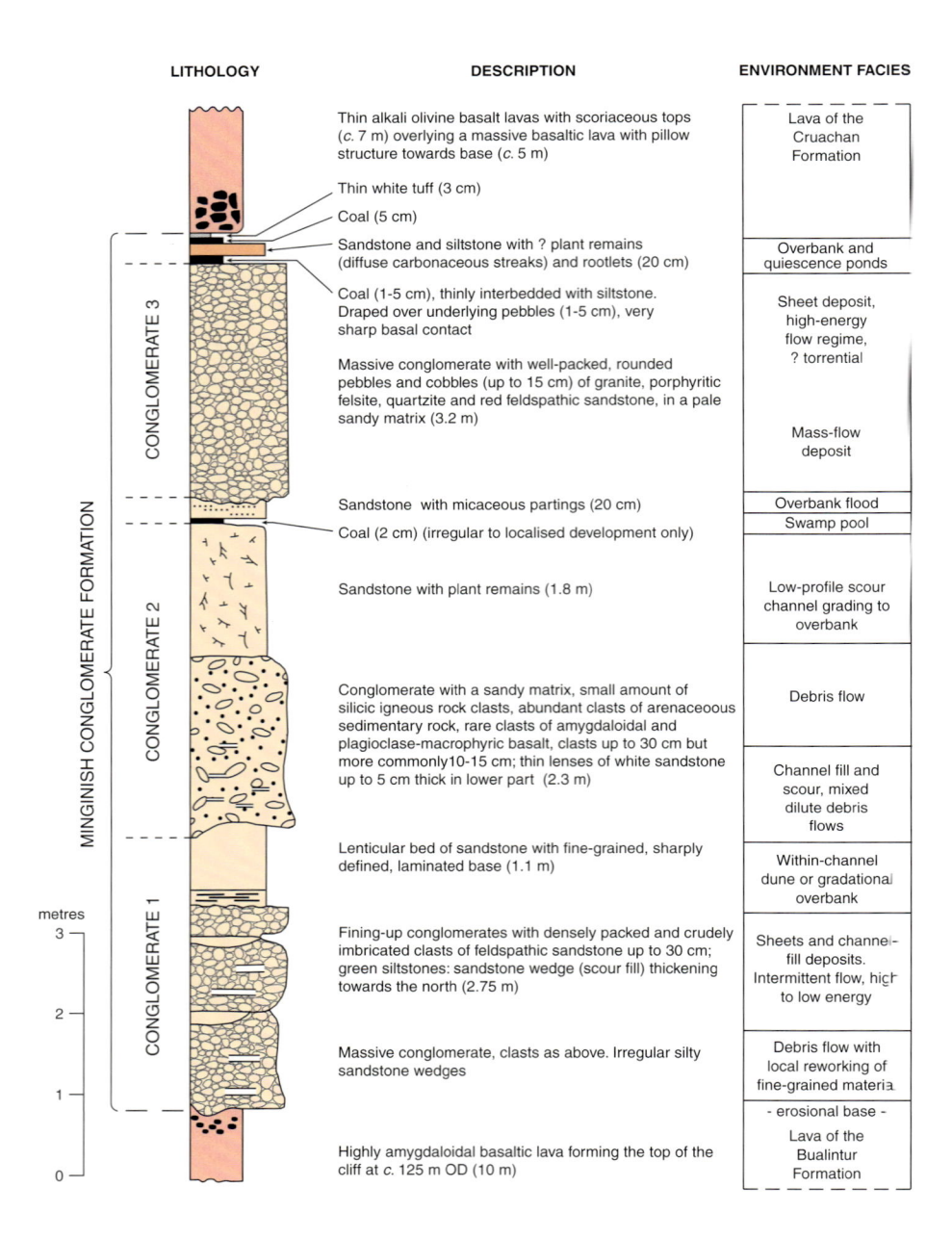

Figure 11 *Stratigraphical section for the Allt Geodh' a'Ghàmhna Member of the Minginish Conglomerate Formation, Skye (based on Williamson and Bell, 1994, fig. 22).*

and may have been erupted more or less simultaneously. In the area south-west of Fiskavaig, flows belonging to the lower part of the Gleann Oraid Formation are intercalated or interdigitated with hawaiites and mugearites belonging to the laterally restricted Loch Dubh Formation, which crops out on the northern and western slopes of Beinn nan Dubh Lochan.

The Preshal Beg Conglomerate Formation, which crops out around the base of Preshal Beg and on the north-east slopes of Preshal More, overlies the Arnaval and Sleadale members of the Gleann Oraid Formation (Plate 16). It comprises a thick, laterally restricted sequence of heterogeneous volcaniclastic material with clasts of local derivation; it is interpreted as talus and alluvial fan deposits that accumulated rapidly within a restricted but broad valley system.

The upper part of the formation is invaded by irregular apophyses (lava pillows, globe breccias and 'neptunian' dykes) of the overlying tholeiitic olivine basalt flows of the Talisker Formation, suggesting that the sediments were not lithified prior to eruption of the lavas.

The Talisker Formation comprises two flows of tholeiitic olivine basalt (Tables 9; 11) that were originally impounded in steep-sided canyons and now form the twin hills of Preshal More and Preshal Beg, south-east of Talisker Bay.

Northern Skye In northern Skye, volcanic activity commenced with the deposition of the Portree Hyaloclastite Formation onto flat-lying Middle Jurasssic strata. The formation comprises palagonitised hyaloclastite breccias (Plate 9), locally developed pillow lavas, and a variety of volcaniclastic sedimentary rocks (sandstone, siltstone and mudstone). These hydromagmatic rocks are exposed around Portree and at other localities on the east side of the Trotternish peninsula (Anderson and Dunham, 1966). Elsewhere in north Skye, for example in the Staffin, Flodigarry, Uig and Loch Bay areas and in the Lealt River of the Trotternish peninsula, there are scattered and poorly exposed occurrences of tuffaceous sandstone and associated strata.

The remainder and vast majority of the Skye Lava Group in north Skye is composed of lavas erupted into a subaerial environment. Pauses in the volcanic activity are marked by the development of interlava sedimentary sequences, ranging from conglomerate and sandstone to plant-bearing mudstone and siltstone. None of the recognised sequences appears to be laterally persistent, leading to problems with correlation and stratigraphical subdivision (England, 1994). However, the scheme set out by Anderson and Dunham (1966) offers a useful and approximate device for the recognition of the overall structure (Table 12). The lavas belong to the alkali olivine basalt–hawaiite–mugearite–benmoreite–trachyte series and are referred to by Thompson et al. (1972) as belonging to the Skye Main Lava Series. No simple correlations between composition and stratigraphical position exist. The evolved flows, comprising about 20 per cent of the total thickness, are intercalated with the more common alkali olivine basalt lavas, possibly becoming more abundant at higher levels in the lava sequence. The entire

Plate 16 Outlier of tholeiitic basalt lava of the Talisker Formation, originating as a valley fill, overlying hawaiite lavas of the Gleann Oraid Formation, Skye Lava Group, Preshal Beg, south-west Skye. Macleod's Tables (also Skye Lava Group) in left middle-distance (P580466).

Table 12 The Skye Lava Group: northern Skye. (Anderson and Dunham, 1966).

Formation *	Thickness (approximate)	Details (The divisions between formations are generally defined by thin layers of conglomerate, sandstone, etc. See text for details)
Osdale	At least 500 m	Flows of alkali basalt, hawaiite and mugearite
Bracadale	120 m	Flows are generally evolved and include porphyritic and non-porphyritic mugearites, benmoreites and trachytes
Beinn Totaig	600 m	Alternating unevolved (alkali basalt) and evolved (hawaiite, mugearite) flows, the latter including composite flows with strongly plagioclase-phyric upper parts (see text)
Ramasaig	760 m (see text)	Predominantly of alkali basalt flows, some of which are olivine- and/or plagioclase-phyric
Beinn Edra	300 m	Predominantly alkali basalt flows, with rarer evolved compositions
Portree Hyaloclastite	30 m	Tuffs containing fragments of glassy and devitrified olivine basalt, accretionary lapilli and larger bombs. Thin amygdaloidal basalt flows, pillow basalts and plant-bearing volcaniclastic sandstones and siltstones also occur

* Termed 'groups' in the original account

preserved cumulative thickness of about 1200 m of lavas gives rise to impressive east-facing escarpments on the Trotternish peninsula, together with inland plateaux and coastal cliffs. The most spectacular sequences comprise the cliffs tracable from The Storr, north to Quiraing, inland from Staffin Bay (Plate 43).

The lowest of the lava units, the Beinn Edra Formation crops out on the Trotternish peninsula in north-east Skye. The most complete section forms The Storr, where at least 24 flows can be identified. Here, and elsewhere on the east side of the outcrop, extensive rotational landslips and rock failures obscure the base of the formation (p.168; 176), although the lower part is seen at Uig on the west side of the peninsula. To the south of Loch Portree, lavas belonging to this formation can be traced south to Tianavaig Bay and Ollach. The top of the formation is defined by the conglomerate–sandstone–siltstone–mudstone sequence in Glen Tungadal, about 7 km east of Bracadale.

The Ramasaig Formation crops out on the Duirinish and Waternish peninsulas of north-west Skye. It overlies Middle Jurassic strata west of Waterstein Head. The formation consists of porphyritic and non-porphyritic olivine basalts. The top of the formation is separated from the Osdale Formation by trachytic tuff, and by sandstones and plant-bearing fissile sandstones exposed in the Hamra River, Glendale.

The Beinn Totaig Formation has a type locality north-east of Loch Harport, which includes Mugeary, the type locality for mugearite (Harker, 1904). Within this formation are composite flows of hawaiite and mugearite; an aphyric portion at the base is separated by a sharp interface from the main plagioclase-macrophyric portion (Harker, 1904). They may correlate with a similar flow close to the presumed base of the Gleann Oraid Formation in west-central Skye (see above). The top of the Beinn Totaig Formation is taken at the base of the Glen Osdale sedimentary sequence, deposited during a pause in the volcanic activity. However, as pointed out by England (1994), the Glen Osdale deposits are about 250 m below the top of the formation as described by Anderson and Dunham (1966) close to the summit of Healabhal

Table 13 The Strathaird Lava Formation of the Skye Lava Group: south-eastern Skye (Almond, 1964).

Member *	Average thickness (m)	Details
Upper Lava	160	Subophitic and ophitic olivine basalts, mostly feldspar-olivine-phyric
Picrite Basalt	15	Upper and lower flows rich in olivine, separated by an olivine-phyric basalt
Ophimottled Basalt	61	Highly ophitic olivine basalts with an intercalation of trachybasalt
Basal Lava	46	Olivine-phyric basalts with subophitic or interstitial pyroxene

* Termed groups in the original account

Mhòr. The Glen Osdale deposits comprise about 8 m of conglomerate containing cobbles of red-weathering Torridonian gritty sandstone, rare cobbles of gneiss, and lesser amounts of granite and porphyritic felsite, which do not, however, match clasts found elsewhere on Skye that were derived from the Rum Central Complex. Clasts of locally derived lava lithologies are also present. Interbedded siltstones were deposited in a lacustrine or marginal floodplain environment and contain well-preserved leaf impressions of oak, hazel and plane (Anderson and Dunham, 1966). The plant-bearing, inter-lava beds at Hamra River, Glen Osdale, Forse River, and Red Burn east of Edinbane were considered by Anderson and Dunham (1966, fig. 13) to be lateral equivalents.

The relatively evolved lavas of the Bracadale Formation crop out north-east of Bracadale. With increasing compositional evolution the lateral extent of individual flows diminishes, with the trachytes being particularly restricted. The flows appear to occupy a broad north-west-trending palaeovalley, with a preserved width of about 15 km and a depth of about 300 m, eroded into the underlying lavas. The valley trends parallel to the Skye Dyke Swarm (England, 1994).

The youngest lavas preserved in north Skye (Osdale Formation) crop out west of Bracadale and form many of the islands in Loch Bracadale. They consist mainly of basalt and mugearite.

Other lavas on Skye

There are large areas of basaltic lava in Glen Drynoch, north of Loch Harport and on either side of Loch Sligachan. They form part of the Skye Lava Group, but their equivalence to particular formations described above has not been established. Likewise, isolated areas of basaltic lava occur around the periphery of the Western and Eastern Red Hills granite centres and on the south-east margin of the Cuillin Centre. Of the latter, the most extensive is the lava succession that overlies Jurassic and thin, impersistent Upper Cretaceous strata east of the Camasunary–Skerryvore Fault on Strathaird. Up to 290 m of predominantly olivine basalt lavas, including picritic flows, are preserved on An Stac, where a lava stratigraphy was established by Almond (1964) (Table 13). The lavas show pervasive hydrothermal alteration and a progressive increase in thermal metamorphism over about 1500 m adjacent to the Cuillin Centre, culminating in a sanidinite-facies assemblage being developed immediately adjacent to its margin, as evidenced by the presence of cordierite, sanidine and possibly mullite in altered boles (Almond, 1964). The lavas and underlying strata occur in open folds concentric about, and related to the emplacement of, the Cuillin Centre. The small area of olivine basalt that forms the summit of Dùn Caan on Raasay is an outlier of the north Skye lava pile.

Small Isles Olivine basalt (alkali and transitional) and tholeiite basalt, together with their evolved differentiates, occur within the lavas of the Canna Lava Formation and are interleaved with fluviatile conglomerate–sandstone sequences, thus permitting local subdivision into a number of recognisable members (Emeleus, 1985, 1997; Figure 10; Table 9). Eruption of the lavas and deposition of the sediments occurred on a hill-and-valley topography, graphically illustrated in north-west Rum. The main drainage system flowed north from a highland area initially created during uplift associated with the emplacement and growth of the Rum Central Complex (Williamson and Bell, 1994; Emeleus, 1997). Flows on Canna and Rum have been dated at 60.0 Ma (Chambers et al, 2005).

Canna and Sanday On Canna and Sanday, the lava pile is subhorizontal, with dips of 3° or less. The lowest flows crop out in south-east Sanday, and the uppermost on the high ground of east and west Canna (Figure 10). The formation comprises at least 200 m of lavas, predominantly of the alkali olivine basalt–hawaiite–mugearite lineage. Individual lavas and sedimentary units show considerable variation in thickness along strike. In general, the stratigraphically higher flows are more evolved and strongly feldspar-phyric basaltic hawaiites; hawaiites form thick, continuous flows on the higher parts of Canna. Lava accumulation was predominantly in a subaerial environment. Many of the flows have brecciated a'a flow tops, a few of which are reddened as, for example on the wave-cut platform between Tarbert and Cùil a' Bhainne in eastern Canna. At Cùil a' Bhainne, lavas were erupted into shallow water with the local formation of pillowed flows and hyaloclastite deposits.

Inter-lava fluviatile breccia–conglomerate–sandstone sequences are a prominent feature of the formation. On Canna, they are thickest on Compass Hill, totalling about 50 m, but thin out westwards along the north coast. Sedimentary breccias at Compass Hill contain basalt blocks up to 2 m across and smaller blocks of red feldspathic sandstone. The more widespread conglomerates contain well-rounded cobbles and pebbles of locally derived lava lithologies, red feldspathic sandstone resembling the Torridon Group on Rum, rare schist and gneiss, and porphyritic rhyodacite and granophyric microgranite identical with rock-types within the Rum Central Complex (Emeleus, 1973). No clasts of Mesozoic rocks are known, although Lower Jurassic limestones occur offshore of north-east Canna (Fyfe et al., 1993). Poorly preserved carbonised plants are found in the sedimentary intercalcations at a number of localities.

Rum The lavas and conglomerates of the Canna Lava Formation in north-west Rum (Figure 10; Table 14) were impounded within a system of steep-sided palaeovalleys. The remains of valleys eroded in Torridon Group sandstones crop out north-east of Fionchra and also on Bloodstone Hill. In the east face of Bloodstone Hill, the cross-section of one valley is infilled with two flows of tholeiitic andesite and underlying conglomerates (Plate 17). One side of a valley eroded in the Western Granite is recognised on the north side of Orval. The presence of up to 60 m of hyaloclastite breccia and pillowed lava at the base of the Upper Fionchra Member on the north side of Fionchra indicates the former presence of a shallow lake impounded in a valley. The underlying sedimentary rocks include siltstones with delicate leaf impressions, from which material for palynological dating has been obtained (Jolley, 1997; see Chapter 5). Carbonised logs up to 1 m in length occur in conglomerate at Maternity Hollow, east of Fionchra. Secondary minerals in the lavas include the well-known bloodstone and banded agate found in the basaltic andesite flows on Bloodstone Hill (p.153; 172).

The conglomerates of north-west Rum provide an excellent example of the palaeogeographical information that may be derived from deposits of this type (e.g. Emeleus, 1973). Four groups of clasts are recognised:

- Lewisian gneisses and Torridon Group sandstones derived from the country rocks
- rocks not found on Rum, including quartz-dolerite and vesicular tholeiitic basalt (?lava), and rare psammites (possibly Moine Supergroup)

Plate 17 *Fluviatile conglomerate overlain by tholeiitic andesite 'icelandite' lava. Fionchra, Rum. Hammer shaft is 30 cm (P580467).*

- rocks from the Rum Central Complex (granophyre and microgranite, porphyritic rhyodacite, bytownite troctolite, gabbro, igneous breccias); the absence of peridotite clasts is attributed to the physical instability of the olivine-rich rocks
- in the younger beds, lava clasts derived from earlier lava flows in the formation

The abundance of clasts derived from the Rum Central Complex provides unequivocal evidence of rapid uplift and unroofing of the central complex during the Paleocene, and its continued vigorous erosion during accumulation of the lavas; the configuration of the deposits shows that they were laid down mainly from fast-flowing streams and rivers in rapidly evolving, steep-sided valleys and, less commonly, in small lakes.

MULL LAVA GROUP

The Mull Lava Group crops out throughout north and west Mull, together with the coastal parts of south Mull (around the margin of the younger Mull Central Complex) and the various offshore islands to the west, including Ulva, Gometra, the Treshnish Islands and Staffa. The successions of flat-lying lavas form substantial mountains, such as Ben More, and commonly develop strong trap featuring (Plate 15). The lavas on the Morvern peninsula, east of Mull, are considered to belong to the Mull Lava Group, as do the lavas east of Ben Hiant, Ardnamurchan, and probably also the isolated outcrops on the south coast of the Ardnamurchan peninsula. Offshore lavas, which extend beneath the Sea of the Hebrides to the Skerryvore Fault, are most likely part of the Mull Lava Group. These flows are probably contemporaneous with flows of the Eigg Lava Formation, which they join in a northern, submarine extension (Fyfe et al., 1993; Figure 9; Chapter 5).

There has been no detailed regional mapping of the Mull Lava Field since the time of the original survey (Bailey et al., 1924). This account therefore relies to some extent on the synthesis of Richey (1961), together with observations made during the last thirty-five years (e.g. Kerr, 1995b). The lava stratigraphy has been rationalised here, in keeping with that used for the other lava sequences (Table 15).

The remains of the Mull Lava Field as presently exposed, cover an area of about 840 km^2 on Mull and Morvern (Emeleus, 1991). The total preserved thickness of lavas on Mull is estimated

Table 14 The Canna Lava Formation of the Skye Lava Group: north-western Rum (Emeleus, 1997).

Member	Thickness (m) (maximum, see text)	Details
Orval	125	Several thick flows of basaltic hawaiite
Guirdil	20	Up to two tholeiitic andesite 'icelandite' flows, each generally underlain by conglomerate (0–2 m thick)
Upper Fionchra	170	Flows of tholeiitic basaltic andesite with up to 60 m of hyaloclastite breccia and pillow lava near the base of the succession. Underlain by up to 40 m of conglomerate, and bedded gritty sandstone and siltstone, the latter plant-bearing
Lower Fionchra	300	Predominantly alkali basalt flows. Basaltic hawaiite and hawaiite flows near top of the succession. Underlain by up to 50 m of conglomerate and bedded gritty sandstone, the latter with plant remains (logs)

Table 15 The Mull Lava Group.

Formation	Approximate thickness (m)	Details
Mull Central Lava	900	South-east Caldera (Bailey et al., 1924) *Central Zone*: non-porphyritic, olivine-poor and olivine-free tholeiitic basalt lavas; pillow-structures absent *Middle Zone*: plagioclase-phyric tholeiitic basalt lavas; pillow-structures common *Outer Zone*: olivine-poor tholeiitic basalts (commonly plagioclase-phyric) interleaved with flows tending towards the olivine basalts of the Mull Plateau Lava Formation; pillow-structures common in the tholeiitic lavas
Mull Plateau Lava	240	Ben More Pale Member olivine basalt, including the Coire Gorm magma-type of Kerr (1995b)
	90	Predominantly flows of mugearite, benmoreite and trachyte, commonly with platy jointing in more evolved flows. Rare plagioclase-macrophyric basalt flows (Big Feldspar Basalt)
	120	Olivine basalt
	450	Ben More Main Member olivine basalt (commonly olivine-phyric) and hawaiite, with rarer picrobasalt, mugearite, benmoreite and trachyte; plagioclase-macrophyric flows (Big Feldspar Basalt)
Staffa Lava		Several flows of tholeiitic basalt, commonly with well-developed columnar jointing. Intercalated plant-bearing sedimentary rocks (Ardtun Conglomerate Member)
	1–6	Gribun Mudstone Member reddish purple to purplish brown to greenish buff massive mudstone and siltstone of lateritic origin (see text)

to be about 1800 m (including an estimated 900 m of olivine-poor tholeiitic basalt lavas now largely within the Mull Central Complex). On the Morvern peninsula, the sequence is about 460 m thick but the lowest lavas thin towards the north (Bailey et al., 1924) and are overstepped by subsequent flows. A thin, laterally continuous sequence of Upper Cretaceous rocks underlies the lavas of Morvern except in the extreme east and north-east of the lava field, where the flows rest directly on pre-Cretaceous rocks (Triassic sandstones and gneisses of the Moine Supergroup). The marginal portion of the lava field was therefore erupted onto, and possibly terminated against, a land surface with a significant topographical relief. The northwest-trending Assapol Fault, in south-west Mull, defines the present-day south-west margin of the lava field.

The volcanic succession on Mull (Table 15) is divided into the Staffa Lava Formation at the base, overlain by the Mull Plateau Lava Formation that consists of the Ben More Main Member and the overlying Ben More Pale Member; these were formerly named, respectively as the Staffa, Main and Pale 'suites', of the Plateau Group (e.g. Bailey et al., 1924; Richey, 1961). The Mull Central Lava Formation crops out principally within the Mull Central Complex, and consists of olivine-poor tholeiitic basalt lavas.

The sequence built up by the effusion of lavas, predominantly from north-west-trending fissures now represented by the Mull Dyke Swarm. Pauses in the volcanic activity were more common during eruption of the lower part of the succession. They are marked by the development of laterally discontinuous sequences of conglomerate, sandstone, siltstone and coal. However, unlike the lava piles elsewhere in the Hebridean Igneous Province, the current subdivisions of the Mull lavas are based on lithological changes in the lavas and do not always depend on the presence of interflow sedimentary sequences.

Zeolite minerals are common throughout the lavas. These have been shown to have a depth-related distribution (Walker 1971; Figure 12), similar to the zonal distribution found in Icelandic lavas. The highest zone is characterised by laumontite and overlies a mesolite zone. Based on comparisons with Icelandic zones, it is estimated that the lavas in the vicinity of Ben More were originally over 2200 m in thickness, of which about 1000 m is preserved. A distinct zone characterised by carbonate minerals occurs in the lavas near Tobermory and in north-west Morvern, becoming wider northwards towards Ben Hiant on Ardnamurchan. It has been tentatively suggested that the carbonate may have been deposited by circulating heated waters from the Ardnamurchan Central Complex (Walker, 1971). The Mull Central Complex is surrounded by a zone of pneumatolysis, which is superimposed on the depth-related zeolite zones (Figure 13).

Staffa Lava Formation

The basal part of the volcanic sequence on Mull and locally on Ardnamurchan is marked by the development of a laterally persistent mudstone, the Gribun Mudstone Member, commonly less than 1 m thick, but thicker at several localities. In Morvern it is known locally as the Beinn Iadain Mudstone Formation (Figure 7). Near Feorlin Cottage at Carsaig, Mull, the mudstone occurs within cavities that developed on the surface of chalk (of possible Turonian age). On the Croggan peninsula south of Loch Buie, at An Garradh, the member consists of about 6 m of buff-coloured calcareous mudstone (marl) and is thought to represent a more aluminous equivalent. These features, together with certain petrographical characteristics (for example the presence of quartz grains), suggest that the mudstone is the product of extreme lateritic weathering of a basaltic ash, which was deposited prior to the effusion of the overlying lavas. The quartz grains were possibly derived from Cenomanian sandstones and introduced either by alluvial or aeolian processes.

At Malcolm's Point, on the south coast of the Ross of Mull, sedimentary rocks may represent a useful lithostratigraphical marker at the base of the Staffa Lava Formation and above an older, distinctive group of non-columnar flows. The sedimentary sequence comprises a thin carbonaceous mudstone overlain by an upward-fining fluviatile conglomerate–sandstone sequence containing rounded flints (derived from the Turonian Chalk deposits) and more angular fragments

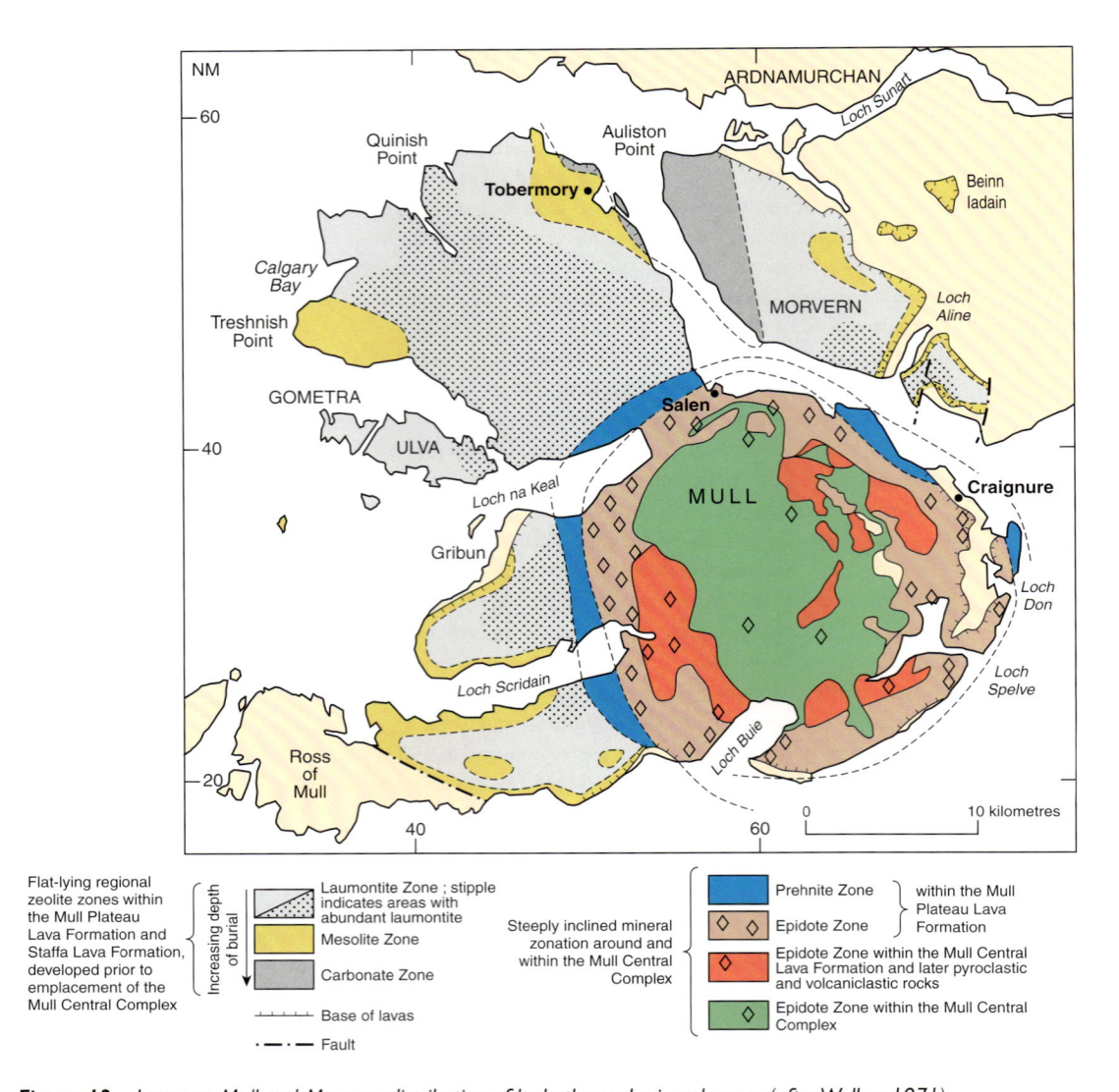

Figure 12 Lavas on Mull and Morvern: distribution of hydrothermal mineral zones (after Walker, 1971).

of basalt. The sequence is capped by a thin fissile mudstone. The main part of the Staffa Lava Formation comprises a distinctive sequence of tholeiitic basalt flows (defining the Staffa Magma-type or sub-type of Thompson et al. 1986; see Chapter 10). The flows crop out around the sea cliffs of south-west Mull, for example on the Ross of Mull at Malcolm's Point and Ardtun, at Burg and The Wilderness on the Ardmeanach peninsula, and on Staffa where they overlie hyaloclastite deposits (Plate 13). Flows of the Staffa Lava Formation also crop out at Bloody Bay, north of Tobermory. Significantly, the formation does not occur everywhere at the base of the lava pile; for example it is absent at Beinn na h-Iolaire in west Mull and throughout the Morvern peninsula (Kerr, 1995b).

Fluvial systems that developed between the eruptions deposited largely clastic sedimentary sequences that are collectively termed the Ardtun Conglomerate Member. These include laterally discontinuous conglomerate–sandstone sequences such as those at Ardtun on the north coast of the Ross of Mull, and laterally equivalent, overbank or lacustrine facies that give rise to siltstone–mudstone–coal–limestone sequences, well preserved on the south coast of the Ross of Mull (Boulter and Kvacek, 1989). The most distinctive field characteristic of flows of the Staffa Lava Formation is the development of typically near-vertical columnar jointing, as

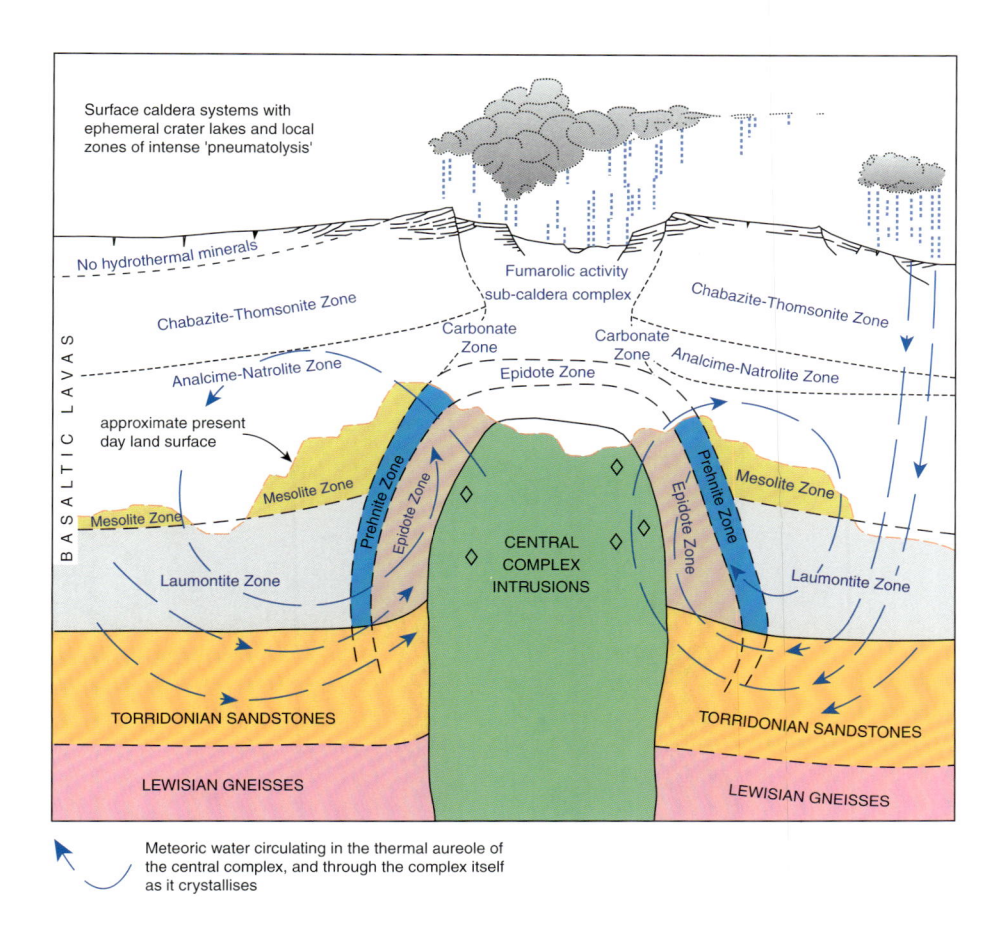

Figure 13 *Hydrothermal circulation in and around the Mull Central Complex.*
Diagrammatic cross-section showing the hydrothermal mineral zones in the lava succession and the superimposed alteration zones developed about the central complex in response to the circulation of heated meteoric water (based on Walker, 1971; Taylor and Forester, 1971; Bell and Williamson, 2002).

exemplified by the Fingal's Cave Flow on Staffa (Plate 13). The locally restricted nature of most flows, with columnar joints that in some instances are near-horizontal, is suggestive of cooling against steep-sided walls of palaeovalleys that cut through a dissected plateau. On the Ardmeanach peninsula, a pillowed facies indicates that the MacCulloch's Tree flow erupted into shallow water. The flow is underlain by interbedded hyaloclastite breccias, together with dark mudstones and siltstones that contain fragments of carbonaceous material (wood), in a fine-grained matrix. The preservation of a cast of an upright tree, 'MacCulloch's Tree' (Plate 18) is the most remarkable feature of the flow, with columnar jointing becoming severely contorted as it approaches the vertical surface of the trunk.

Mull Plateau Lava Formation

The most complete and relatively simplest succession occurs at Ben More, where both the Ben More Main Member and the Ben More Pale Member are present (Table 15). This succession has been the subject of detailed geochemical examination (Kerr, 1995a; Chambers and Fitton, 2000; Chapter 10).

The Ben More Main Member occurs throughout west Mull, as well as on the Morvern peninsula. Reddened, weathered tops to flows, attributable to subaerial weathering, are relatively common, although interflow clastic sedimentary sequences have not been noted to any significant

Plate 18 *MacCulloch's Tree; the cast of a large conifer (Taxodioxylon) encased in columnar basalt lava. Small amounts of charcoal are preserved in places at the margins. Ardmeanach, western Mull (P580468).*

extent. Certain of the reddened deposits on top of flows have been identified as ashy deposits, probably reworked by stream action (Emeleus et al., 1996a). The member is composed of randomly interleaved flows of olivine basalt and hawaiite, together with rarer picrobasalts, mugearites, benmoreites and trachytes. Certain flows, especially the hawaiites and mugearites, are plagioclase macroporphyritic (the Big Feldspar Basalts of Bailey et al., 1924).

Little detailed information is available about the structure and stratigraphy of the Ben More Main Member. Below, two localities are briefly described, in order to indicate some of the complexities and subtleties that are recognised.

The sequence of flows that forms the coastal cliffs at Laggan Bay, near Ulva Ferry, banked up against, and eventually overstepped, a thick heterogeneous accumulation of basaltic ashes, volcaniclastic breccias and debris flow deposits. These most likely mark the site of a vent that penetrated the lava pile relatively early in the development of the Ben More Main Member. The majority of the clasts within the breccia are of basalt and hawaiite; however, rare but conspicuous angular fragments of Moine psammite and Turonian flint can be identified, indicating that material from the subjacent crust has been transported upwards by the magmas or eroded from surface outcrops.

On the west side of the Quinish peninsula, in north Mull, a remarkably well-preserved lava flow crops out for a distance of at least 800 m along the coast between the high and low water lines. The flow preserves both casts (in basalt) and moulds of tree trunks and possibly branches, most of which are flat-lying, and the majority of which trend north-east–south-west. The upper portion of the underlying lava is severely lateritised, representing the soil in which the trees grew. The flow is remarkably well preserved, with surface features and internal structures more typically seen in active volcanic areas, for example, ropy pahoehoe structures, shell-like pahoehoe crusts, and basal or marginal breccias. The remarkable state of preservation of the flow and the presence of fossil trees are taken as clear indication that the Quinish lava was erupted after a significant hiatus in the volcanic activity, and was itself rapidly buried by the succeeding flow.

The Ben More Pale Member crops out around the summit of Ben More, close to the western margin of the Mull Central Complex and, consequently, has been subjected to intense hydrothermal alteration and metasomatism. It overlies the Ben More Main Member, although no sharp boundary has been identified; rather, there is an interdigitation between the two members, over an interval of a few flows, with the increasing preponderance of pale-weathering flows, up sequence. The pale weathering flows are generally of relatively evolved composition, typified by benmoreites (type locality, first identified by Tilley and Muir, 1964). The lower part of the Ben More Pale Member consists predominantly of mugearite, benmoreite and trachyte. The lavas showing the most evolved compositions are typified in the field by a platy jointing and general fissility, and occur in the middle part of the member. The uppermost part of the member marks a return to significantly less-evolved olivine basalt flows (Kerr, 1995a). Intercalated with the lavas are thin sedimentary deposits; on the north side of Ben More, a benmoreite flow is underlain by 30 to 60 cm of fissile black mudstone containing abundant fragments of benmoreite; a similar deposit at the same stratigraphical level occurs on the east side of A' Chioch. South of the summit of Ben More, Bailey et al. (1924) recorded about 3 m of brecciated black mudstone with plant remains overlying a mugearite flow.

Mull Central Lava Formation

The Mull Central Lava Formation consists of tholeiitic basalt flows that crop out in small areas within the Mull Central Complex and possibly around its margins, where they overlie the Mull Plateau Lava Formation. The formation is about 900 m thick, and many of the lavas may have been erupted within a water-filled caldera, forming a thick succession of pillowed flows. This is the South-east Caldera or Early Caldera of Bailey et al. (1924); the structure is about 10 km in diameter, and is thought to have developed as a consequence of the summit collapse of a central vent volcano on a site now occupied by the Mull Central Complex. Subsequent intrusive activity within the central complex has dismembered the pillowed flows, which now occur as isolated exposures a few metres across, preserved as screens between the younger

Plate 19 *Pitchstone lava and ash flow resting on flows of the Eigg Lava Formation, Sgurr of Eigg, Eigg (D1698).*
A site of fossil wood in conglomerate, the 'Eigg pine' (Harker, 1908)
B–B' base of pitchstone where it transgresses several flows of the Eigg Lava Formation

intrusions. Another caldera was also recognised, the so-called North-west Caldera (Bailey et al., 1924, plate III), but flows of the Mull Central Lava Formation within this structure are not pillowed. Throughout the formation, intense hydrothermal alteration has led to severe changes to the primary mineralogy of many of the lavas, with the development of abundant secondary epidote and prehnite, commonly replacing primary minerals, but also in fracture-filling vein systems.

Within the South-east Caldera, Bailey et al. (1924) recognised three zones within the formation (Table 15), but complications due to subsequent structural and intrusive events prevent the thicknesses of these zones being defined. Pillowed flows are restricted to the Outer Zone

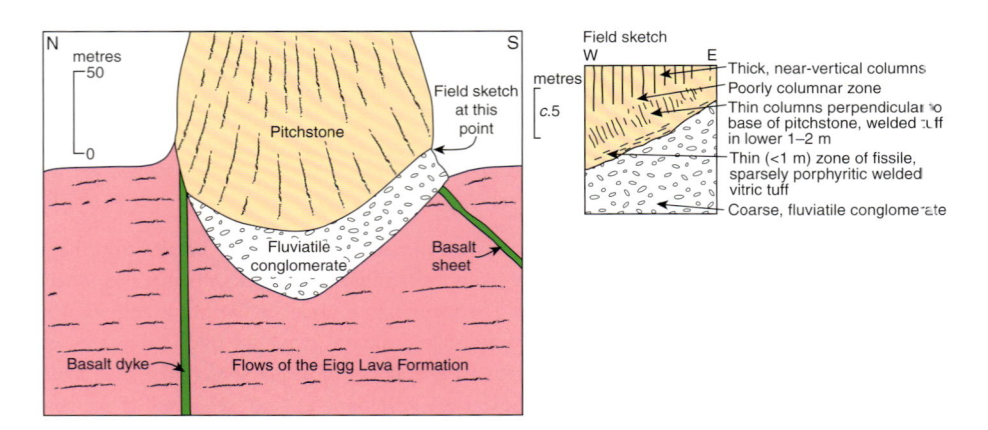

Figure 14 *Sketch of the western end of Sgurr of Eigg pitchstone at Bidean Boideach (from Emeleus, 1997 and based on Allwright, 1980).*

and Middle Zone. On the basis of the significant thicknesses of pillow lavas still preserved, Bailey et al. (1924) argued that there must have been successive subsidence of the caldera floor.

It is important to note that the lavas of distinctive tholeiitic basalt composition were not always contained by the caldera wall. It is envisaged that some were able to flow outwards, presumably down the flanks of the volcanic superstructure, and possibly to interdigitate with flows of the Mull Plateau Lava Formation that had been erupted from fissure systems.

SGURR OF EIGG PITCHSTONE FORMATION

This formation crops out in south-west Eigg and forms the islets of Oigh-sgeir about 15 km west of Rum (Emeleus, 1997). On Eigg, the formation forms the bare, steep-sided ridge that runs north-west from An Sgùrr to Bidean Boideach. The pitchstone lava is up to 120 m thick, of rhyodacitic composition, and is underlain by pockets of fluviatile conglomerate containing plant remains, which locally reach a thickness of 50 m at Bidean Boideach (Figure 14). There, and near the eastern end of An Sgùrr (Plate 19), the steep-sided contacts demonstrate that the pitchstone and the conglomerates occupied a canyon-like, palaeovalley system eroded in the underlying basaltic lavas of the Eigg Lava Formation (Bailey, 1914). The pitchstone is generally characterised by excellent columnar jointing (Plate 14). On Eigg, the attitude of the jointing is extremely variable, due to cooling inwards from the steep-sided valley walls and its development has possibly also been influenced by surface water percolating down into the solidifying mass. The fresh rock is a black, glassy pitchstone studded with small feldspar crystals; flow banding may be developed. Devitrification is common, especially near the base where fragmented, altered pitchstone is mixed with underlying sedimentary material, suggesting that phreatic explosions occurred when the pitchstone flowed over wet sediments in a river bed. At Bidean Boideach, western Eigg, a 3 m-thick zone of welded tuff forms the base of the pitchstone indicating that the initial eruption was an ash flow (Figure 14). Pale, cream-coloured sheets of devitrified pitchstone, a few metres in thickness, are prominent in the cliffs on the south side of the ridge of An Sgùrr. They represent intrusions of more differentiated rhyolitic magma into the main pitchstone mass.

seven

Dykes, dyke swarms and volcanic plugs

Numerous dykes of Palaeogene age are present throughout western Scotland and the Hebrides. The dykes commonly occur in parallel, north-west- to north-north-west-trending regional swarms, becoming more numerous and varied in direction near to and within the central complexes (Speight et al., 1982; Figure 15). Volcanic plugs are much less common and occur principally within or close to the lava fields and central complexes.

Although most of the dykes are a metre or less in thickness, intrusions between 2 and 10 m in width are not uncommon, and thicker examples are known. On Skye, there is evidence that larger dykes may be built up incrementally by numerous small injections of magma (Platten, 2000). The pronounced Minch Magnetic Anomaly to the east of the Isle of Lewis (Figure 15) has been attributed to a dyke that averages about 1.1 km in thickness (Ofoegbu and Bott, 1985; Fyfe et al., 1993) but no outcrop occurs onshore. A notable example of an unusually thick dyke, up to 100 m in width, crosses Muck in several *en échelon* outcrops. This dyke consists of olivine-gabbro and olivine-dolerite, and the gabbroic facies has developed small-scale mineral layering, similar to layering in the sills of north Skye (p.87). Paleocene lavas and sedimentary rocks of Mid Jurassic age intruded by the dyke are altered to high-temperature, sanidinite-facies hornfelses (Emeleus, 1997). The Muck dyke probably acted as a feeder for lava flows; prolonged passage of magma, maintaining high temperatures, is indicated by the intense thermal alteration of the host rocks. In general, however, the dykes produced only slight induration and thermal metamorphism of the immediately adjoining host rocks.

The dykes in and near the central complexes are generally less than 2 m in thickness. However, injection of dykes side by side into the same fissure can result in thick multiple intrusions, which are most common close to the central complexes. On the Strathaird peninsula to the south-east of the Skye Central Complex, 60 per cent of the dykes in the Skye Dyke Swarm are multiple; large numbers of multiple dykes also occur close to the Mull Central Complex and multiple dykes are present along the south coast of Arran. Composite dykes, in which magmas of contrasting composition have been intruded in quick succession, generally without internal chilling, are less common and are also usually restricted to the vicinity of central complexes. On Arran, good examples of composite dykes crop out on the Tormore shore section, north of the composite Drumadoon Sill (p.90). At these localities, initial intrusion of quartz-dolerite has been followed in rapid succession by quartz-porphyry or felsite injected into the centre of the dyke.

Dykes generally have chilled, fine-grained margins against earlier rocks. Glassy selvedges are rare in basaltic dykes, but at Camas na Cairdh, Muck, dark, vitreous tachylyte occurs on the margins of several dolerite dykes and flow-structures in the glass indicate that locally the magma moved nearly horizontally through the dyke fissure. Wholly or largely glassy silicic dykes are, by contrast, comparatively common. Classic examples occur at Tormore on the west coast of Arran, where dark green, flow-banded, glassy pitchstones have intruded Triassic sandstones. The commonest internal structures in basaltic dykes are planar zones of amygdales parallel to the dyke margins; rarely, there is also internal textural variation parallel to the dyke walls (e.g. Drever in Brown, 1969). The dyke walls are typically near-planar, but where dykes have intruded relatively incompetent surroundings the walls may be highly irregular and are bounded by well-defined chilled margins. This is seen, for example, in several thin, lobate, apparently discontinuous basaltic dykes intruded into Cambro-Ordovician dolostones at Camas Malag, Skye (Nicholson, 1985).

Dykes commonly form upstanding, wall-like features, especially where intruded into relatively soft-weathering sedimentary rocks. Many examples are visible at low tide on the foreshore at Kildonnan, south Arran (Plate 20), and on the sea bed between Antrim and Kintyre wall-like

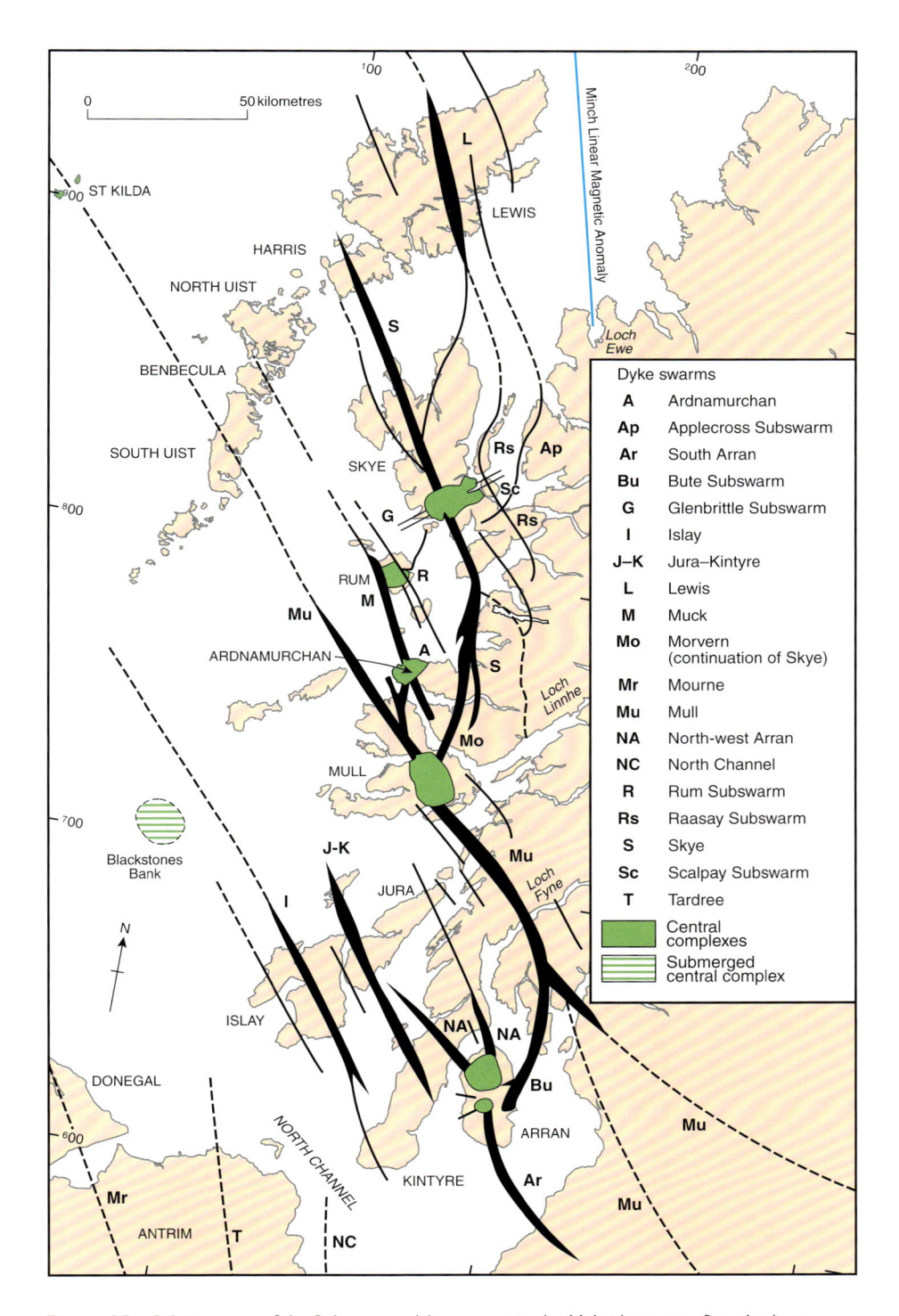

Figure 15 Dilation axes of the Palaeogene dyke swarms in the Hebridean area. Broader lines indicate the main axes of the regional swarms; broken lines indicate less certain axes (based on Speight et al., 1982, fig. 33.5). The Minch Linear Magnetic Anomaly is also interpreted as a dyke or dyke swarm (Ofoegbu and Bott, 1985).

Plate 20 *Dykes of the South Arran Dyke Swarm cutting Triassic sandstones on the foreshore at Kildonnan, Arran (P580470).*

dykes, up to 28 m in height have been recorded (Fyfe et al., 1993). In-weathered dykes are also common, for example at the Bay of Laig, Eigg, where basalt dykes occupy deep trenches between sharply defined, upstanding walls of indurated Mid Jurassic Valtos Sandstone Formation strata (Emeleus, 1997). Columnar jointing in dykes is generally developed perpendicular to the cooling surfaces provided by the country rocks. This helps to identify those dykes that have intruded basalt lavas or dolerite sills (with or without their own columnar jointing), for example at Dippin Head, south-east Arran.

DYKE SWARMS The presence of linear dyke swarms of Palaeogene age has long been recognised in north-west Britain and north-east Ireland (Figure 15). Several regional dyke swarms and localised subswarms have been identified in the Hebridean Igneous Province (Speight et al., 1982) of which the Skye and Mull dyke swarms are the most extensive. The principal features of these dyke swarms, including their dimensions (Table 16), were summarised by J E Richey in earlier editions of this book, and more recently by Speight et al. (1982). The swarms are generally densest in the vicinity of the central complexes, where smaller subswarms of varying trends also occur, and where the dykes commonly exhibit considerable compositional diversity. Intrusion of the dyke swarms has caused pronounced local crustal dilation, the greatest amount estimated for any swarm being 25 per cent to the south-east of the Skye Central Complex (Table 16). In northern Rum, the combination of dykes (and plugs) belonging to the north-west-trending Muck Dyke Swarm and the subsidiary north- to north-east-trending Rum Subswarm simulates a radial dyke swarm focussed on the central complex (Harker, 1908); it is, however, difficult to prove the presence of true radial dyke swarms in the province.

The distribution of the dyke swarms was controlled by a regional north-east–south-west extensional stress field, and the over-riding control on the orientation of the swarms is considered to have been the orientation of lower crustal intrusions that fed the dykes. Locally, subsidiary swarms are orientated approximately north–south, for example in the Outer Hebrides, and between south Skye and Morvern where the subsidiary swarm forms an *en échelon* link

Table 16 Dimensions of selected Palaeogene dyke swarms (based in part on Speight et al., 1982, table 33.1).

Locality	Width of swarm (km)	Number of dykes per kilometre	Percentage crustal stretch due to dyke intrusion
Skye	20	299	25
Mull	20	70	10
Rum	6	40	5
Arran	23	59	12
Islay	15	23	5

Locality details: Skye, south-east of the Cuillin Centre; Mull, south-east of the Mull Central Complex; Rum, coast south-east of the Rum Central Complex; Arran, south coast; Islay Swarm crops out across Islay (Walker, 1960, fig. 1)

between the Skye and Mull swarms (Figure 15). England (1988) suggested that a minor component of dextral shear within the regional stress field controlled emplacement of the north–south dykes on Lewis which cross-cut the pronounced regional 'grain' of the gneisses. Elsewhere, as in Morar, Moidart and Morvern, and on Arran and Ailsa Craig, upper crustal structures may have influenced the orientation of the swarms.

The structure and compositional character of the Mull Dyke Swarm changes as it is traced towards the south-east, away from the central complex. Jolly and Sanderson (1995) examined well-exposed shore sections between Mull and Loch Fyne and found that the multiplicity of thin dykes on and near Mull gives way south-eastwards to more widely spaced, thicker and better ordered dykes (Figure 16). Farther to the east-south-east, from Ayrshire to the extremity of the swarm in the north-east of England, individual thick dykes may be traced over considerable distances. Echelons of the Cleveland Dyke are up to 25 m thick in County Durham and north Yorkshire, some 350 to 400 km from Mull. It has been proposed that the Cleveland Dyke, which is closely similar in composition to tholeiitic basaltic rocks found in the Mull Central Complex, was fed laterally, in a matter of days, by a single pulse of magma that originated in the Mull Volcano (Macdonald et al., 1988). However, Speight et al. (1982) had contended that lateral flow of magma in dyke fissures could only have been very limited and they postulated that the swarms were fed vertically from ridge-like basaltic magma chambers situated at depths of 20 to 50 km. They also suggested that the central complexes developed where elongate magma chambers intersected major lines of crustal weakness such as faults. While the extent of lateral flow remains contentious, recent geophysical observations during active volcanism in Hawaii support the concept (e.g. Walker, 1993a).

The close relationships between the dyke swarms and the central complexes (see above) are evident from Figure 15. The abundance of dykes in the country rocks, including the lava fields, contrasts with their scarcity within many members of the central complexes, although there is much detailed evidence of substantial overlap between emplacement of dyke swarms and the intrusion of successive members of the central complexes. In Skye, many north-west-trending basic dykes intrude both the lavas and the gabbros of the Cuillin Centre from Blà Bheinn in the east to the northern part of the Cuillin near Sligachan; the dykes both cut and are cut by cone-sheets (Chapter 9). By comparison, far fewer north-west-trending dykes intrude the younger Red Hills granites. On Rum, Eigg and Muck, many dykes belonging to both the regional Muck Swarm and the Rum Subswarm intrude earlier members of the Rum Central Complex (Phase 1, see Chapter 9) but the later layered intrusions (Phase 2) postdate the majority of the dykes. On Mull, north-west-trending basic dykes intrude most members of the Mull Central Complex, but appear to be less numerous than in the surrounding lavas. This view is supported by evidence from the shores of Loch na Keal, where up to one-third of the dykes are free from the pervasive pneumatolytic alteration that affects their host lavas around the Mull Central Complex. On Ardnamurchan, only a few dykes cut the intrusions of Centres 2 and 3, although they are common in the country rocks, and a number cut pre-Centre 2 cone-sheets. Members of the north-north-west-trending Arran Dyke

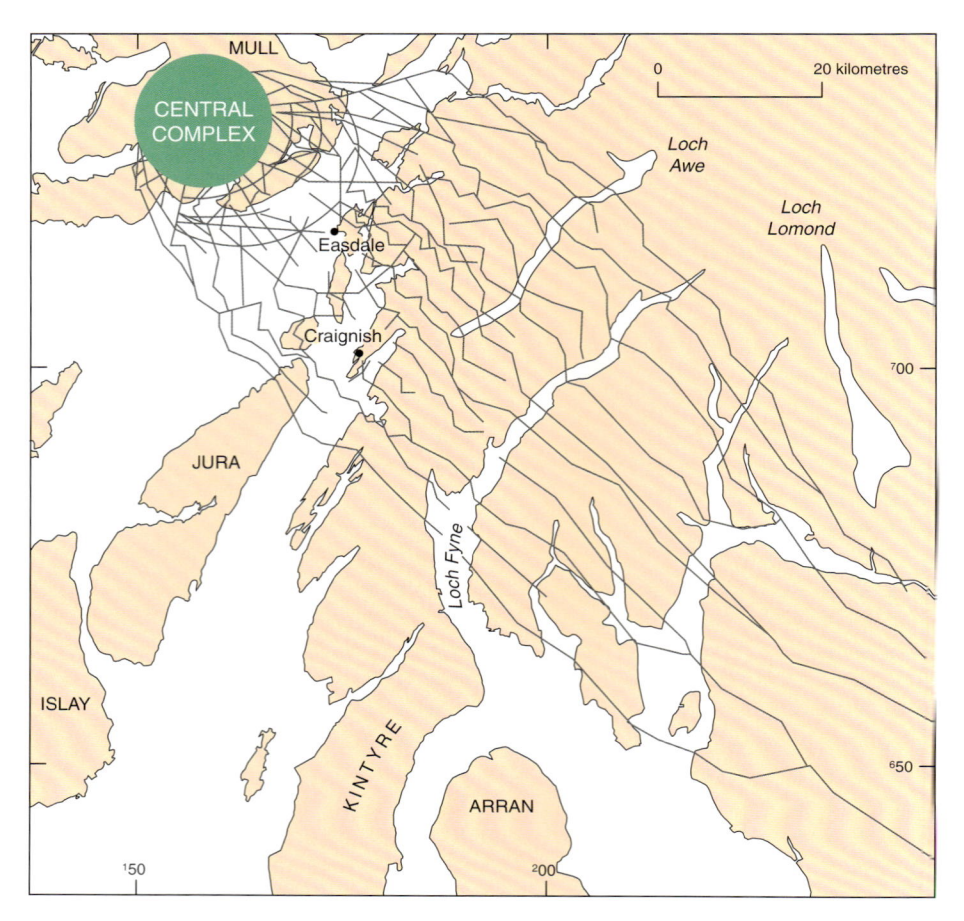

Figure 16 *Diagrammatic representation of the Mull Dyke Swarm, south-east of Mull. South-eastwards, away from the Mull Central Complex, a multiplicity of thin dykes in various orientations gradually gives way to thicker and less numerous dykes that trend predominantly north-west (after Jolly and Sanderson, 1995, fig. 19).*

Swarm intrude the North Arran Granite Pluton, but in fewer numbers than are seen in the country rocks. Few dykes are found in the Central Arran Ring-complex.

With few exceptions (e.g. the Canna Lava Formation in north-west Rum and on Canna and Sanday), the majority of the dykes in the regional swarms and subswarms appear to postdate the Paleocene lava fields surrounding the central complexes. While it is generally assumed that many of the lavas were fed from the dykes, or from plugs that developed on dykes, it is extremely rare to find an example of a dyke feeding a flow. It must be concluded that dyke intrusion was continuous, albeit with diminishing intensity, throughout the growth of the lava fields and the central complexes. In general, the dyke swarms appear to be intimately linked to specific central complexes, although the association of the Islay and Kintyre–Jura dyke swarms with the submarine Blackstones Bank Central Complex cannot be proved (Walker, 1960). Emplacement of a dyke swarm may, therefore, have taken place over a million years or more and the individual swarms may have been separated by even more substantial periods of time. As with the lavas, there was no single period of dyke emplacement in the Hebridean Igneous Province during the Palaeogene.

Dyke composition The majority of the dykes in the linear regional swarms are of basaltic or slightly more evolved composition. In and around the central complexes there are, additionally, dykes of silicic, intermediate and, less commonly, ultrabasic composition (Gibb, 1968, 1969). The basaltic dykes of the regional swarms are predominantly of tholeiitic basalt or tholeiitic olivine basalt, or of mildly alkaline olivine basalt. Dyke compositions may remain fairly uniform over considerable distances, as has been well demonstrated in regional dykes that extend across the Southern Uplands and into north Yorkshire (Macdonald et al., 1988).

Tholeiitic basalts and related, more evolved, lithologies predominate in the Mull Dyke Swarm. However, alkali olivine-dolerites, locally termed 'crinanites', are common as far southeast as Loch Fyne, and the swarm contains silicic and intermediate dykes in the vicinity of Oban and on Mull. The dykes intruding the Skye Central Complex and its immediate surroundings have a wide compositional range (Chapter 10). Sporadic north-west-trending dykes of silicic pitchstone and alkali olivine-dolerite intrude the granites of the Red Hills centres, and abundant earlier dykes of tholeiitic basalt and alkali olivine basalt overlap emplacement of the gabbros and the Coire Uaigneich Granite of the Cuillin Centre. Ultrabasic dykes are principally found close to the central complex, although a few ultrabasic dykes, rich in calcic plagioclase, intrude Paleocene lavas near Bracadale in north Skye (Donaldson, 1977). North-west-trending trachyte dykes occur along the axis of the Skye Dyke Swarm, for example in the vicinity of Drynoch, on Loch Harport.

The dykes on Rum are generally composed of transitional olivine basalt, less commonly of tholeiitic basalt and basaltic andesite, and very rarely of silicic pitchstone (Forster, 1980). Additionally, the Rum layered intrusions are cut by rare picritic basalt dykes, which most likely represent quenched ultrabasic liquids (Upton et al., 2002; see p.143). The dykes of Eigg, which belong to the Rum Subswarm, are principally transitional to mildly alkaline olivine basalts with a few more evolved hawaiites and mugearites, and rare feldspar-phyric dolerites. The dykes of the Muck Dyke Swarm on Muck are generally less evolved than those on Eigg and Rum and include a high proportion of mildly alkaline olivine-dolerites. On Rum, Canna and Sanday, a few north-west- to north-east-trending dykes of olivine basalt and dolerite intrude the flows of the Canna Lava Formation. They are either late members of the Rum Subswarm or, more likely, related to younger activity on Skye. No trachytic dykes are known from the Small Isles.

The Islay and Jura–Kintyre dyke swarms consist of mildly alkaline olivine-dolerites and evolved variants, including hawaiite; on Jura, a small subswarm is of a primitive picrobasalt composition. Along the south coast of Arran, members of the Arran Dyke Swarm exposed in the classic Kildonnan shore section range in composition from alkali olivine-dolerite ('crinanite') to tholeiitic olivine basalt, tholeiitic basalt and quartz-dolerite. The alkali olivine-dolerite dykes of Arran may be earlier than the tholeiitic basalt dykes, since the latter intrude the Central Arran Ring-complex and North Arran Granite Pluton, whereas the former do not. Silicic pitchstones occur in some abundance on Arran where they intrude the North Arran Granite Pluton. There is also a number of composite dykes with quartz-dolerite or tholeiitic basalt margins (commonly containing xenocrysts of quartz and rare alkali feldspar) and quartz-feldspar porphyry or contaminated porphyritic central parts. They are of similar composition to the composite sills (Chapters 8 and 10).

A monchiquite dyke containing xenoliths of mantle and deep-crustal origin, and a variety of megacrysts, crops out at Loch Roag on the Isle of Lewis (Menzies et al., 1987). This dyke is quite unlike any other dyke recorded from the Hebridean Igneous Province (p.144).

VOLCANIC PLUGS

Volcanic plugs are of widespread occurrence in the Hebridean Igneous Province, but are most common in and around the central complexes and in the lava fields. There is a very large concentration on Rum, where more than forty plugs of gabbro, dolerite and peridotite intrude the Torridonian and Triassic sedimentary rocks and most members of the central complex. The gabbro plugs may have fed surface flows but no compositional match can be made with any of the lavas preserved on Rum. The ultrabasic plugs of Rum contain magnesium-rich olivine and calcic plagioclase compositionally similar to the minerals in the Rum layered intrusions (Wadsworth, 1994).

The plugs range from a few tens of metres to over one kilometre in diameter. Their plan view may be nearly circular, but is more commonly oval or elongate. Where elongate, the long axes of the plugs are generally aligned in the direction of the regional dyke swarms as, for example, at 'S Airde Beinn and other localities in northern Mull. However, on Rum, elongate plugs appear to radiate from the central complex. On Islay, a north-north-west-elongated olivine-dolerite plug, or boss, at Cnoc Rhaonstil is considered to be the distended head of a dyke in the Islay Dyke Swarm (Hole and Morrison, 1992). Several of the thicker dykes in the Hebridean Igneous Province are of limited lateral extent and may actually be elongate plugs, possibly related to volcanic necks, as in Antrim. Examples include a dolerite mass that intrudes granite on the north-east shoulder of Beinn na Caillich, Skye, and the olivine-dolerite of Gualainn na Sgurra that is intruded into lavas near the east end of the Sgurr of Eigg.

The majority of the plugs are basaltic in composition, but a few of trachyte composition have been recorded. On Mull, a large trachyte plug intrudes lavas and is, itself, intruded by north-west-trending dykes at Druim Buidhe, south-east of Tobermory. On Ardnamurchan, a plug of biotite trachyte intrudes psammites of the Moine Supergroup east of Ben Hiant. The dolerite plug at Cnoc Rhaonstil, Islay, is unusual in exhibiting considerable internal variation in composition and texture, with marked alkali enrichment (Hole and Morrison, 1992; Preston et al., 1998b; Chapter 10).

Basalt or dolerite plugs intruded into the lava fields may easily escape detection unless they form distinct topographical features. This may account for their apparent absence from the Skye Lava Field. However, they may not occur there as they have not been identified among the adjacent and relatively well exposed Mesozoic sedimentary rocks. Several plugs intrude the lavas of northern Mull (e.g. Kerr, 1997). They include the olivine-dolerite at 'S Airde Beinn which has produced a pronounced thermal aureole in the surrounding lavas and was almost certainly a feeder for lava flows (Cann, 1965), as were a number of the other plugs on Mull. Several plugs of fairly Mg-rich tholeiitic dolerite have, however, no compositional equivalents among the lavas and probably acted as feeders for flows subsequently removed by erosion. The lavas of Eigg and Muck are cut by rare dolerite plugs; on Muck the only plug is an alkali olivine-dolerite.

Sedimentary and metasedimentary rocks intruded by plugs are commonly thermally metamorphosed. On Rum, high-grade thermal alteration of the Torridonian feldspathic sandstones surrounding the gabbroic plugs involved partial melting and the formation of tridymite around larger quartz grains; limited fusion of sandstone, to form buchite, is also present. The mineralogy of the altered sandstone adjoining a gabbroic plug in Kinloch Glen indicates that the thermal alteration occurred when the sedimentary rocks were buried under as little as 500 m cover of strata (Holness, 1999). By contrast, alteration of sandstone around the peridotite plugs of Rum is generally restricted to induration and discoloration of the sedimentary rocks with evidence of only limited melting; since the peridotites were emplaced carrying large amounts of suspended olivine crystals, their heat capacity and ability to affect their surroundings was apparently much reduced. Several olivine-dolerite plugs intrude Moine metasedimentary rocks east of the Ardnamurchan Central Complex where, in Glenmore, a plug has caused a limited amount of fusion of the feldspathic country rocks to form buchite. Alteration of the Dalradian metamudstone around the Cnoc Rhaonstil plug on Islay is slight; however, the Dalradian metasiltstone surrounding a plug of coarse olivine-dolerite at Sithean Sluaigh, south of Strachur on Loch Fyne, is intensely altered and its foliation has been deformed to parallel the margins of the plug. Here, partial melting has resulted in granophyric veins and refractory residual material containing mullite, magnetite, corundum and spinel (Smith, 1969).

eight
Sills and sill-complexes

Sills and sill-complexes are intimately associated with the lava fields and central complexes of the Hebridean Igneous Province. In general, the sills occur within the Mesozoic sedimentary sequences below the lavas, although on Mull the Loch Scridain Sill-complex also intrudes the overlying Paleocene lava field. Thus, the depth of emplacement of the upper parts of the sill-complexes is relatively shallow, most likely less than 1 km. Individual sills can vary from less than 1 m thick up to several tens of metres, and are exceptionally over 100 m. Bifurcations, transgressions and terminations are common. Locally, the country rocks may be disturbed and chilled margins are generally well developed, but thermal alteration of country rocks is typically slight. It is difficult to estimate the total thickness of each of the sill-complexes due to a lack of continuous vertical sections and the fact that many of them continue below sea level. However, it is likely that they are at least a few hundreds of metres thick.

Each sill-complex has its own distinctive compositional signature, unrelated to the typically older lava sequences. Alkali olivine basalt and tholeiitic basalt magmas are represented, together with their fractionation products (trachyte and rhyolite, respectively). Magma mixing between basaltic and silicic liquids can be recognised in some sills and may be a significant process in the evolution of some sill-complexes. Intra-sill variation in mineral proportions may be due to a variety of processes, including crystal settling or flotation and flow differentiation. Crustal contamination of the sill magmas may be deduced from whole-rock isotopic characteristics. Xenoliths are rare, except within the Loch Scridain Sill-complex.

The main sill-complexes are:

- the Little Minch Sill-complex, of alkali olivine basalt affinity on north Skye, Raasay and the Shiant Isles
- the Loch Scridain Sill-complex, of tholeiitic basalt affinity on south-west Mull

In south Arran there are several unrelated groups of sills of very different compositions. These include:

- the Holy Island and Dippin sills, of alkaline affinity
- microgranite/porphyritic rhyolite sills in southern Arran

Certain groups of sills show clear spatial and temporal relationships with the central complexes. These include composite sills (basaltic andesite–rhyolite) associated with the Mull Central Complex and the Eastern Red Hills Centre of Skye. Other groups of composite sills show no obvious link to central complexes, for example those in south Arran (Rogers and Gibson, 1977) and south Bute. The thick microgranite sill in southern Raasay is most likely related to the nearby granitic intrusions of the Western Red Hills Centre on Skye.

Ultrabasic sills are much less common and are generally associated with central complexes. Typically they are modally layered, with alternating olivine- and plagioclase-rich layers, as exemplified by the Gars-bheinn Sill on the southern margin of the Cuillin Centre of Skye (Weedon, 1960). Cryptic layering may also occur.

LITTLE MINCH SILL-COMPLEX

The Little Minch Sill-complex crops out on the Trotternish peninsula of north Skye, and it forms almost all of the Shiant Isles, with more restricted outcrops on the Waternish and Duirinish peninsulas (north Skye) and the west coast of Raasay (Figure 17). Significant submarine outcrops occur within the Little Minch, between Skye and Harris. The country rocks

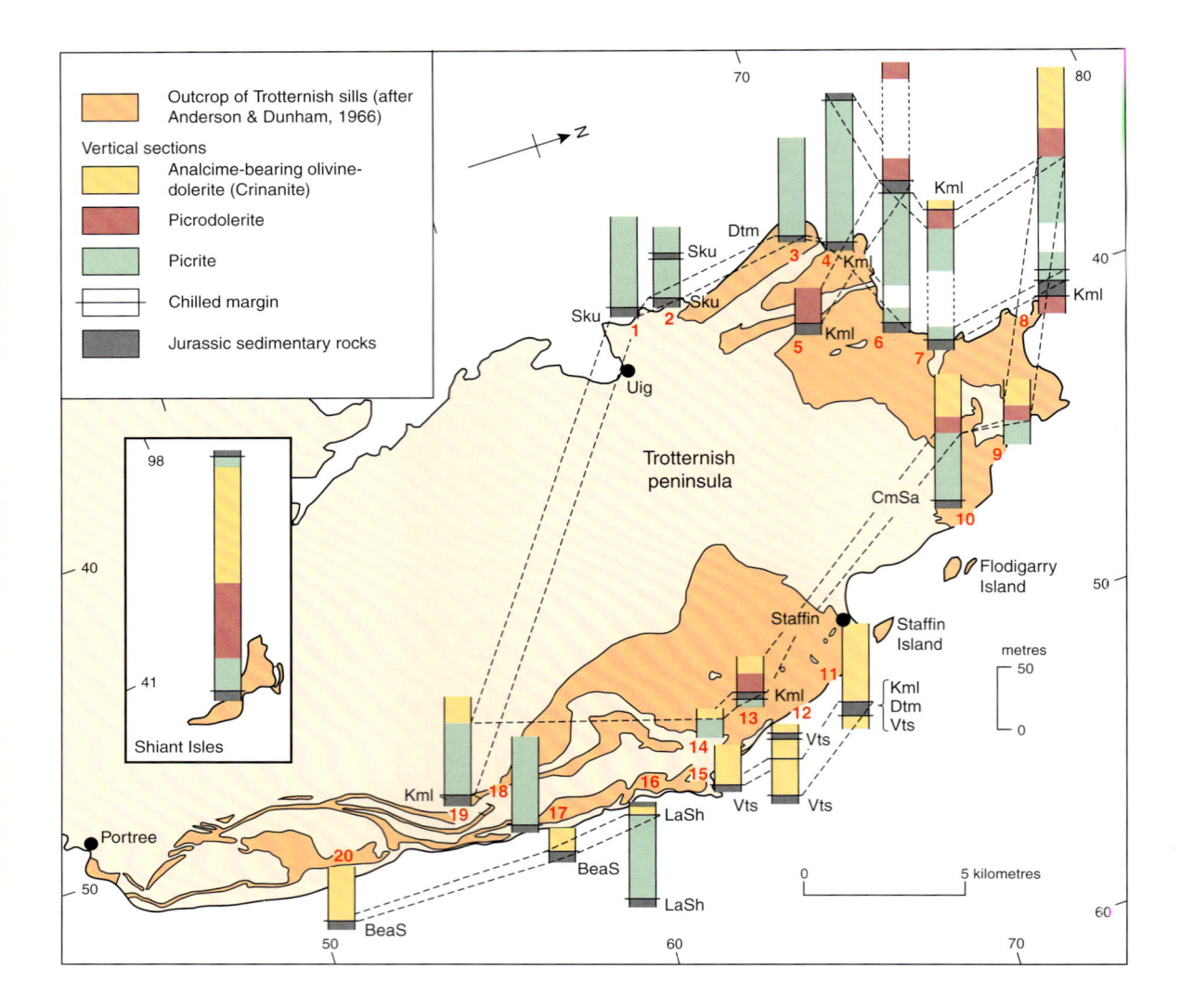

Figure 17 The Little Minch Sill-complex: the stratigraphy of the sills at selected localities on the Trotternish peninsula and on the Shiant Isles is shown. The Jurassic sedimentary rocks in contact with the sills are also indicated (after Gibson and Jones, 1991).
Key to localities: **1** Skudiburgh; **2** Kilbride Point; **3** Sgeir Lang; **4** Bornesketaig; **5** Heribusta; **6** Osmigarry; **7** Cregan Iar; **8** Meall Tuath; **9** Balmaqueen; **10** Flodigarry; **11** Staffin; **12** Loch Mealt; **13** Dùn Raisaburgh; **14** Culnaknock; **15** Dùn Connavern; **16** Inver Tote; **17** Leac Tressirnish; **18** Rigg; **19** Tottrome; **20** Bearreraig Bay
Jurassic sedimentary rocks: **BeaS** Bearreraig Sandstone Formation; **LaSh** Lealt Shale Formation; **Vts** Valtos Sandstone Formation; **Dtm** Duntulm Formation; **Kml** Kilmaluag Formation; **Sku** Skudiburgh Formation; **CmSa** Carn Mor Sandstone Member

are typically Jurassic sandstones and mudstones. Studies by Drever and Johnston (1965) and Anderson and Dunham (1966) were concerned predominantly with the field relationships and petrography of the sills. These investigations identified their alkaline nature, ranging in composition from olivine-rich picrite and picrodolerite, through to more-evolved analcime-bearing olivine-dolerite ('crinanite'). Simkin (1967) developed a model involving flow differentiation and gravity settling or flotation processes to explain the mineralogical variation and layering within certain of the sills. More recently, the sills on the Shiant Isles and Skye have been investigated in order to understand their mode of emplacement and petrogenesis (Gibson, 1990; Gibson and Jones, 1991; Foland et al., 2000; Henderson et al., 2000 and references therein).

The Little Minch Sill-complex postdates the Skye Lava Group and has a shallow dip towards the west. On the Trotternish peninsula, where the columnar-jointed sills are responsible for much of the impressive coastal scenery (Plate 21b), the sill-complex is transgressive, invading progressively younger Jurassic strata towards the north. The sill-complex has an aggregate thickness of at least 250 m, with individual sills from 10 to over 100 m in thickness. The sills can either be simple (one lithology), multiple, or composite involving some combination of picrite, picrodolerite and alkali olivine-dolerite. Where sills are composite, the lower unit tends to be the least evolved. The internal contacts between different units can show evidence of their relative age, based on the development of chilled margins. Isolated rafts of country rock may also occur at internal contacts in the multiple sills. This material is thermally altered, ranging from mild baking which involved recrystallisation and reduction in porosity, through to the formation of very fine-grained or glassy buchites where melting has occurred; the buchites provided material for Mesolithic stone tools at Staffin.

Layering is common in sills of basaltic compositions; within the Little Minch Sill-complex good examples are seen at Duntulm Castle (Plate 21a), Eilean Flodigarry, Meall Tuath and the Ascrib Islands, all on or just offshore of the north of the Trotternish peninsula. Layering tends to occur mostly within picritic units and is most commonly found adjacent to contacts with the country rocks. It is typically on a centimetre scale, involving variations in texture and in the proportions of olivine, clinopyroxene and plagioclase. However, such layering can also occur within a sill adjacent to rafts of sedimentary rock, implying that heat loss from the magma was important in the development of layering. Furthermore, there is no field or petrographical evidence to suggest that crystal settling has taken place to any significant extent. Thus, an in-situ crystallisation model is preferred to account for the layering (Gibson and Jones, 1991).

Gibson and Jones (1991) deduced that the parental magma was of alkali olivine basalt composition, with about 10% MgO. Fractionation occurred at relatively shallow depths, and batches of the three main magma compositions (picrite, picrodolerite and alkali olivine-dolerite) were then tapped from the processing chambers and emplaced at shallow levels as sills. The most primitive melt, which formed the picrites, carried chrome-spinel, olivine and plagioclase prior to emplacement. During emplacement, flow differentiation was a significant process close to the margins of the sills, whereas relatively late-stage filter pressing yielded various pegmatite sheets. The chilled margins to the sills are typically contaminated by country-rock mudstone and sandstone.

LOCH SCRIDAN SILL-COMPLEX, MULL

The Loch Scridain Sill-complex intrudes the Moine basement, the Mesozoic sedimentary sequence and the overlying Paleocene lavas in south-west Mull, particularly on the Ross of Mull and on the north side of Loch Scridain. The sills are typically between 0.5 and 6 m in thickness but exceptionally exceed 10 m, and commonly have well developed chilled margins. They are of tholeiitic affinity, ranging in composition from tholeiitic basalt, through andesite and dacite, to rhyolite (including glassy variants). Basaltic compositions are the most common, with about 20 per cent of the intrusions being of intermediate composition, and a relatively small number of rhyolitic composition. Many of the sills are xenolithic, with cognate and accidental (upper crustal) material represented; two of the best examples are to be found at Killunaig and Kilfinichen Bay, on the south and north sides of Loch Scridain, respectively (Plate 22). Composite sills occur, commonly involving crystalline and glassy rock types of significantly different compositions as in, for example, the classic xenolithic sill at Rudh' a' Chromain on the south side of the Ross of Mull. The sills may be related to the initial stage of the development of the Mull Central Complex (Centre 1 or Glen More Centre, see p.126; Dagley et al., 1987).

The sills were described in detail by Bailey et al. (1924). These early studies were concerned mainly with the complex high-temperature mineral assemblages that developed within crustal xenoliths prior to and/or during sill emplacement. More recently, Brearley (1986) described the melting reactions of the Moine basement materials, and Kille et al. (1986) inferred that the sill magmas were actively convecting during emplacement, maintaining high temperatures at the contacts, which resulted in localised thermal erosion of the pelitic Moine wall-rocks.

Plate 21 Dolerite sills of the Trotternish peninsula, northern Skye.
a   Rhythmically layered structures in the picrodolerite sill at Duntulm Castle. The darker layers are rich in olivine and clinopyroxene, the lighter layers are relatively plagioclase rich (P580471). The cliff is approximately 20 m high.
b   Columnar jointed olivine-dolerite sill intruding Valtos Sandstone Formation at the Kilt Rock. The sandstone is cut by another sill just above sea level. The cliff is approximately 100 m high (Locality 12, Figure 17) (P580472).

The Loch Scridain sills may be divided into three distinct geochemical groups (Preston et al., 1998a; Chapter 10). The basic sills (Group 1) are markedly xenolithic, containing both cognate and crustal types (Plate 22). The cognate xenoliths are of ultrabasic and basic composition, and are most likely of cumulate origin (Preston and Bell, 1997). There are two broad groupings of the crustal xenoliths (Preston et al., 1999):

- siliceous xenoliths derived from psammites of the Moine Supergroup or, less commonly, as in the Rudh' a' Chromain Sill, sandstone and conglomerate xenoliths from the local Mesozoic country rocks
- aluminous xenoliths, of which there are three types, each dominated by glass produced by melting of the xenolith, hence the term buchite

The aluminous xenoliths include:

- mullite buchites — a mass of mullite needles ($3Al_2O_3.2SiO_2$) set in clear glass and pale lilac in hand-specimen
- cordierite buchites — small crystals of cordierite and mullite needles set in clear glass and virtually black in hand specimen
- plagioclase-rimmed mullite buchites — a core of mullite-rich glass surrounded by a thick rim of white-weathering, coarse-grained plagioclase. (An_{87-60})

Clear blue corundum (sapphire) also occurs. Pockets of quenched, isotopically contaminated basic glass with skeletal plagioclase and clinopyroxene occur trapped between the plagioclase crystals. The highly aluminous composition of the buchite glasses is consistent with their derivation from a clay-rich sediment or its metamorphic equivalent, most likely the pelites of the Moine Supergroup (Dempster et al., 1999).

HOLY ISLAND AND DIPPIN SILLS, ARRAN

Holy Island, on the east side of Arran, is dominated by a sheet of sanidine-microphyric trachyte about 250 m thick, emplaced into Permian sandstones. This peralkaline intrusion is riebeckite-

Plate 22 Quartzite xenolith in basalt sill, Loch Scridain, south-western Mull. Hammer shaft is 30 cm (P580473).

bearing, similar to the microgranite intrusion that forms Ailsa Craig (p.138). Other related intrusions are sills or transgressive sheets of analcime-bearing olivine-dolerite which crop out around Lamlash Bay. These inclined sheets may have a focal point at depth in the vicinity of Lamlash Bay (Tomkeieff, 1961), akin to the focal point of cone-sheets associated with the central complexes (Figure 31).

The Dippin Sill is 40 m thick at Dippin Head and forms a transgressive sheet emplaced into Triassic mudstones and siltstones (Gibb and Henderson, 1978). It is a composite sill, with a thick central unit of analcime-bearing olivine-dolerite ('crinanite') bounded top and bottom by analcime-dolerite ('teschenite'). Pegmatitic facies occur sporadically throughout the sill. The parental magma was of alkali olivine basalt type that fractionated within a magma chamber at a deeper level than the sill. The first phase of intrusion involved relatively evolved magmas from the upper part of the chamber, followed by intrusion into the central part of the sill of less evolved magmas carrying suspended olivine, with or without plagioclase crystals from the lower part of the chamber. As with the Little Minch sills, flow differentiation of the magmas occurred during emplacement, together with assimilation of material from the conduit walls. The numerous pegmatitic segregations are the product of late-stage filter pressing within the crystallising body. A number of subsolidus processes are also recognised, including decomposition of magmatic nepheline (± alkali feldspar) to produce analcime. Circulation of hydrothermal (meteoric) fluids caused the formation of secondary serpentine, chlorite, iron oxides, and natrolite and other zeolites (Dickin et al., 1984b).

MICRO-GRANITIC AND RHYOLITIC SILLS, SOUTH ARRAN

These thick silicic sills form prominent crags at Brown Head south of Drumadoon Bay and inland for some distance to the east (Tyrrell, 1928; Figure 31). They are most likely linked to dykes of similar composition, for example those that crop out south of Corrygills Point on the east coast. In some respects these sills are gradational with the composite sills described below.

Most of the sills are conspicuously porphyritic, with phenocrysts of quartz and alkali feldspar, up to 1 cm and 1.5 cm across, respectively, set in a fine-grained groundmass. Sparse phenocrysts of sodic plagioclase are also present. Commonly, the phenocrysts are anhedral, suggestive of some form of magmatic corrosion by the groundmass 'liquid', akin to the types of reaction envisaged for the composite intrusions (see below).

COMPOSITE SILLS OF ARRAN AND SOUTH BUTE

These composite sills are up to 80 m thick, and form some of the higher ground in south-east Arran, and the prominent cliffs at Bennan Head and at Drumadoon on the west coast (King, 1982) (Plate 23). Typically, they comprise a volumetrically dominant, central silicic unit flanked by relatively thin basaltic or doleritic units, but some consist of a central doleritic unit, flanked by marginal felsites (Rogers and Gibson, 1977; Macgregor, 1983). The Arran composite sills and dykes generally postdate the emplacement of the Holy Island and Dippin alkaline sills.

The internal contacts between the basic and the silicic units are typically gradational over a few tens of centimetres and 'xenoliths' or pillow-like masses of basic material may occur within the silicic unit. Within the xenoliths are xenocrysts of feldspar and/or quartz of the type seen in the silicic unit. In places, marginal felsites vein the central dolerite. These intimate relationships are suggestive of mixing processes, initially prior to intrusion, explaining the xenocrysts in the basic units, and then during intrusion, explaining the gradational contacts and the xenoliths of basic material in the silicic material (Rogers and Gibson, 1977).

TIGHVEIN INTRUSION-COMPLEX, ARRAN

Augite diorite in gradational contact with underlying xenolithic quartz-dolerite is intruded conformably into Triassic strata at Tighvein, in south Arran (Figure 31). Intrusions of microgranite, some granophyric, cut the augite diorite, which is also extensively veined by microgranite. The quartz-dolerite and augite diorite form a sill, but the microgranitic intrusions are steep-sided and may be parts of a ring-dyke (Herriot, 1975). Other sheets and dykes of felsite and porphyritic rhyolite cut the whole 'complex', together with dykes of pitchstone and tholeiitic basalt. The xenoliths in the quartz-dolerite are a fine-grained basic rock in various stages of digestion.

RAASAY SILL

The Raasay Sill is of granitic composition, crops out over an area of about 12 km² in southern Raasay, and is emplaced transgressively into Triassic and Lower Jurassic sedimentary rocks. Despite the sill being at least 30 m thick, its thermal effects on the adjacent country rocks are minor. Texturally, it varies from coarse to fine grained, with granophyric and porphyritic facies, and it contains alkali amphibole. The sill also occurs on the nearby small peninsula of An Aird on Skye, suggesting a link with the granites of the Western Red Hills Centre. One possible correlation is with the Maol nam Gainmhich Granite which, like the Raasay Sill, is distinctly alkaline in composition and contains riebeckite.

GARS-BHEINN ULTRABASIC SILL, SKYE

On the southern slopes of Gars-bheinn, adjacent to the Cuillin Centre of Skye, a layered ultrabasic sill, 80 m thick, has been emplaced into hornfelsed lavas. It is composed of feldspathic peridotite, the uppermost 15 m of which comprises alternating plagioclase-rich and olivine-rich layers. Pegmatitic textures are common within the plagioclase-rich layers. The base of the sill appears to be connected with a dyke of similar composition, and material from the chilled margin of this dyke suggests a parental magma of picritic composition. The source of these intrusions has been identified as the marginal gabbro of the adjacent Cuillin Centre. Weedon (1960) interpreted the layering as a consequence of crystal–liquid fractionation, with settling out of the olivine crystals. By contrast, Bevan and Hutchison (1984) concluded that the plagioclase-rich layers arose from the injection of 'sills' into already-crystallised olivine-rich material.

OTHER SILLS

Other dolerite sills of probable Paleocene age occur at a number of localities on the periphery of the Hebridean Igneous Province. Most notable are a massive, easterly dipping sill, which intrudes gneisses at Lochmaddy, North Uist (Mackinnon, 1974), and the Prestwick–Mauchline Sill-complex north of Ayr, which is petrographically similar to analcime-bearing olivine-dolerite sills on Arran, and has yielded a K-Ar mineral age of 58 Ma (De Souza, 1979).

Plate 23 Columnar jointed, composite (quartz porphyry–dolerite) Drumadoon Sill, Arran. Exposures in the foreground are in the composite dyke at Cleiteadh nan Sgarbh. The top of the cliff is about 50 m above sea level (P580474).

nine

Central complexes

The Hebridean Igneous Province is one of the World's classic areas for the development of central complexes (Figure 1). Each central complex may contain many, or few, igneous rock types in intimate relationship with various metamorphic and sedimentary country rocks. The igneous rocks exhibit a variety of modes of intrusion and crop out within restricted, approximately circular areas, up to about 15 km in diameter. These represent the eroded roots of central volcanoes and, where reliable age data are available, individual central complexes generally appear to have been relatively short-lived. From initiation to decay, and subsequent Palaeogene unroofing, possibly took less than 1 Ma in some instances. Most of the central complexes contain clear evidence of surface volcanism at an early stage in their evolution, and extrusive igneous rocks are commonly preserved within them. Conversely, the associated and relatively young, coarse-grained intrusions cooled and crystallised slowly at some depth (probably from 2 to 3 km). The coarse-grained intrusions were most likely emplaced beneath an ever-growing volcanic superstructure (e.g. Skye, Rum, Mull), but some probably formed beneath a thick cover of country rocks and may not have experienced associated surface volcanism (e.g. the North Arran Granite Pluton).

The central complexes have been foci of intense, very localised magmatic activity dominated by basic and/or ultrabasic magmas derived from the Earth's upper mantle. These primitive magmas were usually accompanied by silicic magmas of mixed origin:

- those generated through partial melting of the continental crust as this was traversed by the basic magmas
- those originating by fractionation of the basic magmas as they ponded in the upper crust (Chapter 10)

Most of the central complexes are sited on ridges of pre-Mesozoic rocks, at the intersections of major north–south-trending lineaments with older cross-cutting (generally north-east–south-west) faults (Figure 3). The continental crust within the Hebridean Igneous Province is generally thinner than in adjoining areas (Meissner et al., 1986), which may have focussed the igneous activity (Thompson and Gibson, 1991). The surrounding country rocks comprise older Paleocene lava fields, as for example on Skye and Mull, and/or basement lithologies, for example Lewisian gneisses and Torridonian sedimentary rocks on Rum, Dalradian metasedimentary rocks plus younger sedimentary strata on Arran, and Moine psammites and pelites on Ardnamurchan. These basement lithologies are commonly overlain by attenuated basin-margin sequences of Triassic, Jurassic and Cretaceous strata.

The geometry of the intrusions depends on the lithologies involved, and five main associations are described below.

i   Coarse-grained ultrabasic and basic rocks, such as peridotite, troctolite and gabbro, commonly form layered, relatively flat-lying sequences that may be several hundreds of metres thick, and possibly dip at a shallow angle towards a focal point at a few kilometres depth. Alternatively, these rocks may form sets of concentric conical intrusions (cone-sheets) with a focal point at depth (see below). Ring-dyke intrusion models that were previously invoked to explain the arcuate plan of certain large basic intrusions (e.g. Bailey et al., 1924; Richey, 1932) are not now considered to be valid for a number of reasons. In particular, the central 'block' would have to drop to unrealistic depths along steeply inclined, outwardly dipping fractures in order to create the space necessary for intrusions that commonly have ring widths in excess of 1 km.

ii   Fine-grained basic rocks, for example dolerite and basalt, occur in sets of cone-sheets and other sheet-like intrusions emplaced into older coarse-grained ultrabasic and basic intrusions and the adjoining country rocks; they are related to dyke emplacement during growth of the central complexes (see below).

iii   Intermediate, hybrid and granitic rocks typically occur in nested plutons comprising small, stock-like intrusions, usually with steep-sided, outwardly inclined margins, which are for the most part younger towards the core of the intrusive centre. Narrow ring-dykes, commonly of mixed-magma composition (basic to silicic), may also be present.

iv   Cone-sheet emplacement and its relationship with the linear dyke swarms have been examined by Walker (1975b, 1993b). Sets of cone-sheets comprising basalt or dolerite form conical intrusions, each typically less than 5 m thick, with a common focal point at depth. They are generally more numerous in two opposing quadrants of the central complex, for example north-west and south-east, and are related to a dyke swarm that, in this example, would trend north-west–south-east (Figure 18a). Thus, the central complex appears to have acted as a focal point for cone-sheet and dyke emplacement. Detailed mapping of the cone-sheets and dykes indicates that the emplacement of both sets of intrusions overlaps in time and space. Where crustal dilation is slow, then magma excursions cannot be accommodated as dykes, but rather will be emplaced as cone-sheets. Intrusion of sets of cone-sheets can cause vertical thickening and uplift of the volcanic edifice by as much as 1500 m (e.g. Le Bas, 1971), similar to the amount of horizontal crustal dilation resulting from the intrusion of some dyke swarms.

v   Ring-dyke emplacement occurs when typically silicic magma is intruded along steep, outwardly dipping fractures resulting from the subsidence of a central, cylinder-like block. The fractures may either reach the Earth's surface, forming a caldera and initiating surface volcanism, or be truncated along a subsurface, planar and essentially horizontal fracture, resulting in a cauldron subsidence and little or no surface volcanism (Figure 18b). Associated with the central subsidence process, the silicic magmas may create space by stoping and by gas brecciation during their ascent. Ring-dykes commonly occur in nested sets with complicated intrusive relationships, as exemplified by the Mull Central Complex (e.g. Richey, 1932).

Evidence for surface volcanic activity during the development of the central complexes takes the form of volcaniclastic materials (tuffs, agglomerates, breccias etc) and lavas, covering a wide range in composition, that are commonly caught up as screens between younger intrusions. The amount of such evidence varies, but this may simply reflect the relative level of erosion. For example, at the present day, the rocks exposed in the Central Ring Complex of Arran are predominantly lavas and volcaniclastic rocks with minor amounts of gabbro, granite and other coarse-grained rocks, but it is likely that with further erosion these proportions would be reversed and eventually the volcanic rocks would be removed. Surface venting would appear to have been more common during the earlier stages of development of individual complexes.

   Intrusion of magma into the upper, water-saturated crust resulted in the development of convecting hydrothermal systems that circulated meteoric water through both the intrusions and the adjacent country rock (Figure 13). This resulted in pervasive low-grade alteration that is estimated to have taken place at about 400°C, and caused alteration of the mafic minerals generating 'serpentine' from olivine, amphibole and chlorite from pyroxene and biotite. Alteration of silicic minerals resulted in turbidity and clay-mineral formation in feldspars, and fluid inclusion development in quartz. When fluid flow was concentrated along channels, secondary precipitation of minerals such as chlorite, epidote, prehnite and carbonate occurred in the myriad of tensile fractures that developed in reponse to both cooling of the intrusions and fracture by fluid ingress. The heated meteoric waters also imparted distinctive stable isotopic signatures to the altered rocks; these are useful in determining the origin of the water and in tracing the circulation patterns (e.g. Taylor and Forester, 1971; Forester and Taylor, 1976, 1977). The pervasive alteration haloes extend laterally for as much as 20 km from the margins

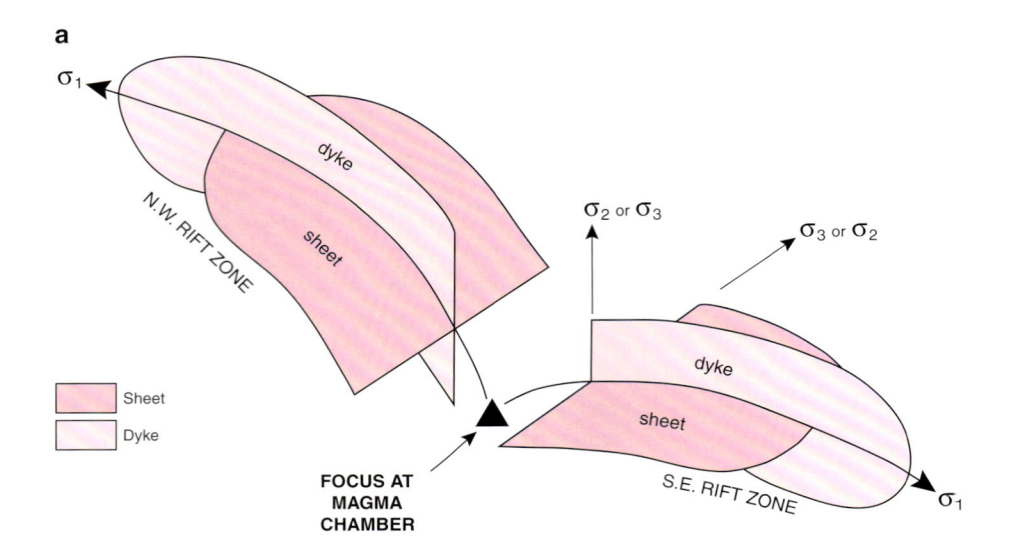

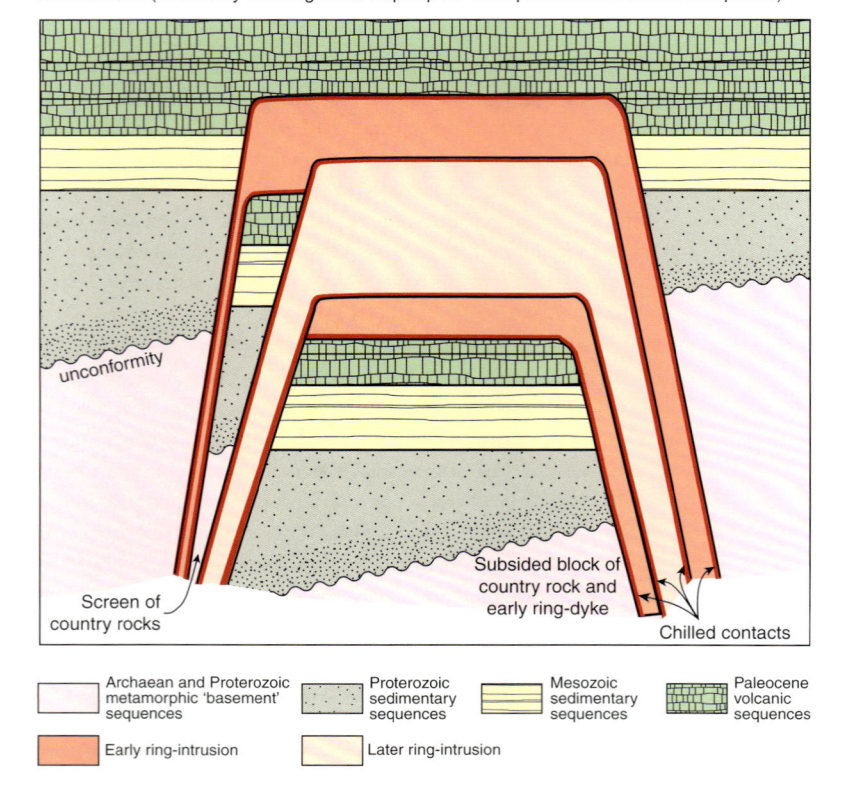

Figure 18 *Dyke swarms, cone-sheets and ring-dykes.*
a Relationship of the dyke swarm and cone-sheets to the stress field in the Cuillin Centre. Dykes are injected when the principal deviatoric stress axis σ_3 is horizontal and sheets are injected when σ_3 is vertical (based on Walker, 1993b, fig. 9).
b Form and mode of occurrence of ring-dykes (after Richey, 1932, fig. 2)

of the central complexes, as on Mull, but the most intense alteration that was caused by temperatures in excess of 800°C is restricted to a zone less than 1 km in width (Figure 27). The mineral-filled fractures are more localised, either within the central complexes or in narrow zones in the surrounding country rocks that are commonly north-west-trending, parallel to the regional dyke swarms. Within the lava piles intruded by the Mull and Skye central complexes, the narrow concentric zones of secondary mineral growth that are attributed to the emplacement of the central complexes are superimposed on flat-lying zeolite and other mineral zoning that resulted from burial metamorphism of the lavas (Walker, 1971; Figures 12; 13). The pervasive alteration of rocks in and surrounding the central complexes by circulating hydrothermal systems is a potential source of change to the rock chemistry and careful sampling is necessary when fresh material is required for geochemical and isotopic studies (e.g. Pankhurst et al., 1978; Walsh et al., 1979; Dickin and Exley, 1981).

The deformation of the surrounding country rocks during emplacement, and possibly as a consequence of central uplift and gravitational collapse, has led to the development of spectacular annular folds (both anticlines and synclines), which extend outwards for distances of up to 5 km from the central complexes and produce significant upper crustal shortening (Chapter 11). The most complete example occurs around the margin of the Mull Central Complex, where folding of pre-Palaeozoic country rocks may have commenced even prior to the eruption of the Mull Lava Group, which itself predated the central complex (Walker, 1975a). Within the area of the central complex, the lavas rest on metasedimentary basement rocks of the Moine Supergroup or on Triassic sedimentary rocks. In contrast, at the margins of the updomed area, they overlie Jurassic and Cretaceous sedimentary rocks.

Central uplift of several hundreds of metres occurred during emplacement of the layered ultrabasic and basic units and the predominantly basic cone-sheet sets, for example on Skye and Mull. This uplift may have been very rapid and possibly led to very early unroofing of some of the complexes, for example on Rum. Emplacement of granites also caused central uplift, as in north Arran, where there are steep-sided domes. However, where fracture systems penetrated to the contemporaneous land surface, as on Rum, emplacement of silicic magmas was commonly accompanied by central subsidence, with caldera formation and associated volcanic eruptions. Central collapse, possibly by several hundreds of metres, would have accompanied the emplacement of ring-dykes such as at Loch Bà on Mull, at Marsco in the Western Red Hills Centre of Skye, and the mixed-magma ring-dyke of Centre 2, Ardnamurchan, but these examples do not preserve direct evidence of associated volcanicity.

Gravity and magnetic anomalies are associated with the central complexes. All of the complexes are characterised by positive Bouguer gravity anomalies which are generally large, very localised and characterised by steep gradients (Figure 19). The anomalies are attributed to the presence of steep-sided bodies of dense rock, such as gabbro or peridotite, extending to depths of 15 km or more (McQuillin and Tuson, 1963; Bott and Tuson, 1973; but see also Emeleus et al., 1996b). Based upon these geophysical data, it is inferred that where granites dominate the present-day surface outcrop, they are thin and do not extend to depth. For example, on Mull, Bott and Tantrigoda (1987) estimated that granitic rocks form as little as 10 per cent by volume of the entire complex and are less than 2 km thick. Supporting evidence comes from seismic investigations carried out over the Western Red Hills Centre on Skye, where it is inferred that the granites are about 2 km thick and overlie interleaved basic and silicic sheets lying at depths of between 2100 and 2400 m (Goulty et al., 1992). The strong, comparatively localised magnetic anomalies over the complexes probably reflect the predominance of gabbroic rocks; such magnetic anomalies are due to the presence of high modal amounts of iron-titanium oxides, which are generally only sparsely present in rocks more primitive than gabbro (e.g. peridotite and troctolite).

Some 25 other igneous centres or central complexes have been identified throughout the continental shelf north and west of the Outer Hebrides from geophysical surveys (Figure 2). Some are exposed on the sea bed and have been sampled, but others are concealed by younger strata. Their features are summarised by Fyfe et al. (1993) and by Stoker et al. (1993).

CHAPTER NINE: CENTRAL COMPLEXES

The onshore complexes, numbered I to VIII in Figure 1, are described from north to south. Complexes IX to XI, Slieve Gullion, the Mourne Mountains and Carlingford are not included in this district, but the submerged Blackstones Bank is described.

ST KILDA The St Kilda archipelago comprises a group of precipitous islets consisting entirely of igneous rocks that form a central complex close to the edge of the continental margin (Harding, 1966, 1967; Harding et al., 1984; Stoker et al., 1993; Figure 1, 1). Although no Pre-Paleocene rocks are found above sea level, Lewisian gneisses crop out on the surrounding sea bed. Detrital minerals characteristic of regional and thermal metamorphic rocks occur in stream sediments on Hirta, washed out of drift deposits (Chapter 13). The central complex is the site of a strong positive Bouguer gravity anomaly and the igneous rocks are reversely magnetised, most likely during the magnetic interval C26R. The principal intrusive units comprising the varied geology of Hirta, the largest island, are described below in order of their emplacement (Figure 20).

The **Western Gabbro** is a texturally variable olivine-gabbro. An equigranular variant exhibits streaky banding which reflects variation in modal mineral concentrations. A second type is characterised by large black augite crystals (about 2 cm across) containing numerous small white plagioclase crystals, which impart a characteristic speckled appearance to weathered surfaces. A third variety has a particularly marked textural range; in addition to containing elements of the first two types, there are pyroxene-rich and porphyritic facies and the rocks are commonly layered due to variation in both texture and mineral proportions. The dip of this layering, while locally variable, is general easterly to north-easterly, at about 45°. These gabbros may originally have formed part of a large layered intrusion. Gabbro identical to some of the variants in the Western Gabbro occurs (with other gabbros) as blocks in the extensive breccias described below, which therefore postdate the Western Gabbro. On the north end of Hirta, a concentration of fine-grained, olivine-dolerite sheets intruding the Western Gabbro is termed the Cambir Dolerite.

Breccias of gabbro and dolerite crop out at Glacan Mor on Hirta, and on Boreray and Soay; they also form numerous islets and stacks. These spectacular breccias are made of large blocks of coarse-grained gabbro, commonly sheared, which are fragmented, veined and enclosed by dolerite. Similar brecciated gabbros have been recovered from the sea bed between the main islands.

The **Glen Bay Gabbro** is an olivine-gabbro that crops out either side of Glen Bay and differs chemically from the older Western Gabbro. The age relationship is established by coarse-grained gabbroic veins, most likely derived from the Glen Bay Gabbro, which cut sheared parts of the Western Gabbro. On the eastern side of Glen Bay, the Glen Bay Gabbro has a black, splintery chilled margin against the brecciated gabbros and dolerites of Glacan Mor, in which glass extends for about 10 mm from the contact.

The **Glen Bay Granite** is medium grained, granophyric in places, and contains microphenocrysts of zoned plagioclase. Ferroaugite and ferropigeonite are variably replaced by amphibole. The rare-earth-bearing, titanium-rich silicate chevkinite is an accessory mineral. The granite is fine grained (probably chilled) adjacent to the Glen Bay Gabbro, and both have been affected by shearing.

The **Mullach Sgar Intrusion-complex** is one of the most distinctive rock assemblages in the St Kilda Central Complex. It comprises an extremely complicated assemblage of microgranite, microdiorite and dolerite sheets, dykes and igneous breccias. At least four generations of dolerite and microgranite intrusion are recognised and there is abundant evidence for the co-existence and mixing of silicic and basic magmas. Some of the most spectacular examples of

Figure 19 Gravity anomaly image of western Scotland: Bouguer anomalies onshore and free-air anomalies offshore. Colour shaded-relief image is illuminated from the north-west: red = high, blue = low. Superimposed contours are at 10 mGal intervals. The Paleocene central complexes are characterised by strong circular positive anomalies up to 50 km in diameter, indicating the presence of localised masses of dense ultramafic or mafic rocks high in the Earth's crust.

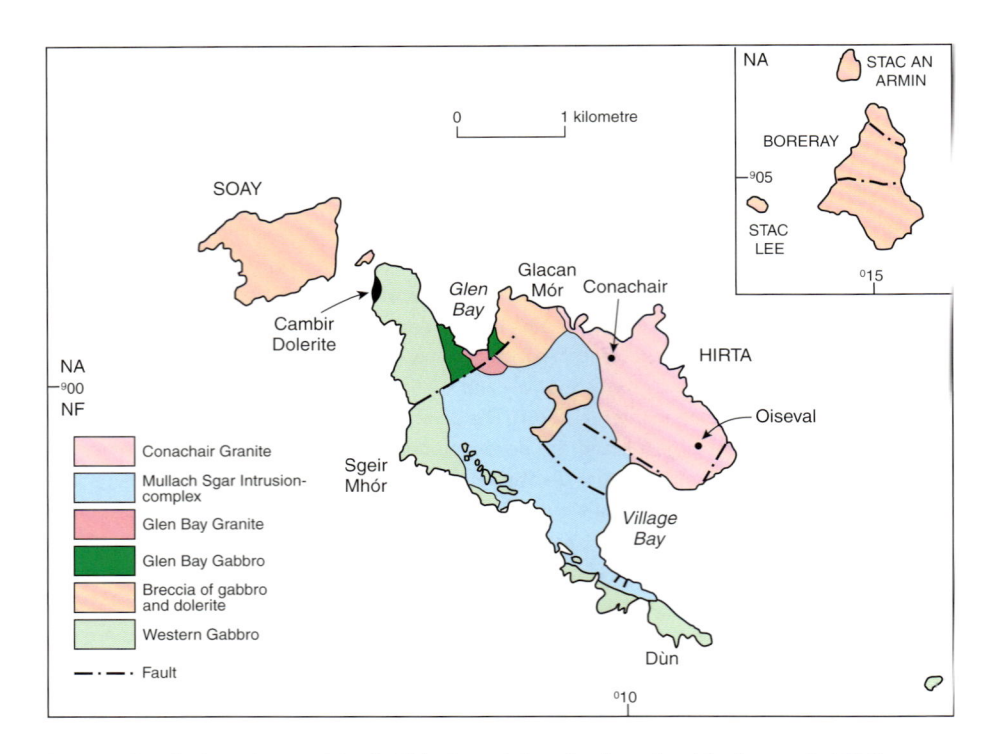

Figure 20 *Geological map of the St Kilda Central Complex (based on Harding et al., 1984).*

this mixing phenomenon in the Hebridean Igneous Province are provided by the pillow-like masses of dolerite and basalt enclosed by microgranite, against which the basic rocks have developed fine grained, glassy selvedges. Members of the Mullach Sgar Intrusion-complex intrude the Western Gabbro and enclose blocks derived from it; similar material intrudes the Glen Bay Granite and the gabbro and dolerite breccias in north Hirta.

The **Conachair Granite** is the youngest major intrusion on Hirta. It forms the peninsula containing the hills of Conachair and Oiseval. The granite is a somewhat drusy, leucocratic, medium-grained rock, and its exposures have a distinctive slabby appearance caused by prominent horizontal and vertical jointing. The rock consists largely of intergrown quartz and microperthitic alkali feldspar and albite, with small amounts of calcic amphibole and biotite. Chevkinite, sphene and zircon are accessory minerals. The granite is in sharp, but not noticably chilled contact with the Mullach Sgar Intrusion-complex and with the gabbro and dolerite breccias, both of which are altered near the contacts.

Numerous sheets and dykes of basic and silicic composition, and composite (basic–silicic) sheets cut all the plutonic rocks, being most conspicuous in the pale-weathering cliffs formed of Conachair Granite. The dykes and inclined sheets appear to converge on a common focus at depth, east of Hirta.

SKYE

The Skye Central Complex comprises the Cuillin (oldest), Srath na Creitheach, Western Red Hills and Eastern Red Hills (youngest) centres (Bell and Harris, 1986; Figure 21).

Cuillin Centre

The Cuillin Centre consists of gabbroic and peridotitic ring-intrusions, some of which have well-developed mineral layering and lamination (Plate 24). These structures attracted the attention of early geologists, who attributed layering in the gabbros of Druim Hain to intrusion of a heterogeneous magma (Harker, 1904). The intrusive contacts between units are typically well exposed. These coarse-grained rocks are cut by numerous centrally inclined, fine-grained sheets of tholeiitic basalt and

Plate 24 *Well-developed mineral layering in the Outer Bytownite Gabbros (Figure 21, C) of the Cuillin Centre, Skye Central Complex, Meall na Cuilce. Scale: hammer shaft is 30 cm (P580475).*

dolerite, and are also invaded by volcaniclastic breccia pipes. All of the units of the centre are cut by north-west-trending basaltic dykes. The margin of the centre is steeply inclined inwards, as are many of the internal contacts between various basic and ultrabasic intrusions, resulting in an overall nested disposition. In general terms, the intrusions are older around the margin of the centre, younging progressively inwards. The north-east quadrant of the centre is missing as a consequence of the emplacement of the younger granites of the Srath na Creitheach and Western Red Hills granite centres.

The country rocks are predominantly lavas, which have been thermally metamorphosed up to pyroxene hornfels facies (Almond, 1964). Along the south-east margin, the contact is with Torridonian sandstones and siltstones, which have been baked and, locally, partially melted. Small outcrops of Jurassic limestone at Camasunary, along the south-east margin of the centre, are altered to high-grade calc-silicate hornfels.

The constituent units of the Cuillin Centre are shown on Figure 21 and are described below in order of intrusion.

The **Outer Gabbros** (Figure 21, A1 to A3) consist of a rather heterogeneous group of intrusions comprising layered and unlayered gabbro and bytownite gabbro ('eucrite'), which crop out along the margin of the centre from Sgurr nan Gillian in the north to Loch Scavaig in the south. At their inner contact against younger intrusions, the Outer Gabbros are altered and thermally metamorphosed, and take on a dull, matt-black appearance (like the gabbro at Lochan an Aodainn, Ardnamurchan; see below). One particularly distinctive intrusion is the fine-grained gabbro (Gars-bheinn type; A3), which crops out in the southern part of the centre on and around the summit of Gars-bheinn. This intrusion has a distinctive green coloration and is particularly altered along its northern contact with younger intrusions.

The **Outer Bytownite Troctolites** have, at their outer margin, a fluxioned, xenolithic tholeiitic dolerite (Figure 21, B1). This contains xenoliths of bytownite gabbro and grades inwards into a massive coarse-grained, plagioclase-rich bytownite troctolite (B2) of striking pale coloration that is commonly referred to as the 'White Allivalite' (Hutchison, 1968). The bytownite troctolite immediately adjacent to the marginal tholeiitic dolerite has a distinctive, steeply inclined, inward dipping 'wispy banding'.

EASTERN RED HILLS CENTRE

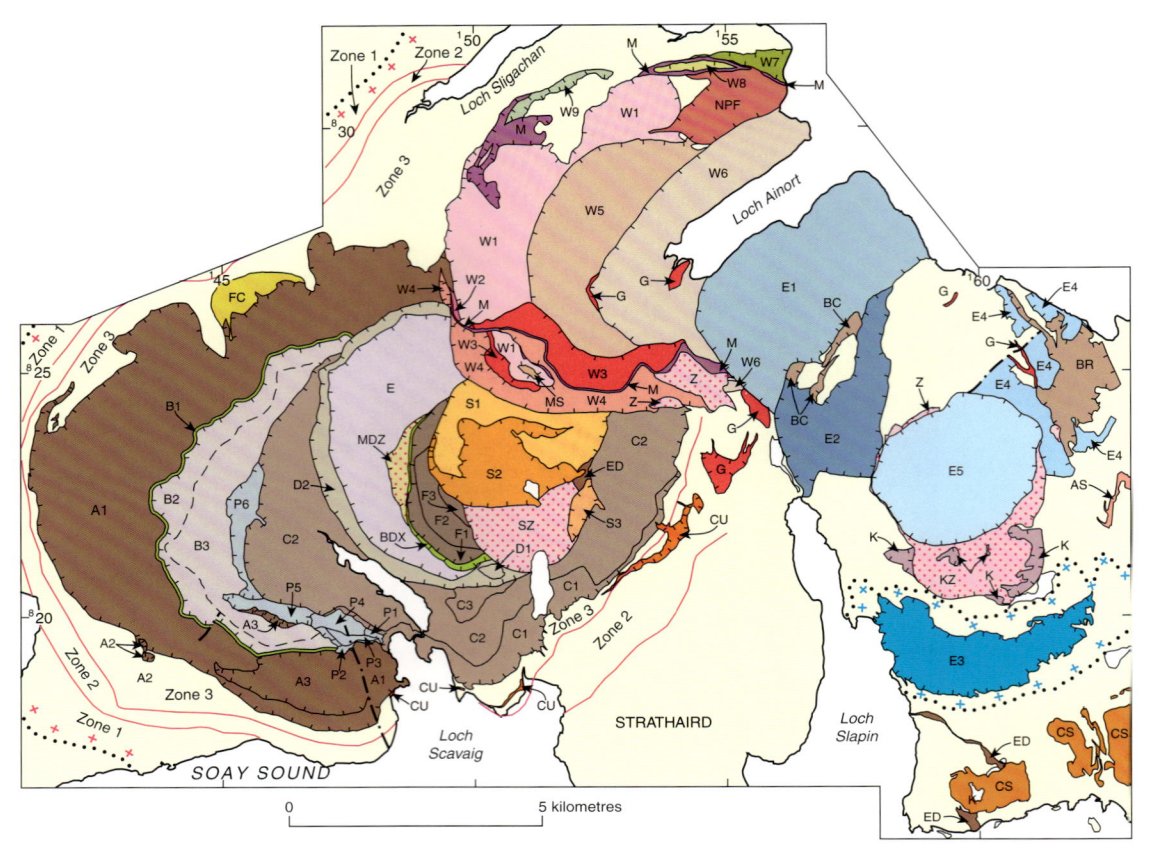

- **E5** **Beinn na Caillich Granite:** granophyric-textured, relatively coarse-grained, macroporphyritic with alkali feldspar and quartz phenocrysts, hornblende- and biotite-bearing. Miarolitic, with quartz, alkali feldspar and fluorite. Chilled marginal facies locally, with microphenocrysts of ferrohedenbergite and fayalite
- **E4** **Creag Strollamus Granites:** granophyric-textured, variably coarse- to medium-grained, macroporphyritic with alkali feldspar and quartz phenocrysts, amphibole- and biotite-bearing
- **E3** **Beinn an Dubhaich Granite:** granophyric-textured, relatively coarse-grained, macroporphyritic with alkali feldspar and quartz phenocrysts, amphibole- and biotite-bearing. Rare rounded basaltic inclusions. Alkali pyroxene-bearing facies in vicinity of skarns at margin
- **E2** **Beinn na Cro Granite:** porphyritic with phenocrysts of alkali feldspar and quartz, amphibole- and biotite-bearing
- **E1** **Glas Bheinn Mhor Granite:** granophyric-textured, macroporphyritic with plagioclase phenocrysts, amphibole- and biotite-bearing. Rare rounded basaltic inclusions

Other intrusions and volcaniclastic deposits associated with the Eastern Red Hills Centre

- **BR** **Broadford Gabbro:** variably medium- to coarse-grained, unlayered, typically hydrothermally altered
- **BC** **Beinn na Cro Gabbro:** variably medium- to coarse-grained, unlayered, typically hydrothermally altered. Composed of sheet-like masses
- **K** **Kilchrist Hybrids:** leucocratic to mesocratic, medium-grained rocks. Contain irregular crystals of quartz fringed by amphibole and/or pyroxene and resorbed crystals of alkali feldspar. Common rounded basaltic inclusions. Flow banding common in marginal facies
- **KZ** **Kilchrist Breccias:** predominantly unbedded coarse breccia with clasts of basalt, hawaiite, quartzite, Torridonian sandstone and fissile mudstone, Ordovician dolostone, Jurassic limestone, sandstone and siltstone. Matrix of hydrothermally altered volcaniclastic siltstone and finely comminuted fragments of clast lithologies

WESTERN RED HILLS CENTRE

- **W9** **Eas Mor Granite:** microgranite with phenocrysts of microperthite, amphibole-bearing
- **W8** **Meall Buidhe Granite:** microgranite with phenocrysts of plagioclase; fayalite- and ferrohedenbergite-bearing
- **W7** **Maol na Gainmhich Granite:** relatively coarse-grained, arfvedsonite-bearing
- **W6** **Loch Ainort Granite:** granophyric-textured, glomeroporphyritic with anorthoclase phenocrysts; ferrohedenbergite-, fayalite- and amphibole-bearing, pegmatitic in parts
- **W5** **Beinn Dearg Mhor Granite:** granophyric-textured, glomeroporphyritic with anorthoclase phenocrysts; ferrohedenbergite-, fayalite- and amphibole-bearing
- **W4** **Marsco Granite:** granophyric-textured, relatively coarse-grained, miarolitic in part, fluxioned in part, ferrohedenbergite- and fayalite-bearing
- **W3** **Southern Porphyritic Granite:** granophyric-textured, macroporphyritic with alkali feldspar and quartz phenocrysts, brecciated in part
- **W2** **Glen Sligachan Granite:** macroporphyritic with plagioclase and alkali feldspar phenocrysts, minor ferrohedenbergite, fayalite, amphibole and biotite
- **W1** **Glamaig Granite:** amphibole- and biotite-bearing, with abundant rounded basaltic inclusions
- **NPF** **Northern Porphyritic Felsite:** granophyric-textured, macroporphyritic with alkali feldspar and quartz phenocrysts, ferrohedenbergite- bearing in part. Abundant rounded basaltic inclusions. Flow-banded and brecciated in part
- **M** **Marsco Hybrids:** a composite ring intrusion of 'marscoite' (a hybridised, medium-grained, intermediate rock with xenocrysts of quartz, alkali feldspar and plagioclase), *ferrodiorite* (with or without phenocrysts of plagioclase), *'glamaigite'* (a heterogeneous, hybridised, coarse-grained intermediate rock) and *porphyritic felsite* (granophyric-textured with phenocrysts of alkali feldspar and quartz)
- **MS** **Marsco Summit Gabbro:** unlayered, olivine-poor, brecciated in part

Figure 21 Geological map of the Skye Central Complex (based on 1:50 000 geological map sheets 70 Minginish and 71W Broadford (2000, 2002).

CHAPTER NINE: CENTRAL COMPLEXES

Key to Figure 21 *continued*

SRATH NA CREITHEACH CENTRE

- **S3** **Blaven Granite:** granophyric-textured with phenocrysts of perthite, hornblende-bearing
- **S2** **Ruadh Stac Granite:** granophyric-textured, relatively coarse-grained with sparse phenocrysts of perthite, arfvedsonite-bearing
- **S1** **Meall Dearg Granite:** granophyric-textured, relatively coarse-grained, ferrohedenbergite- and fayalite-bearing. Cut by sheets of hornblende-bearing granophyric microgranite with xenoliths of gabbro

SRATH NA CREITHEACH VOLCANICLASTIC ROCKS

- **SZ** Volcaniclastic breccia, lapilli tuff and tuffaceous sandstone. Layered in part, matrix-supported. Includes large slabs of hydrothermally-altered bytownite gabbro

CUILLIN CENTRE

- **BDX** Intrusion breccia of basalt, dolerite and tuffisite
- **MDZ** Volcaniclastic breccia pipe of Meallan Dearg
- **CU** **Coire Uaigneich Granite:** granophyric-textured with quartz ?paramorphs. Heterogeneous with partly digested xenoliths of sandstone and siltstone
- **F** **Inner Gabbros:**
  - F3 Olivine-gabbro, laminated with rare layers of anorthosite and magnetitite, commonly pegmatitic
  - F2 Olivine-gabbro, distinctly laminated with layers of dunite, anorthosite and magnetitite, xenolithic
  - F1 Olivine-gabbro, poorly laminated, xenolithic
- **E** **Inner Bytownite Troctolites:** laminated, massive, xenolithic
- **D** **Inner Cross-cutting Bytownite Gabbro:**
  - D2 Unlayered olivine-bytownite gabbro, ophitic, commonly pegmatitic
  - D1 Unlayered olivine-rich bytownite gabbro, red-weathering, xenolithic
- **C** **Outer Bytownite Gabbros:**
  - C3 Layered olivine-bytownite gabbro with megacrysts of calcic plagioclase
  - C2 Layered olivine-bytownite gabbro with lenses of anorthosite and dunite, megacrysts of calcic plagioclase and xenoliths of hornfelsed basaltic rock
  - C1 Layered olivine-bytownite gabbro, pegmatitic in parts, granitic schlieren. Notably xenolithic at inner margin. Unlayered, xenolithic and hybridised at outer margin
- **P** **Layered Peridotites:**
  - P6 Layered peridotite and bytownite troctolite, xenolithic
  - P5 Layered peridotite and bytownite troctolite
  - P4 Xenolithic and brecciated peridotite and bytownite troctolite
  - P3 Layered peridotite and bytownite troctolite
  - P2 Fluxioned peridotite and bytownite troctolite
  - P1 Dunite with chrome-spinel layers
- **B** **Outer Bytownite Troctolites:**
  - B3 Layered clinopyroxene-bytownite troctolite
  - B2 Clinopyroxene-bytownite troctolite
- **B1** B1 Tholeiitic dolerite, fluxioned and xenolithic
- **A** **Outer Gabbros:**
  - A3 Microgabbro (Gars-bheinn type)
  - A2 Olivine-microgabbro
  - A1 Olivine-gabbro

FIONN CHOIRE FORMATION

- **FC** Trachyte and rhyolite lavas, trachytic and rhyolitic pyroclastic rocks, volcaniclastic sedimentary rocks and minor intrusions along the margin of the Cuillin Centre

OTHER ASSOCIATED INTRUSIONS

- **AS** **An Sithean Granite:** granophyric-textured, variably coarse- to medium-grained, macroporphyritic with alkali feldspar and quartz phenocrysts, hornblende- and biotite-bearing. Brecciated in parts
- **CS** Composite sheets of rhyolite and basaltic andesite
- **G** Granite and microgranite sheets, unclassified
- **ED** Gabbroic and doleritic intrusions, unclassified
- **Z** Volcaniclastic breccia, unclassified

OTHER ROCKS

- Intrusive boundary, ticks on side of younger rock
- Fault
- Internal, gradational boundary
- Outer margin of contact metamorphic zones in lavas around the Cuillin Centre, + towards intrusion
- Margin of zone within aureole

Outside Zone 1, lavas have a regionally developed hydrothermal mineral assemblage typified by partial development of chlorite and fibrous amphibole in the groundmass

Zone 1: Olivine pseudomorphed by saponite or talc; secondary groundmass epidote after olivine and/or clinopyroxene

Zone 2: Olivine variably replaced by saponite and chlorite; clinopyroxene pseudomorphed by aggregates of amphibole and chlorite; plagioclase variably replaced by aggregates of albite + sericite + calcite

Zone 3: Intensely hornfelsed with recrystallised granoblastic aggregates of plagioclase + olivine + clinopyroxene + magnetite ± orthopyroxene

- Margin of contact metamorphic aureole around Beinn an Dubhaich Granite, + towards intrusion

Zones within the aureole, not shown here, are characterised by the following index minerals in carbonate rocks (Holness 1992):

Zone 1: talc (outermost zone)
Zone 2: tremolite
Zone 3: diopside
Zone 4: olivine
Zone 5: periclase (innermost zone)

The layered bytownite troctolites (Figure 21, B3) accumulated against the massive bytownite troctolites (B2). The nature of the layering is quite variable, comprising both plagioclase-pyroxene and plagioclase-olivine cumulates. The inward dipping attitude of the layering suggests that the Outer Bytownite Troctolites were formed within a funnel-shaped chamber (Hutchison and Beavan, 1977).

Six units are identified in the **Layered Peridotites** (Claydon and Bell, 1992; Figure 21, P1 to 6). These comprise various layered olivine-rich cumulates, ranging from dunite through to feldspathic peridotite (Weedon, 1965). The structurally lowest unit (P1) is dominated by dunite and contains abundant layers rich in chrome-spinel (Bell and Claydon, 1992). Other units contain plagioclase, both cumulus and intercumulus, and have excellent modal layering. Xenoliths of various ultrabasic rock-types are common and typically accentuate the layering. Unit 4 (P4) is a heterogeneous breccia comprising plagioclase-dominated blocks in a matrix dominated by olivine, and olivine-dominated blocks in a matrix dominated by plagioclase. Block types vary from anorthosite through to dunite. Such breccias, which are similar to breccias in the Central Intrusion of the Rum Central Complex (see p.117), are most likely to be of intrusive origin, and formed by the upward forceful intrusion of aluminous, ultrabasic magma into ultrabasic rocks, which were consequently disrupted and brecciated.

A number of unusual textural varieties of peridotite, as well as structures involving various ultrabasic rocks, are preserved within the Layered Peridotites; similar features are described from the Rum layered intrusions (see p.117). For example, within Unit P5 on the south side of An Garbh-choire, structures that are approximately hemispherical and range in size from 15 to 200 cm, comprise dendritic intergrowths of poikilitic intercumulus plagioclase enclosing orientated cumulus olivine. Such structures most likely formed within a crystal mush from a hydrous, aluminous, ultrabasic magma. Also within Unit P5, along the interface of peridotite and troctolite layers, 'fingers' of peridotite penetrate upwards into troctolite (Plate 25). These structures are interpreted as evidence that hot ultrabasic magma invaded pre-existing solid troctolite and eroded it by melting. From these and other observations it is now generally accepted that thick sequences of ultrabasic layered rocks, such as those which form the Cuillin Centre and the Rum layered intrusions, were formed through the combined operation of a number of processes including crystal settling, in-situ crystallisation, and the intrusion of sheets of magnesium-rich magma into pre-existing solid material.

The **Outer Bytownite Gabbros** (Figure 21, C) crop out around a significant sector of the centre. These gabbros are subdivided into three units, based upon obvious field characterisitics, such as layering (Plate 24), modal mineralogy and the presence of xenoliths. Where bytownite gabbros are in contact with the Layered Peridotites, they contain abundant xenoliths of peridotite (Weedon, 1961). Elsewhere, for example on the south side of Sgurr na Stri, the Outer Bytownite Gabbros have an unlayered marginal facies in steep, inward-dipping contact with country-rock lavas.

Interior to the crescent-shaped outcrop of the Outer Bytownite Gabbro is the **Inner Cross-cutting Bytownite Gabbro of Druim nan Ramh** (Figure 21, D). This unlayered, steep-sided ring-intrusion is about 200 m wide and might represent some form of marginal unit to the layered rocks of the Inner Bytownite Troctolites and the Inner Gabbros, although this interpretation has not been proved.

The **Inner Bytownite Troctolites** (Figure 21, E) form a thick sequence of layered rocks in an outcrop concentric with and inside that of Unit D. They consist of plagioclase-pyroxene and plagioclase-olivine cumulates, commonly with a fine mineral lamination. The layering dips typically at about 25°, towards a point below Meall Dearg at the southern end of Glen Sligachan.

The **Inner Gabbros** (Figure 21, F) are interpreted as the youngest sequence of layered rocks preserved within the centre, and crop out on the ridge of Druim Hain, north-east of Loch Coruisk. These gabbros are separated from the Inner Bytownite Troctolites by a 50 m-wide vertical zone of intrusive breccias, consisting of shattered tuffaceous rocks invaded by numerous basalt and dolerite sheets and containing xenoliths of various gabbros and peridotites. The Inner Gabbros have an arcuate distribution, with modal layering inclined towards a point below Meall Dearg. Layering is poorly developed immediately adjacent to the bounding intrusive breccias, whereas close to the top of the preserved sequence the layering is well preserved and dips at up to 70°. Mineral lamination is well developed in places, with magnetite enrichment at the base of individual layered units, passing upwards into plagioclase-enriched cumulates. Xenoliths of basalt and dolerite are common throughout, their flattened shapes suggesting plastic deformation of hot material soon after incorporation into the cumulate pile.

Late-stage intrusions of tholeiitic basalt and dolerite

The various coarse-grained basic and ultrabasic rocks of the Cuillin Centre are cut by myriad basic minor intrusions, which may be divided into two types:

i   Multiple intrusions of tholeiitic basalt and dolerite, some xenolithic, which crop out in the western part of the centre. Inner and outer sets are recognised, with all of the components

Plate 25 *Replacement 'finger' structures of more olivine-rich peridotite in feldspathic peridotite, Unit P5, Cuillin Centre, Skye Central Complex (P580476).*

dipping towards a focal point in the middle of the Cuillin Centre (Hutchison, 1966). The inclination of the sheets generally increases inwards. Members of the inner set are further complicated by a volcaniclastic facies that includes welded breccias and xenolithic intrusions. Brecciation may have been caused by the release of volatiles during emplacement of individual sheets.
ii Cone-sheets, of tholeiitic basalt and dolerite, are most concentrated in the north-west and south-east quadrants of the centre, coinciding with the north-west-trending regional dyke swarm (Figure 18a). A number of features suggest that the final stages in the growth of the lava field overlapped in time with the formation of the central basaltic volcano that is now represented by

the Cuillin Centre. These include the distinctive and near-identical composition of the cone-sheets, the dykes from the axial portion of the regional dyke swarm, and flows of the Talisker Lava Formation (Preshal More type) at the top of the lava succession in west-central Skye (Chapter 6).

Volcaniclastic pipes

More than forty pipe-like structures containing volcaniclastic breccia pierce various rocks of the Cuillin Centre; only the largest example is shown on Figure 21. They are typically less than 100 m across, circular or oval in plan, and are narrower at lower structural levels, suggesting an overall funnel shape. They are composed of blocks of coarse-grained basic or ultrabasic lithologies of local derivation set in a matrix of basalt, dolerite or tuffaceous material. Crush zones are common. The pipes formed when volatiles, expelled from upward-penetrating minor intrusions, caused initial brecciation of country-rock lithologies, and were also responsible for further fragmentation, rounding and sorting of the breccia clasts. Final emplacement of the breccias occurred when the fluidised mass collapsed as a consequence of insufficient pressure gradient to drive the system.

Coire Uaigneich Granite

This intrusion crops out around the south-east margin of the Cuillin Centre (Figure 21, CU), between Coire Uaigneich on the east side of Blà Bheinn, where it is dyke-like with fine-grained margins, and the Sgurr na Stri peninsula, where it appears to be a sill, emplaced between Torridonian sedimentary rocks and Paleocene lavas. It is a narrow, steeply inclined intrusion, and space for it appears to have been made largely by horizontal crustal dilation. The granite is heterogeneous, both fine and coarse grained, with partially digested xenoliths of sandstone (most likely Torridonian). The most conspicuous mineral is hypersthene, set in a pale granitic groundmass in which quartz paramorphs after tridymite have been recognised. The field characteristics, mineralogy and geochemistry indicate that it is the product, at least in part, of partial melting of Torridonian sedimentary rocks (Chapter 10). Melting most likely occurred at a shallow depth, the heat being supplied by the Cuillin Centre magmas. The granite is cut by cone-sheets associated with the Cuillin Centre, whereas the granites of the Srath na Creitheach and Western Red Hills centres, farther north and east, are not.

RED HILLS GRANITE CENTRES

Granites form the Red Hills of Skye, which extend from Glen Sligachan to Strath, near Broadford. There are three distinct centres: from oldest to youngest they are the Srath na Creitheach, Western Red Hills and Eastern Red Hills centres (Figure 21). Each centre contains a number of discrete intrusions, varying from steep-sided bosses through to segments of ring-intrusions. Field relationships suggest that certain of the intrusions are true ring-dykes, where space was made available by central subsidence of a block of older country rocks (which may in part have comprised older granites, e.g. Figure 18b), but this style of intrusion is by no means ubiquitous. More commonly, the outcrop patterns of the granites suggest broad, curved, dyke-like masses. Emplacement of the granites does not appear to have resulted in appreciable disturbance of the earlier rocks. Stoping and brecciation of the country rocks is generally absent, although there may be peripheral folding of country rocks on the northern slopes of Glamaig and in the vicinity of the Beinn an Dubhaich Granite (pp.108, 149).

Evidence for volcanic activity, possibly from calderas that developed as a consequence of central subsidence, is at best circumstantial. Silicic pyroclastic rocks are preserved locally, for example within the sequences of volcaniclastic deposits in the Kilchrist area of the Strath district (see below). In addition, within the offshore sedimentary basins of the North Sea and west of the Shetland Isles, thin beds of silicic pyroclastic rocks are intercalated with marine sedimentary rocks of Paleocene age, and are considered to have been derived from the volcanic centres of either the Hebridean Igneous Province or even East Greenland (e.g. Knox and Morton, 1988).

Srath na Creitheach Centre

This centre is located at the southern end of Glen Sligachan and comprises three granitic intrusions: Meall Dearg (S1), Ruadh Stac (S2) and Blaven (S3) (Jassim and Gass, 1970; Figure 21). Its western and eastern sides intrude gabbros of the Cuillin Centre. Apophyses of the Meall Dearg Granite, in the form of metre-wide dykes of spherulitic rhyolite, invade the Inner Gabbros on Druim Hain and attest to the younger age of the granite. Along the northern margin of the centre, the Srath na Creitheach granites are cut out by younger granites of the Western Red Hills Centre. To the south, the granites are in contact with the older Srath na Creitheach volcaniclastic rocks.

The white-weathering Meall Dearg Granite (Figure 21, S1) crops out on the upper slopes of Meall Dearg and Ruadh Stac and overlies the younger dome-shaped pale brown-weathering Ruadh Stac Granite (S2). Screens of gabbro are present along the contact. On Ruadh Stac, the Meall Dearg Granite comprises interleaved sheets of two distinct types: a fine-grained, hornblende-bearing granite and a younger, coarse-grained pyroxene (hedenbergite)-bearing granite with granophyric texture. The Ruadh Stac Granite contains an iron-rich, sodic amphibole that varies between arfvedsonite and ferrorichterite. The Blaven Granite (S3) crops out on the lower western slopes of Blà Bheinn where the light-coloured granite underlies and intrudes dark gabbros in the Outer Bytownite Gabbros of the Cuillin Centre. Its age relative to the two other granites is not known.

Western Red Hills Centre

The granites of this centre have an outcrop area of about 35 km^2 in rounded, scree-mantled hills that contrast with the adjacent rugged Cuillin (Plate 26). Detailed mapping has identified ten granites, together with a composite ring-dyke (Marsco Hybrids), various masses of explosion breccia and a gabbro that was emplaced during the main granite intrusion events (Thompson, 1969). The margin of the centre is generally steeply inclined, against a variety of country rocks. Crush zones are recognised, in particular against hydrothermally altered Paleocene lavas and Torridonian strata.

The earliest event attributed to the Western Red Hills Centre is the formation of various masses of explosion breccia (Figure 21, Z), now preserved on Belig and Meall a' Mhaoil in the south-east and north-east parts of the centre, respectively. These comprise rounded to subangular blocks of pre-granite lithologies, such as Torridonian and Mesozoic sedimentary rocks, basalt and bytownite gabbro, together with blocks of silicic igneous rocks, including granite, felsite and quartz-porphyry. The breccias were formed by the release of volatiles during subvolcanic emplacement of the granitic magmas. Juvenile material is absent and, consequently, the breccias represent the action of gas streaming through, and fluidising of, the country rocks (J D Bell, 1966).

Three distinct structural groupings are identified in the Western Red Hills granites.

i   A set of intrusions broadly elongated in a north-east–south-west direction, for example the Glamaig (Figure 21, W1), Beinn Dearg Mhor (W5) and Loch Ainort (W6) granites.

ii   In the north, a set of intrusions elongated in an east–west direction, the Northern Porphyritic Felsite (NPF), and the Maol na Gainmhich (W7), Meall Buidhe (W8) and Eas Mor (W9) granites.

iii   In the south, a set of intrusions also elongated east–west, the Southern Porphyritic (W3), Glen Sligachan (W2) and Marsco (W4) granites.

Members of (i) were emplaced sequentially inwards, whereas members of (ii) and (iii) were emplaced sequentially outwards. All of the intrusions are approximately co-focal around a point east of Loch Ainort (Figure 21).

Plate 26 The northern ridge of Blà Bheinn and the eastern Red Hills, Skye. Brown-weathering Outer Bytownite Gabbros (Figure 21, C) forming the ridge of Clach Glas in the foreground are cut by west-dipping (right to left) dolerite cone-sheets. The pale, scree-covered granite hills of the Eastern Red Hills Centre, in the distance, are (left to right) Glas Bheinn Mhor (Figure 21, E1), Beinn na Crò (Figure 21, E2) and Beinn na Caillich (Figure 21, E5). (Photograph: Michael Macgregor.)

These field relationships may be explained in terms of a cauldron subsidence model, as follows. After the formation of the explosion breccias and the emplacement of the Marsco Summit Gabbro (MS), which forms a resistant cap on the summit of Marsco, the first major silicic intrusion to be emplaced was the Glamaig Granite (W1). This intrusion has a steep-sided outer western margin against lavas and a steep inner margin against the younger Beinn Dearg Mhor Granite (W5). The Beinn Dearg Mhor Granite was, in turn, intruded along its eastern margin by the Loch Ainort Granite (W6), resulting in three north-east-trending, nested plutons. Emplacement of the east–west sets of intrusions involved subsidence of the block of crust formed by the three earlier granites. Jamming of the block, possibly in the vicinity of Glen Sligachan, caused it to rotate about an east–west-trending axis. According to this model, each successive tilt of the block would lead to similar, though not necessarily identical, intrusions being emplaced to the north and south (Thompson, 1969).

The Glamaig Granite (Figure 21, W1) is characterised by abundant inclusions, apparently cognate in origin. Many small, rounded to subangular, basic inclusions (5 to 50 mm) typically constitute up to 5 per cent of the rock. The larger inclusions have relatively sharp edges, whereas smaller inclusions have the appearance of clots of mafic minerals. These inclusions may have been introduced into the granite during an early stage, involving the mixing and homogenisation of volumetrically dominant granitic magma and a basaltic magma (Chapter 10). Rounded inclusions, up to 50 cm across, of a microgranite are also present.

Further evidence of bimodal magma mixing is provided by the Marsco Hybrids (M). One of the classic field locations in central Skye is the gully on the north-west side of Marsco at the southern end of Glen Sligachan that is now referred to as Harker's Gully, in deference to Alfred Harker who first described the locality (Figure 22). Within the accessible lower reaches of the gully are rocks that provide unambiguous evidence for the mixing of two magmas, one a differentiated iron-enriched tholeiitic basalt (crystallising as ferrodiorite) and the other of rhyolitic composition (Chapter 10). In the upper, less accessible part of Harker's Gully, the centre of the Marsco Hybrids ring-dyke is occupied by ferrodiorite flanked by a homogeneous hybrid, 'marscoite', and a porphyritic felsite successively on each side; however, in the lower exposures the ring-dyke is apparently assymetrical since the south-western side has been removed by intrusion of the later Marsco Granite. The ferrodiorite encloses a few large xenoliths of partially melted Lewisian gneiss.

The course of the ring-dyke composed of Marsco Hybrids is visible on the eastern slopes of Marsco. It is not exposed on the floor of Glen Sligachan, but it crops out to the north, on the

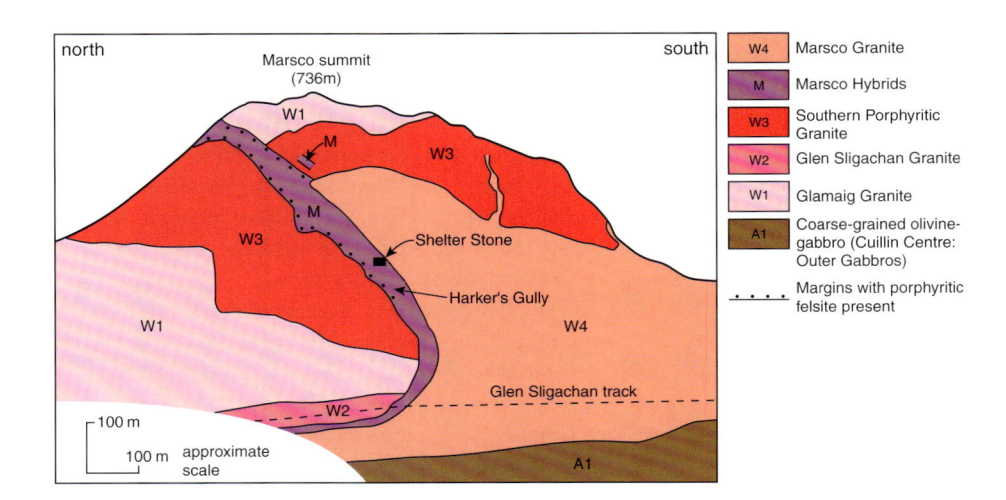

Figure 22 The Marsco Hybrids ring-dyke at Harker's Gully, Marsco, Western Red Hills Centre, Skye Central Complex (after Brown, 1969, fig. 13).

south-west flank of Glamaig and on the coast of Loch Ainort, north of Moll. On Glamaig, more complex magma mixing processes are recognised in the form of a considerably more heterogeneous hybrid, given the local name 'glamaigite' and consisting of dark, rounded cognate inclusions, generally less than 2 cm in diameter, set in a paler matrix, with xenocrysts of plagioclase in both components.

Eastern Red Hills Centre

This centre covers a larger area than either of the two other granite centres but, in contrast, has a significant proportion of country rocks preserved between the intrusions (Figure 21).

The oldest major intrusion of the Eastern Red Hills Centre is the Glas Bheinn Mhor Granite (Figure 21, E1), which cuts across the eastern margin of the Western Red Hills Centre and thus provides clear evidence of their relative ages. Within and to the east of the Glas Bheinn Mhor Granite is a granite comprising several discrete outcrops: Beinn na Cro (E2), Beinn an Dubhaich (E3), Creag Strollamus (E4) and Allt Fearna (also E4); these are collectively referred to as the Outer Granite. The most studied of these is the Beinn an Dubhaich Granite which, uniquely, was emplaced into Cambro-Ordovician (Durness Group) dolostones. Thermal metamorphism of the dolostones has produced a well-developed aureole, which can be subdivided into a number of zones based upon the appearance of distinctive index minerals (in terms of increasing temperature: talc, tremolite, diopside, olivine, periclase; Holness, 1992). Locally, magnetite-rich skarns have formed at the granite contacts. The altered dolostones form a commercially exploited marble (pp.174–5).

The dolostones are folded into large, open, annular structures, the largest of which is the Broadford Anticline (p.150). In addition, many of the dykes of the regional swarm intruded into the dolostones preserve evidence of deformation during emplacement. However, these deformation events occurred prior to the emplacement of the granite, which appears to have utilised, but not caused, the main fold structures. Similar metamorphic and deformation events are identified within the Cambro-Ordovician dolostones farther north, in the area west of Broadford, but here the older Broadford Gabbro may be responsible for the present-day mineralogy and structure of the dolostones.

The Broadford Gabbro (Figure 21, BR) has an outcrop area of about 3 km^2. It has both faulted and intrusive margins. On Creag Strollamus, the gabbro is in sharp contact with bedded pyroclastic rocks, which in turn overlie a weathered surface of the Allt Fearna Granite. The granite is thought to predate the gabbro (Johnston, 1996). The shape of the intrusion is not readily deduced from its field relationships, although its north-west extension is clearly dyke-like, with a width of about 200 m. Unlike the gabbros of the Cuillin Centre, it is unlayered and devoid of olivine. In general, it has been subjected to hydrothermal alteration, with the resultant development of secondary amphibole, chlorite and epidote.

The geographical middle of the Eastern Red Hills Centre is occupied by the Beinn na Caillich (or Inner) Granite (Figure 21, E5). This boss-shaped intrusion has a near-perfect circular outcrop pattern and, locally, has well-developed marginal facies composed of fayalite- and ferrohedenbergite-bearing porphyritic felsite. The marginal rocks are well exposed in the gorge of the Allt Slapin, on the south-west side of the intrusion, where there is also evidence of severe, local deformation of the Mesozoic country rocks.

Also associated with, and generally to the east of, the centre is a suite of composite sills (Figure 21, CS) and dykes, with a basaltic andesite marginal facies and a central felsic facies. The sills and dykes crop out over a broad arc exterior to the granites and co-focal with the centre of the Beinn na Caillach Granite (Figure 21). The sills dip inwards at about 15° towards the Beinn na Caillach Granite, and are typically emplaced into Mesozoic strata. The upper and lower basaltic andesite portions are typically 0.5 to 2.5 m thick, whereas the central felsites are considerably thicker, up to 50 m. The dykes of the suite may have acted as feeder conduits to the sills. Many of the intrusions are symmetrical and the field evidence suggests that the basaltic andesite portions were emplaced first, since there are fragments of basaltic rock within the central felsites, and veins of felsite in the marginal basaltic andesites. However, the presence of gradational contacts between basaltic andesite and felsic rocks, as found for example at Rubh' an Eireannaich, north of Broadford, suggests that the two magmas were available (i.e. liquid) at the same time (Harker, 1904). Evidence for mixing of basaltic and rhyolitic magmas also comes from a detailed analysis of the petrography and geochemistry of the intrusions (Bell and Pankhurst, 1993; Chapter 10). The composite sheets of the Broadford area have many features in common with the composite sheets of Arran (p.90).

Volcaniclastic rocks of the Central Complex

Two large areas of volcaniclastic rocks occur within the Skye Central Complex, in Srath na Creitheach, at the northern end of Loch na Creitheach, and near Kilchrist, in the district of Strath. Unlike the other volcaniclastic breccias described above, these two breccias do not have any obvious association with the main intrusive events.

Srath na Creitheach

The volcaniclastic breccias and associated pyroclastic deposits that crop out in Srath na Creitheach have an outcrop area of about 2 km^2 (Figure 21, SZ) The deposits were formed after the emplacement of the Cuillin Centre, since they contain large slabs of bytownite troctolite, and prior to emplacement of the granites of the Srath na Creitheach Centre, since the northern margin is intruded by the Ruadh Stac Granite. The southern margin of the volcaniclastic breccias is an arcuate ring-fault. The dominant lithology is an unsorted breccia consisting of abundant subangular fragments, mainly of basalt, dolerite and gabbro, set in a tuffaceous matrix. Little, if any, juvenile magmatic material is recognised. Within these deposits are at least twelve large slabs of bytownite troctolite, ranging in size from 40 to 900 m across, which were derived from the Cuillin Centre. Layers of fine-grained, bedded tuff and tuffaceous sandstone help to define the overall structure and stratigraphy. The formation of all these deposits most likely involved the opening of a volcanic conduit to the surface. Fragmentation and incorporation of material derived from the vent walls produced the main breccia mass, whereas some form of reworking process, considered to be subaerial, produced the bedded tuffs and tuffaceous sandstones (Jassim and Gass, 1970).

Kilchrist

The volcaniclastic breccias and associated rocks preserved within the Kilchrist area, west of Loch Cill Chriosd, are extremely heterogeneous and comprise a wide variety of rock types (Bell, 1985; Bell and Harris, 1986; Figure 21, KZ).

The main mass of volcaniclastic breccia crops out over an area of about 2 km^2 and is surrounded and invaded by mixed-magma intrusions, the Kilchrist Hybrids (Figure 21, K). These hybrid rocks, comprising five separate intrusions, contain dark irregular enclaves of basaltic material up to 3 cm across, commonly with diffuse edges, set in a paler granitic matrix. The basaltic rocks contain xenocrysts of quartz and alkali feldspar, which provide further evidence of magma mixing. Three of the hybrid intrusions have steeply inclined outer margins against Cambro-Ordovician (Durness Group) dolostones. These field relationships are consistent with some form of cauldron subsidence, with the volcaniclastic breccias subsiding to create space for the hybrid ring-dyke. Intruded into the volcaniclastic breccias are small stocks and sheets of brecciated porphyritic rhyolite.

The volcaniclastic breccias are composed of subangular to rounded blocks of various pre-Paleocene rocks (Torridonian sandstone and siltstone, Cambro-Ordovician dolostone and quartzite and Jurassic sandstone, limestone and mudstone) together with fragments of basalt, dolerite, gabbro, rhyolite, tuff (including ignimbrite), granite, pitchstone, and blocks of older volcaniclastic breccia. The breccias are unstratified, very poorly sorted and may be either block-supported or matrix-supported. Intercalated with the volcaniclastic breccias are numerous thin basaltic volcaniclastic sandstones with eroded tops, and rhyolitic tuffs less than 2.5 m thick. Three thin ignimbrites are also intercalated with the volcanic breccias and two large masses of ignimbrite are preserved (Bell, 1985; Bell and Harris, 1986). A layer of lateritised breccia, about

0.5 m thick, which crops out north of Loch Cill Chriosd in the Allt Coire Forsaidh, and a thin hyaloclastite layer north of Cnoc nam Fitheach are further evidence of surface processes.

It is evident that the very heterogeneous assemblage of volcanic and subvolcanic magmatic and volcaniclastic rocks preserved at Kilchrist is the product of a variety of surface and near-surface processes. The ignimbrites and other silicic tuffs appear to be in situ extrusive deposits intercalated within the dominant volcaniclastic breccias. The heterogeneous nature of the volcaniclastic breccias, with little or no stratification and very poor sorting, suggests they were formed within the upper part of a vent system, exposed from time-to-time to the atmosphere (giving rise to lateritised material) and invaded by water (resulting in hyaloclastite deposition). The stocks and sheets of brecciated rhyolite attest to the important role played by silicic magmas and may be related to the ignimbrites and other silicic tuffs.

RUM

The Rum Central Complex was emplaced into sandstones of the Neoproterozoic Torridon Group which, together with Lewisian gneisses, form the ridge (or horst) that separates the Sea of the Hebrides and the Inner Hebrides Mesozoic sedimentary basins (Binns et al., 1974; Fyfe et al., 1993; Figure 5). There were at least three distinct phases in the growth and decay of the Rum Central Complex (Emeleus, 1997). *Phase 1* was dominated by silicic magmatism and the development of an arcuate system of faults, termed the Main Ring Fault (Figure 23). Basic and ultrabasic magmas were intruded mainly during *Phase 2*, when numerous minor intrusions, and gabbros and layered ultrabasic rocks were emplaced. *Phase 3* was marked by deep subaerial erosion and unroofing of the central complex, with intermittent burial of the developing topography by basaltic lavas of the Skye Lava Field (p.67), and by the coarse detritus eroded from the central complex and its surroundings.

The basic and ultrabasic intrusions of Phase 2 were emplaced at about 60.5 Ma (Hamilton et al., 1998; Figure 8). All of the intrusive and extrusive rocks are reversely magnetised and it is probable that emplacement of the central complex occurred over about 1 million years, during magnetic anomaly C26R. The central complex is the site of a pronounced positive Bouguer gravity anomaly.

Phase 1

The Northern Marginal Zone, the Southern Mountains Zone and the Western Granite were all formed during Phase 1 (Dunham, 1968; Emeleus, 1997; Figure 23). Initial activity involved uplift within the Main Ring Fault, causing Lewisian gneisses and basal members of the Torridon Group to be elevated by as much as 2 km inside the fault system. The uplift was accompanied by tilting, deformation and doming of the Torridon Group rocks that surrounded and overlay the central complex. At this time the two major masses of Torridonian sandstone that form Mullach Ard and Welshman's Rock (Figure 23), each now underlain by low-angle faults inclined at about 35° to the east and north-north-east, respectively, are thought to have slid off the developing dome. The Torridonian rocks south of the Main Ring Fault in southern Rum most likely comprise a similarly displaced block (SB, Figure 23).

Subsidence within the Main Ring Fault next led to caldera formation, during which quartz-feldspar-phyric rhyodacitic magmas were intruded near to and along the faults, leading to explosive, surface eruptions. These eruptions formed the ignimbrites and crystal-vitric tuffs exposed in Dibidil, and on Cnapan Breaca and Meall Breac (Figure 24). The porphyritic rhyodacite ignimbrites commonly show a distinctive streaky banding resulting from the welding of flattened and attenuated fiamme (Plate 27). In addition to quartz and plagioclase, they also contain small, generally altered phenocrysts of augite, pigeonite, rare orthopyroxene and iron-titanium oxides. The ignimbrites overlie or, less commonly, are interbedded with coarse-grained volcaniclastic deposits in the form of largely unbedded breccias. The breccias consist of angular to subrounded blocks and megablocks of Torridonian sandstone; clasts of gneiss, basalt and dolerite also occur but are rare, as are scoria fragments. Examples of megablocks of Torridonian sandstone, tens of metres across, occur in Dibidil. Finer grained, bedded tuffaceous deposits containing sandstone clasts, thin

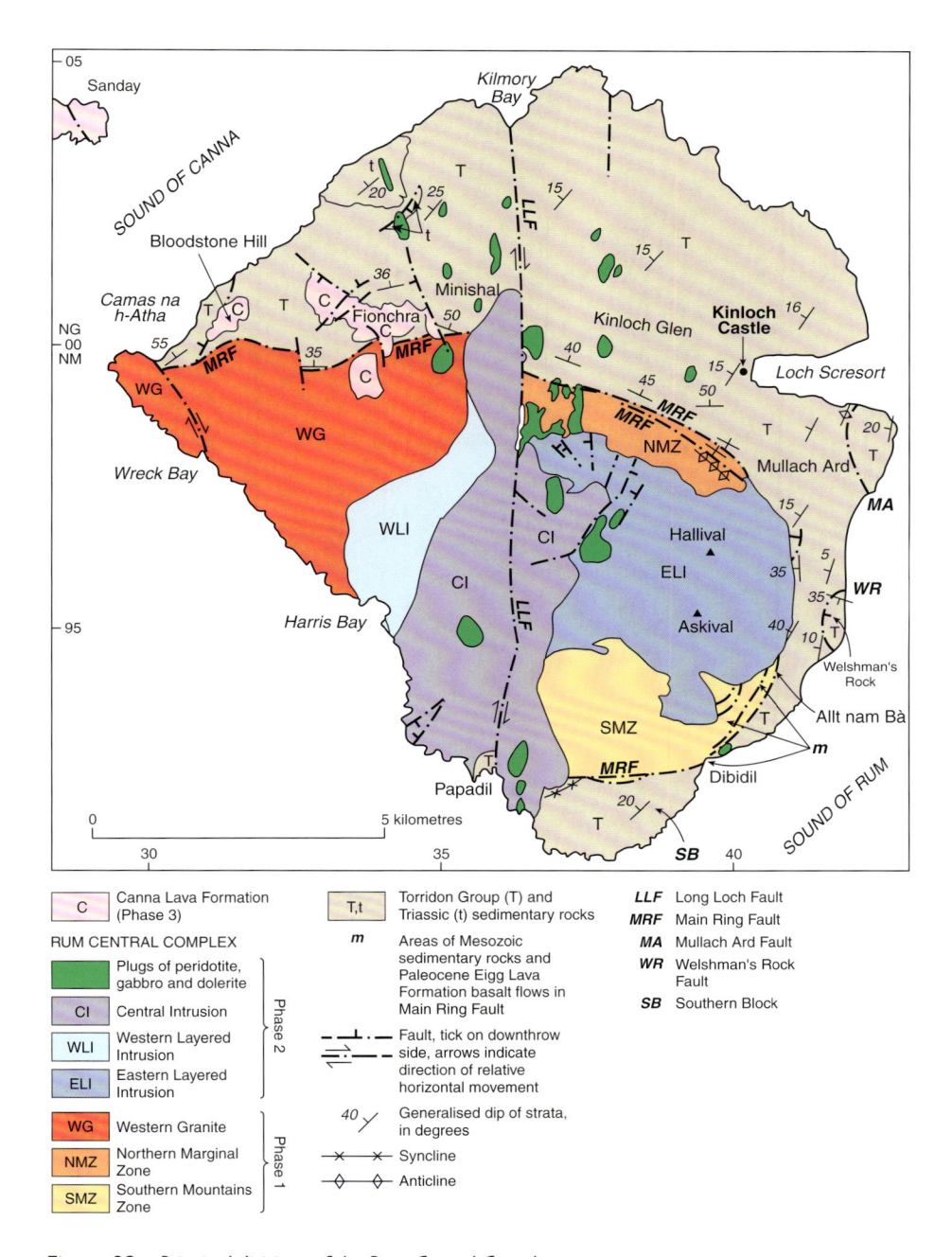

Figure 23 *Principal divisions of the Rum Central Complex.*
At Mullach Ard, Welshman's Rock and the Southern Block, major blocks of Torridon Group sandstone have been displaced over the central complex

layers of crystal tuff and impersistent beds of coarse, gritty sandstone are interbedded with and immediately underlie the ignimbrites. Dykes and veins of porphyritic rhyodacite cut the breccias, as do rare dykes of breccia and tuff which contain lobate inclusions of rhyodacite, plagioclase crystals derived from rhyodacite and rounded fragments of baked sandstone.

These deposits, termed the Coire Dubh Breccias, are interpreted as caldera-fill materials (Troll et al., 2000). The coarse, poorly sorted breccias are interpreted as having formed when the over-steepened caldera walls collapsed onto the caldera floor, and the thin beds of gritty

CHAPTER NINE: CENTRAL COMPLEXES

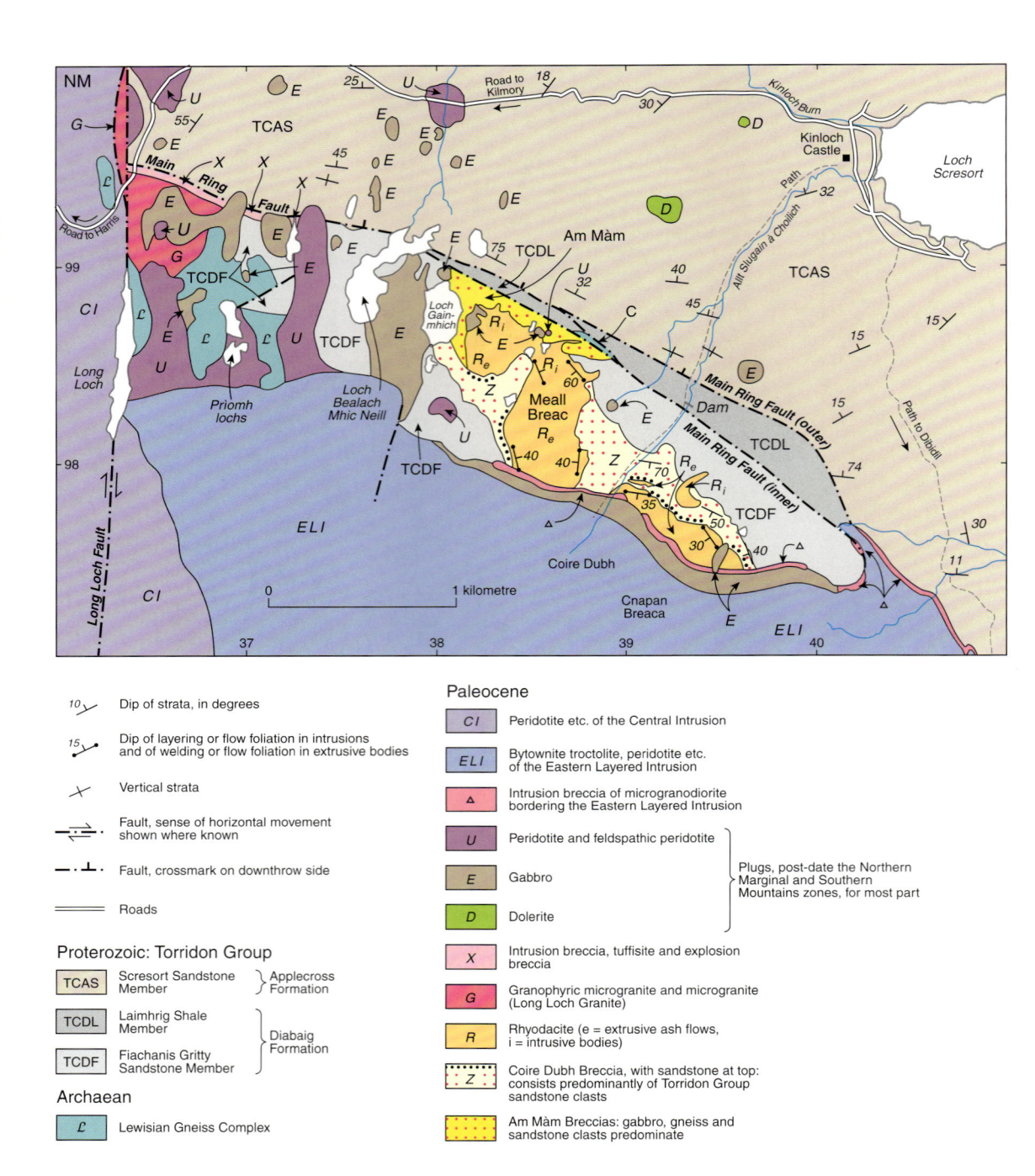

Figure 24 *Geological map of the Northern Marginal Zone, Rum Central Complex.*

sandstone represent materials washed out of the coarser deposits. The porphyritic rhyodacites formed as ignimbrites, erupted from feeders situated on the Main Ring Fault, their eruption being preceded by several ash-fall eruptions. On Meall Breac and in Dibidil, the steep-sided, intrusive rhyodacite bodies most likely pass laterally into ignimbrites (Donaldson et al., 2001). In Coire Dubh and in Dibidil, certain of the intrusive porphyritic rhyodacites contain numerous irregular, rounded masses of basic material, indicating that there was

Plate 27 Fiamme deformed beneath a lithic block in a porphyritic rhyodacite ash flow, Cnapan Breaca, Northern Marginal Zone, Rum Central Complex. Scale: coin is 24 mm diameter (P580477).

thorough mixing of co-existing silicic and basic magmas beneath the central complex during Phase 1, and suggesting that some of the eruptions were of mixed magma. (Troll et al., 2004) Subsurface explosive disintegration of the partially crystalline rhyodacite magma also shattered the sandstone surroundings, the mixture of fragments forming the rather rare tuff dykes as well as the more extensive areas of explosion breccia exposed south of Dibidil.

Breccias of a different kind occur in the Northern Marginal Zone. These rocks, termed the Am Màm Breccias, comprise blocks and megablocks of gabbro, blocks of thermally altered gneiss and sandstone, and rare feldspathic peridotite fragments. The matrix is a medium-grained, hybrid rock of quartz-dioritic or granodioritic composition. The breccia is intruded by the porphyritic rhyodacite but its relationship to the other breccias described above is uncertain; it is regarded as an early member of Phase 1. The clast content of the Am Màm Breccias is important since it shows that plutonic basic and ultrabasic rocks had formed at a very early period in the evolution of the central complex, long before the rocks of Phase 2, described below.

On the north-east side of Dibidil and on the eastern slopes of Beinn nan Stac, slivers of Lower Jurassic (Broadford Beds) limestone, siltstone and sandstone, and of altered basaltic lava derived from the Eigg Lava Formation, owe their preservation to subsidence within the Main Ring Fault. However, the final event of Phase 1 involved further uplift on the Main Ring Fault system, when Torridon Group sandstones were reverse-faulted over the Mesozoic sedimentary rocks and Paleocene lavas on the eastern slopes of Beinn nan Stac in eastern Rum. In western Rum, granite was similarly faulted over sandstones of the Torridon Group near Bloodstone Hill. This late uplift is tentatively attributed to the forcible emplacement of the silicic magma that formed the Western Granite.

The Western Granite and similar smaller granite bodies near the Long Loch and Papadil postdate the porphyritic rhyodacites. The well-jointed Western Granite is magnificently exposed in cliffs and raised-beach platforms in south-west Rum, although inland exposure is poor. The granite is microporphyritic and of similar composition to the porphyritic rhyodacites. It contains phenocrysts of plagioclase together with ferroaugite and ferropigeonite, both of which are generally partially altered to amphibole. A small area of granophyric granite containing microphenocrysts of fayalite and ferroaugite, as well as plagioclase, crops out on the south-east slopes of Sròn an t-Saighdeir.

Phase 2 Basaltic and picritic magmatism predominated throughout Phase 2. Initially, numerous basaltic dykes were intruded, accompanied by many inclined sheets (cone-sheets), which are present throughout the Northern Marginal Zone and Southern Mountains Zone. The cone-sheets focus on a point at depth beneath upper Glen Harris; unlike the cone-sheets of the Cuillin Centre on Skye, those on Rum are fine grained, possibly reflecting emplacement at a shallower level. The principal trend of the basaltic dykes is north-west, and they merge with the regional Muck Dyke Swarm (Chapter 7). Additionally, there are many dykes that trend north–south or north-north-east so that, overall, the dykes define a radial pattern, as was noted by Harker (1908). In south-west Rum, several generations of approximately north-west-trending dykes intrude the Western Granite but few extend into the layered intrusions.

The greatest development of basic and ultrabasic rocks is in the layered intrusions, which crop out over much of the central and southern parts of the island (Figure 23; Plate 28). These intrusions postdate the cone-sheets and the majority of the dykes. Many of the layered structures are identical with, and of similar origin to, those in the Cuillin Centre of Skye (see above). In eastern Rum, the layered rocks are intruded by numerous semiconcordant sheets of gabbro, for example in Atlantic Corrie and on Askival where they accentuate the bedded appearance of the exposures (Volker and Upton, 1990). The layered rocks are also cut by plugs of gabbro and feldspathic peridotite, as are earlier members of the central complex and the country rocks (see also below).

The layered structures in the ultrabasic rocks and bytownite gabbros (the latter were formerly known as 'eucrites') vary in scale from tens of metres to millimetres (Plates 28; 29).

Plate 28 Layered ultrabasic rocks of the Eastern Layered Intrusion on Hallival, Rum Central Complex. Bytownite troctolite ('allivalite') forms the resistant layers, the easily weathered layers are made of feldspathic peridotite. Westward-dipping Torridonian strata are visible in the middle distance (P580478).

They are defined principally by varying proportions of anorthite-rich plagioclase, forsteritic olivine and subsidiary chrome-spinel and diopsidic clinopyroxene, although in some instances, textural features define the layers. Predominance of olivine gives rise to dark, easily weathered layers of feldspathic peridotite, whereas the resistant layers of light-coloured bytownite troctolite (or 'allivalite', so-called after the mountain Hallival on Rum; Harker, 1908) contain abundant plagioclase. In addition to their bedded appearance, many of the layered rocks contain structures closely resembling those found in clastic sedimentary rocks. These include size-graded and density-graded bedding, cross-bedding, slump structures, flame structures, drop-stones with associated deformation of small-scale layering, and coarse breccias (Wadsworth, 1992; Plates 30; 31; 32). Flat-lying zones of shearing and dislocation, as evidenced by deformation of small-scale structures, indicate that there has been much internal movement during accumulation of the layered rocks prior to their lithification. In south-west Rum, layering in bytownite gabbros at Harris and in feldspathic peridotites on Ard Mheall is accentuated by the presence of elongate olivine crystals which appear to have grown upwards from a succession of planar surfaces in the magma chamber. Rocks with this distinctive structure (Plate 33) were termed 'harrisites' by Harker (1908).

The microscopic textures, particularly in the ultrabasic rocks, suggest an origin by sedimentation of crystals from cooling magmas, with crystal sorting by density and, to a lesser extent, by size, followed by 'cementation' by later-crystallising phases. Feldspathic peridotite contains abundant, rounded to well-formed olivine and small chrome-spinel crystals enclosed by anhedral plagioclase and pyroxene, whereas the bytownite troctolites contain tabular plagioclase with pronounced parallel orientation or lamination, interstitial pyroxene and only rare olivine. The term 'cumulates' was coined to describe these and other layered rocks with distinctive textures presumed to reflect the accumulation of crystals precipitated from a magma (Wager et al., 1960; Wager and Brown, 1968). However, many of the textures now preserved are not wholly primary. It is now recognised that the rocks have achieved a high degree of

Plate 29
Photomicrograph of the boundary between an anorthosite at the top of Unit 11 (below) and a typical feldspathic peridotite at the base of Unit 12 (above), Eastern Layered Intrusion, Rum Central Complex, northwest Hallival. A 2 mm-thick chromite-rich seam occurs at the boundary. View is under crossed polarised light (P580479).

textural equilibration: during prolonged cooling and consolidation their crystal shapes were modified by resorption and reprecipitation, processes that were aided by hot residual fluids percolating through the pile of layered rocks (e.g. Hunter, 1996). Composition was also modified during recrystallisation, as is evident from detailed studies of mineral compositions across the boundaries of layers on Hallival (Dunham and Wadsworth, 1978; Tait, 1985) and where late-stage veins cut the layered rocks (Butcher, 1985). Some of the larger-scale structures are attributable to replacement, for example where olivine-rich rocks cut across apparently undisturbed layered troctolites, as on the north-west shoulder of Hallival, and where olivine-pyroxene-rich 'finger' structures cut across undisturbed small-scale layering in bytownite troctolites (Butcher et al., 1985; compare with Plate 25).

The hot picritic magmas (1100° to 1300°C; Chapter 10), parental to the ultrabasic and basic rocks were emplaced into rhyodacites, sandstones, granites and gneisses. Since these host rocks generally have relatively low melting points (less than 1000°C), they underwent complete or partial melting and remobilisation (rheomorphism). The silicic melts back-veined and shattered the chilled margins of the basic and ultrabasic rocks, giving rise to the spectacular zones of intrusion breccia generally found at the margins of the layered intrusions. Intrusion breccias are common throughout the central complexes of the Hebridean Igneous Province, where felsic and *later* mafic rocks

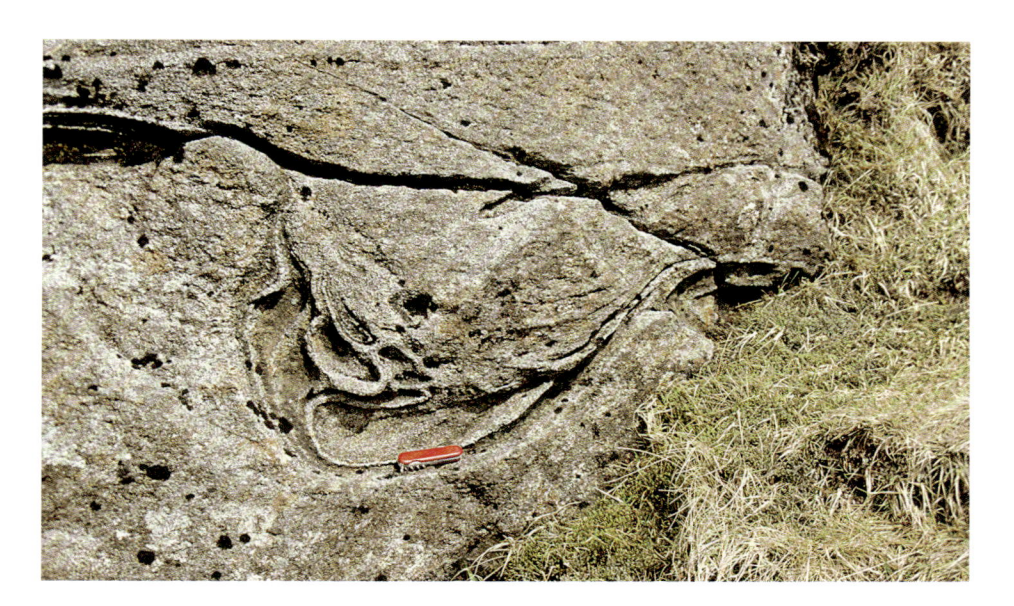

Plate 30   Slumping in bytownite troctolite of Unit 15, Eastern Layered Intrusion, Rum Central Complex on Askival. Scale: pen-knife is 10 cm (P580480).

Plate 31   Peridotite block ('dropstone') with deformed layering in underlying bytownite troctolite of the Central Intrusion, Rum Central Complex. North of the Long Loch, Rum. Scale: hammer shaft is 30 cm (P580481).

are juxtaposed; on Rum some of the most accessible and varied examples of this phenomenon occur at Harris Bay (Plate 34). At Harris, and elsewhere on Rum, there is also evidence that rheomorphic silicic melts, together with basic and rare ultrabasic blocks from the intrusion breccias, may have become detached from their sources and intruded for quite considerable distances into the layered suite. Such an origin could provide an explanation for the Am Màm Breccias of the Northern Marginal Zone (see above), although these would have been related to a much earlier period of intrusion of hot basic and ultrabasic magmas. Elsewhere, rocks in contact with the

Plate 32 *Block of layered bytownite troctolite and feldspathic peridotite in a breccia of bytownite troctolite. Note the 'finger-structures' within the block. South of the Long Loch, Central Intrusion, Rum Central Complex. Scale: hammer shaft is 30 cm (P580482).*

layered intrusions are thermally metamorphosed. At Allt nam Bà, in eastern Rum, impure limestones of Early Jurassic age are altered to calc-silicate hornfels containing tilleyite, spurrite and other high-temperature (sanidinite-facies) minerals, and there are also small amounts of peralkaline hybrid rock (Hughes, 1960). East of Hallival, xenoliths of fine-grained pyroxene-plagioclase-olivine rock in the Eastern Layered Intrusion are probably fragments of metamorphosed country-rock basalt; in these rocks, small, diffuse areas a few centimetres or so in diameter contain fassiaite and hydrogarnets and may represent metamorphosed amygdales (Faithfull, 1985).

Three major intrusions of basic and ultrabasic rock, the Eastern Layered Intrusion, the Western Layered Intrusion and the Central Intrusion, form the core of the Rum Central Complex (Figure 23). The Eastern Layered Intrusion (Brown, 1956; Wager and Brown, 1968) contains the classic layered sequences of feldspathic peridotites and bytownite troctolites that are at least 700 m thick. The Western Layered Intrusion (Wadsworth, 1961) consists of layered bytownite gabbro at Harris, overlain by a thick succession of layered feldspathic peridotite on Ard Mheall. In both areas, the layering is commonly accentuated by layers containing

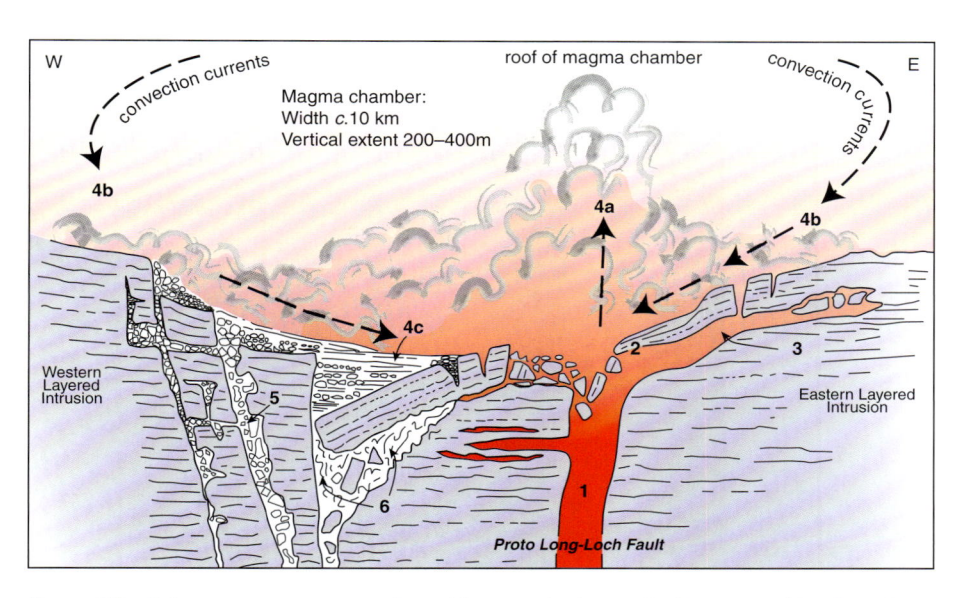

Figure 25 Schematic representation of possible events leading to the formation of the Central Intrusion, Rum Central Complex (after Emeleus et al., 1996b, fig. 21).
Periodic replenishments of picritic magma (1) rejuvenated the magma chamber, causing disruption, sliding and slumping of earlier layered sequences (2), and intruded laterally into earlier cumulates (3). Magma fountaining into the magma chamber (4a) subsequently flows off the roof and down the sides as crystal-laden, gravity-driven currents (4b), dislodging crystal mushes as they move. The gravity currents spread across the floor, reworking cumulate debris and depositing this material and primary crystals on the floor (4c). Movement on faults was accompanied by magma injection, thermal erosion of earlier rocks and their fragmentation to form breccia zones (5). Slides of large, coherent masses of cumulates across partly liquefied cumulates led to spectacular slump mélanges (6) (Plate 32).
Solidified or semi-solidified cumulates are shaded.

harrisitic olivines (compare with Plate 33). The layered succession is about 500 m thick. The Central Intrusion (McClurg, 1982; Volker and Upton, 1990) intrudes both of the Eastern and Western layered intrusions and truncates layered structures in each. The Central Intrusion is characterised by narrow zones of complex ultrabasic breccias enclosing areas of normally layered troctolites and peridotites. The breccias consist essentially of blocks and megablocks derived from earlier parts of the layered intrusions, including blocks which themselves exhibit complex internal relationships (Emeleus, 1997, Plate 32). The Central Intrusion is considered to have been the feeder zone for all of the layered intrusions (Emeleus et al., 1996b; Figure 25). Successive batches of ascending ultrabasic and magnesium-rich basaltic magmas were focussed into an approximately north–south zone of weakness on the proto-Long Loch Fault (p.150), spreading laterally when they encountered the density trap at the Lewisian gneiss–Torridonian sandstone boundary and crystallising as layered rocks. Later batches intruded and brecciated layered rocks formed earlier and, in turn, spread laterally over the earlier, denser, and now largely crystalline material to build the Eastern and Western layered intrusions.

Over 40 small plugs of feldspathic peridotite, gabbro and dolerite intrude the central complex and the surrounding country rocks. Where the plugs intrude the Torridon Group and Triassic sandstones, the sedimentary rocks are generally bleached, closely jointed and thermally metamorphosed, commonly with the formation of tridymite (preserved as quartz paramorphs); rarely, there is also evidence of melting. The altered feldspathic sandstones are commonly spotted, with pale rounded patches (a few millimetres in diameter) that have a

Plate 33 *Elongate olivine crystals ('harrisitic texture') in bytownite gabbro of the Harris Bay Member, Western Layered Intrusion, Rum Central Complex. Harris Bay, Rum. Scale: coin is 20 mm diameter (P580483).*

spherulitic microstructure. Thin, linear zones of bleaching and, less commonly of brecciation, traverse the sedimentary rocks north of the central complex. Termed fissure breccias, the altered rocks are thermally metamorphosed in a similar manner to the sandstones adjoining the peridotite and gabbro plugs. However, the only spatially associated igneous rocks are rare stringers and veinlets of fine-grained basalt.

Phase 3 During phases 1 and 2, a considerable volcanic edifice must have built up over the Rum Central Complex. At the end of Phase 2, with the cessation of all igneous activity within the central complex, this massif was subjected to intense erosion. At the same time, lavas of the Canna Lava Formation, derived from sources outwith Rum, were ponded in the developing valley and canyon system on its flanks, where they became interbedded with the coarse detritus being stripped off the Rum massif. These features, which comprise Phase 3, are discussed more fully in Chapter 6.

ARDNAMURCHAN The Ardnamurchan Central Complex is situated on the mainland north of Mull. Three distinct centres of activity have been recognised and numbered sequentially from oldest to youngest as Centre 1, Centre 2 and Centre 3 (Richey and Thomas, 1930; Gribble et al., 1976). There is no obvious spatial progression as the central complex developed: Centre 2 lies to the west of Centre 1, but Centre 3 lies to the north-east of Centre 2 (Figure 26a).

The Ardnamurchan Central Complex was emplaced into a metamorphic basement of relatively low-grade Moine psammites and pelites, overlain by Triassic breccias, conglomerates and sandstones, Lower Jurassic sandstones, limestones and mudstones, and Paleocene lavas. The proportion of country rocks to intrusive rocks in the vicinity of Centre 1 is high, whereas centres 2 and 3 are largely devoid of remnants of Pre-Palaeogene country rock between the various major intrusions. The Mesozoic strata and Paleocene lavas have been domed, and dip

Plate 34 *Intrusion breccia at the contact between granophyre and later gabbro of the Central Intrusion, Rum Central Complex. East side of Harris Bay, Rum (P580484).*

away from the central complex at about 30°. The updoming occurred at an early stage, since it predates cone-sheet emplacement (see below). No marginal annular folds are preserved.

Centre 1 and the Ben Hiant volcaniclastic rocks

Along the south-east margin of the central complex, on Ben Hiant, a small remnant of the Mull Lava Field is preserved, overlain by about 200 m of volcaniclastic conglomerates and breccias. There are no direct links between these volcaniclastic rocks and the nearby ring-intrusions, which define Centre 1, but they are described here because of their spatial association with Centre 1.

The volcaniclastic rocks are heterogeneous, with matrix-supported clasts derived mainly from the older lava field. Fragments derived from the basement Moine rocks and the Mesozoic cover sequence are rare. One fragment type that cannot be matched with in-situ material is of rhyodacitic ignimbrite with a well-developed eutaxitic fabric. The ignimbrite is clearly the product of explosive activity, although the vent(s) from which the magma was erupted has not been recognised. The clasts within the conglomerates and breccias range in size from several metres, with rare megablocks of several tens of metres, down to sand and silt grade. Bedding is relatively uncommon, but where present, it is typically close to horizontal and tends to be defined by thin layers of tuffaceous siltstone and sandstone. Fragments within these finer grained deposits are predominantly of lava lithologies, especially trachyte, together with Moine psammite and pelite, and Mesozoic sedimentary rocks. The Ben Hiant conglomerates and breccias are interpreted as the products of debris flow and avalanche deposition (Brown, 2003).

Emplaced into the Ben Hiant volcaniclastic rocks are sheets of andesitic pitchstone ('inninmorite'), typically with well-developed columnar jointing. These have previously been interpreted as lavas, although in at least one example on the south-east side of Ben Hiant, the intrusive nature is particularly obvious, with columnar jointing fanning away from the leading edge of the intrusion where it cooled in contact with possibly water-saturated conglomerates and breccias. The volcaniclastic rocks are also invaded by a variety of plagioclase-phyric and non porphyritic dolerite sheets of unclear association. However, the main intrusive sheets, forming the prominent summit of Ben Hiant, are of a distinctive quartz-dolerite and belong to the Centre 1 set of cone-sheets with a focal point at depth, about 1 km west of Meall nan Con (Figure 26). The outer members of the set of cone-sheets are inclined inwards at relatively shallow angles (15° to 20°), whereas those towards the interior are inclined at up to 40°. There are also a few multiple and composite (dolerite plus felsite) cone-sheets.

On the north coast of the Ardnamurchan peninsula, around Achateny and Kilmory, are further outcrops of conglomerate and breccia, interpreted as having a similar origin to those at Ben Hiant (Brown, 2003). These rocks are cut by the Centre 1 cone-sheets and were previously referred to as the products of the 'Northern Vents'.

Other intrusions possibly associated with Centre 1 are shown on Figure 26. All are either sheet- or dyke-like and appear to predate the cone-sheets of Centre 1.

Centre 2

Centre 2 was formed by four recognisable phases of intrusive activity, all with a focal point at depth, below Achosnich (Figure 26). The first phase involved the emplacement of an outer (older) set of basic cone-sheets, emplaced into the Mesozoic cover sequence and the underlying Moine psammites and pelites. These cone-sheets are of similar tholeiitic composition to the Centre 1 set, are typically 1 to 5 m thick, and are inclined at 35° to 45° towards the focal point. In the vicinity of Kilchoan, the distribution density of the cone-sheets is very high (Plate 35). Cone-sheets were emplaced into cone-sheets, and in the relatively uncommon instances where country rock is preserved as screens, it has undergone extreme thermal metamorphism. Multiple and composite members of the set occur, but are not particularly common. Emplacement of these outer cone-sheets produced a relative central uplift of about 1300 m (LeBas, 1971), although a degree of central subsidence may also have been involved. In the critical area, near Camphouse, where the outer (older) set of Centre 2 cone-sheets and the Centre 1 cone-sheets occur together, exposure is poor, and it is not possible to establish the order of intrusion. The two sets of cone-sheets are geochemically indistinguishable (Holland and Brown, 1972; Geldmacher et al., 1998).

The first major ring-intrusion of Centre 2 is a hypersthene-gabbro (Wells, 1954) (Figure 26b, 2a). It is clearly younger than the outer (older) set of cone-sheets, which it cuts at several places along its southern margin, particularly on the south side of Beinn nan Codhan. The hypersthene-gabbro shows excellent, if uncommon, mineral layering. Layers rich in pyroxene, plagioclase and iron-titanium oxides dip inwards towards the focal point but cannot be traced far. Augen structures and upward-penetrating finger structures occur in the section north of Sanna Bay and suggest that significant post-cumulus modifications have taken place. The thermal effects of the hypersthene-gabbro are substantial, with the formation of various rocks of rheomorphic origin from a variety of basement and Mesozoic lithologies. At Glebe Hill, strongly hornfelsed rocks contain a high-grade thermal metamorphic mineral assemblage, with sapphire, spinel and plagioclase, produced from a highly aluminous parent, possibly a Paleocene claystone, a Jurassic mudstone, or pelitic material from the Moine basement.

The gabbro that crops out at Lochan an Aodainn close to Sonachan is a thoroughly recrystallised olivine-gabbro, severely shattered throughout its mass and with plagioclase crystals that are clouded due to the presence of abundant inclusions of opaque minerals; the clouding has been attributed to thermal alteration. The outcrop pattern is largely controlled by the surrounding younger intrusions and, consequently, its ring shape did not result from its mode of emplacement (Figure 26b, 2b). The degree of fracturing and alteration present clearly indicate that it was emplaced early in the development of Centre 2.

The quartz-dolerite of the Sgurr nam Meann Ring-dyke is a hybrid intrusion, consisting of dolerite veined by microgranite and felsite. It crops out in an arc over a distance of about 6 km south and south-east of Sanna Bay (Figure 26b, 2c). This intrusion, most likely a steep-sided, outward-dipping ring-dyke, was formed by the turbulent mixing of basic and silicic magmas (Blake et al., 1965; Skelhorn and Elwell, 1966). Typically, the silicic material either forms veins (millimetre to centimetre wide) within large masses of dolerite, resulting in relatively angular outlines to the dolerite 'clasts', or acts as 'matrix' to more rounded masses of dolerite (Plate 36). In the latter situation, thin, irregular

123

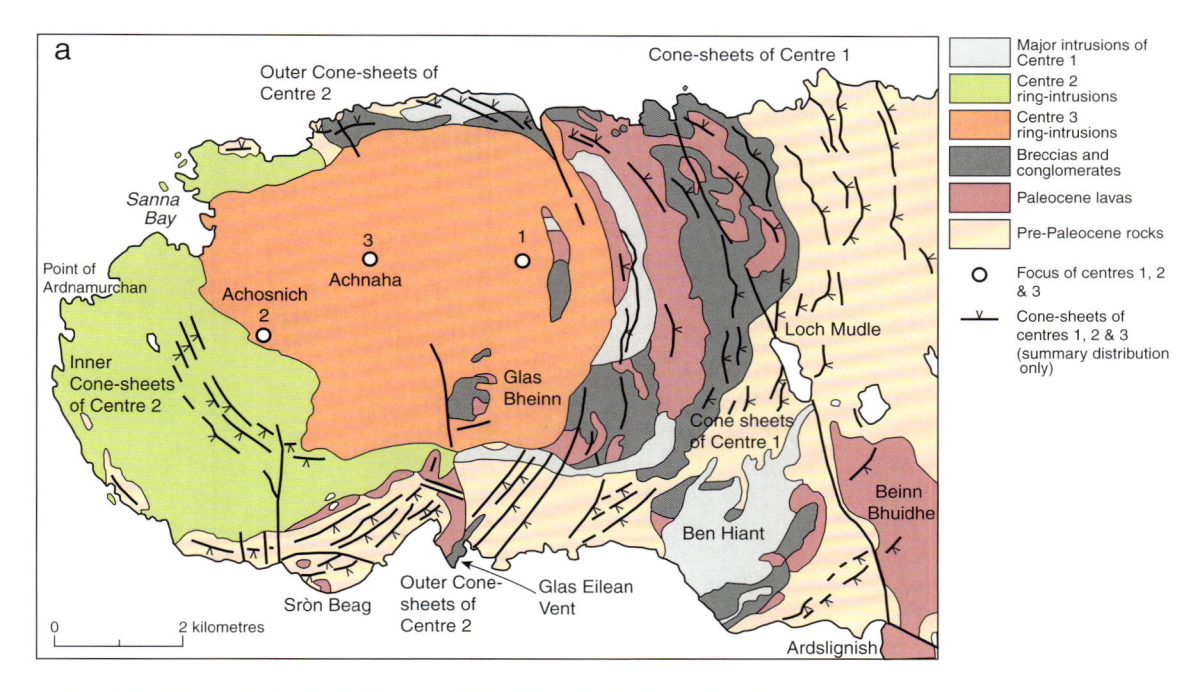

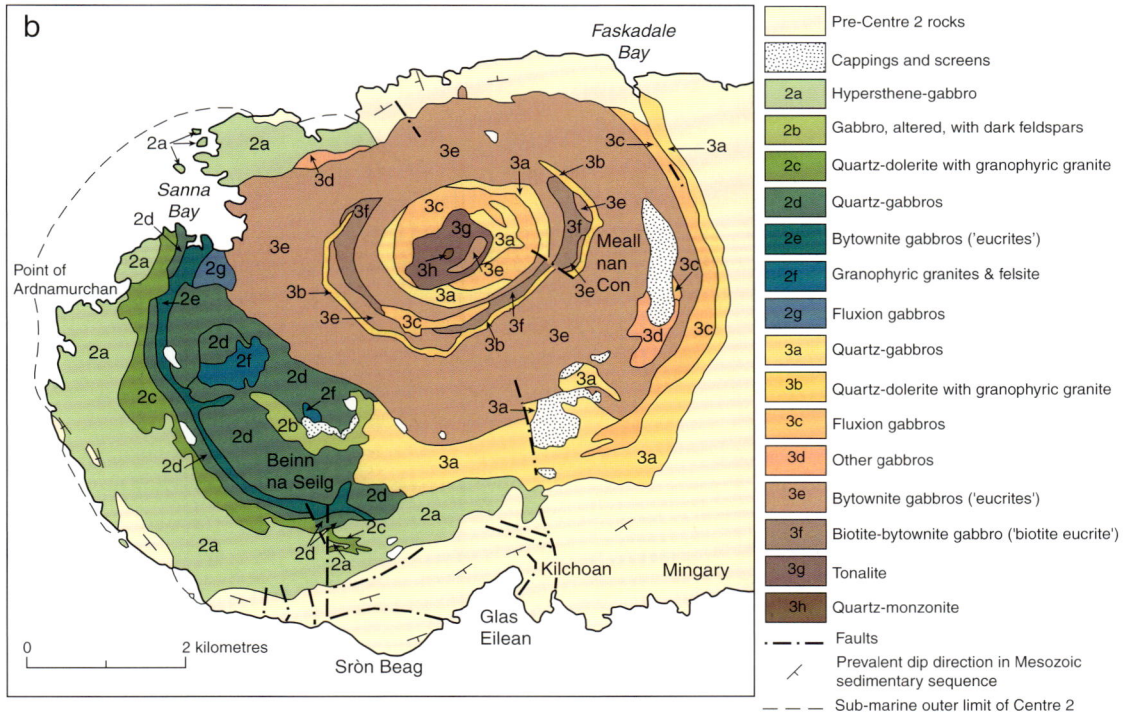

Figure 26 Principal components of the Ardnamurchan Central Complex.
a   Centres, cone-sheets and country rocks
b   Ring-intrusions of centres 2 (2a–g) and 3 (3a–h)
(based on Bell and Williamson, 2002, fig.14, 15)

CHAPTER NINE: CENTRAL COMPLEXES

Plate 35 *Inclined dolerite cone-sheets intruding Moine metasedimentary rocks surrounding the Ardnamurchan Central Complex. The sheets are from 1.5–2 m in thickness; they focus on Centre 2, but predate the major gabbroic intrusions in this centre (forming the more distant hills), west of Mingary pier, Ardnamurchan (P580485).*

(crenellated) chilled margins to the dolerite indicate that the basic material was liquid when it came into contact with the silicic magma. Thus, on mixing, the hot basic magma underwent rapid cooling and solidification against the cooler but still liquid silicic magma, forming 'pillows' of dolerite with chilled, tachylitic margins. Contraction cracks formed on the edges of the 'pillows', and silicic magma was then injected into these and beyond, resulting in veining and fragmentation of the dolerite.

Within Centre 2 there are several other less complete ring-intrusions that predate the inner basic cone-sheets (see below). They comprise quartz-dolerites and gabbros, a granite and various felsites.

The third phase of intrusive activity associated with Centre 2 was the emplacement of the inner (younger) basic cone-sheets. These intrusions are steeply inclined inwards, at up to 70°, towards the focal point below Achosnich. They are typically plagioclase-phyric dolerites.

The final phase of intrusive activity was the emplacement of a further set of incomplete ring-intrusions, mainly of quartz-gabbro, some of which have a fabric due to the alignment of plagioclase (the so-called fluxion structure, or lamination). Whether certain of these intrusions are late-stage members of Centre 2 or early members of Centre 3 is not clear.

The formation of the Glas Eilean Vent (Figure 26a), cropping out on the east side of the narrow promontory and tidal island on the east side of Kilchoan Bay, was later than the intrusion of the outer basic cone-sheets, but contemporaneous with the development of the younger intrusions of Centre 2. The outcrop of the vent material is partly fault-bounded, with Moine basement rocks along the south-east margin and basaltic lavas to the west. The fragmental material within the vent consists of blocks of Moine psammite and pelite, Jurassic sandstone and limestone, and various basic igneous rock-types (porphyritic and non-porphyritic dolerite and basalt, some perhaps derived from the outer cone-sheets). Much of the fragmental material appears to have been brecciated almost in situ, and not subsequently transported any significant distance. The matrix that binds the blocks together forms vein-like features, up to 0.5 m across, with fine-grained margins against the blocks. It is composed of comminuted block material together with pumiceous and shard-like fragments of devitrified (silicic?) glass. The

Plate 36 *Intrusion breccia of quartz-dolerite, net-veined by granophyric microgranite in the Sgurr nam Meann Ring-dyke of Centre 2, Ardnamurchan Central Complex. Opposite Eilean Carrach, Ardnamurchan. Scale: lens cap is 60 mm diameter (P580486).*

vein-like features form an anastomosing network, and were probably formed when pyroclastic material was emplaced during rapid volatile escape. The devolatilisation process may have been responsible, to a large extent, for the initial and subsequent brecciation of the blocks.

Centre 3 The youngest intrusive centre within the Ardnamurchan Central Complex comprises annular gabbroic intrusions with near-complete ring geometries. The dips of the inter-intrusion contacts are most likely inwardly inclined and give rise to a nested set of funnel-shaped intrusions. Fabrics due to plagioclase alignment ('fluxion structures') help to distinguish between some of the intrusions. Medium-grained intermediate and silicic intrusions associated with the centre are of hybrid origin, containing components derived by fractional crystallisation of basic magmas, together with partial melts of country rock (Gribble et al., 1976).

The focal point of Centre 3 is 1 km east-north-east of Achnaha, directly below the compositionally evolved intrusions (Figure 26). The centre is slightly ovoid, measuring about 7 km east-north-east and 6 km north-north-west. The margin of the centre is defined by various relatively small gabbros with outcrops that define sectors of ring-intrusions. However, the largest of the Centre 3 intrusions and the one which helps to define its overall geometry and give it topographic expression, is the

Great Eucrite (Figure 26b, 3e), a bytownite olivine-gabbro, which forms an arc of high ground (Plate 37). The outcrop width of the Great Eucrite is about 1 km and the external diameter of the intrusion is between 4.5 and 6 km. Thus, if it were a true ring-dyke, dipping steeply outwards (compare with Figure 18b), unrealistically large amounts of central subsidence would have been required. More likely, the Great Eucrite, together with other gabbroic intrusions of Centre 3, constitute a composite intrusion of lopolithic or funnel-like shape. One of the most remarkable features of the Great Eucrite is that, given its relatively large volume, it is largely devoid of internal structure; layering, mineral fabric, xenoliths and pegmatites are uncommon, implying that the intrusion had a very simple emplacement history.

Within the interior of Centre 3 is a hybrid ring-dyke that was formed by the partial mixing of basic and silicic magmas (Figure 26b, 3b). This narrow, steep-sided intrusion crops out over about 270° of arc, suggesting off-centre subsidence of the central block or some form of rotational or 'trap door' subsidence mechanism for its emplacement. Certain of the gabbroic ring-intrusions within the inner part of Centre 3 crop out over the same sector as this ring-dyke.

The innermost part of Centre 3, directly above the focal point, is composed of a small area of amphibole-rich tonalite and a central mass of quartz-monzonite with large biotite crystals. Given their position within Centre 3 and their hybrid mineralogical and compositional characteristics, it is likely that these evolved rocks are the product of reaction between fractionating basic magmas of the centre and partially melted roof rocks. Radiometric age determinations on the tonalite and Great Eucrite give ages of about 59 Ma (Figure 8).

There are very few minor intrusions associated with Centre 3. They are limited to sparse dolerite cone-sheets cutting the older members of the centre in the south, and a few north-north-west-trending dykes of the regional swarm that cut the younger ring-intrusions.

MULL

Our understanding of the order of intrusive events within the Mull Central Complex is still largely due to Bailey et al. (1924). Subsequent work has been restricted to a small number of studies of some of the main intrusions and a brief summary and field guide by Skelhorn and Longland (1969). Three centres are recognised and numbered sequentially 1, 2 and 3; the first and last are considered to have been related to the development of calderas: Centre 1 to the Early Caldera or Glen More Centre and Centre 3 to the Late Caldera or Loch Bà Centre. Centre 2, the Beinn Chaisgidle Centre, is composed of various cone-sheet and ring-dyke intrusions.

During formation of the Mull Central Complex, there was a gradual shift of activity from Centre 1 through to Centre 3 (Figure 27). Movement was in a south-east to north-west direction, by a few kilometres, parallel to the trend of the regional dyke swarm. Large annular folds surround the central complex (see p.150).

Centre 1, the Glen More Centre

Within the Mull Central Complex and partly acting as country rock to the intrusions, are remnants of pillowed basaltic lavas. The relationship between these lavas and the main lava field on Mull (p.75; Table 15) is unclear, although Bailey et al. (1924) concluded that the pillowed material constitutes the stratigraphically youngest part of the lava field and formed within a caldera, hence the Glen More Centre is also referred to as the Early Caldera (Figure 28). The pillowed flows are referred to as being of the 'Non-Porphyritic Central Magma Type' or the 'Central Mull Tholeiites' (Chapters 6 and 10).

Early granites

The oldest of the main intrusions of the Glen More Centre are the steep-sided granites of Glas Bheinn and Derrynaculean, which possibly form parts of ring-dykes or steep-sided stocks. Emplacement of these granites was, in part, controlled by ring-faults, with central collapse. Brecciation, due either to gas escape or ring-faulting, is common throughout the granites, especially in the Derrynaculean mass. In addition, the Glas Bheinn intrusion was emplaced into the core of the somewhat imperfectly developed marginal Loch Spelve Anticline. Both granites

Plate 37 *Aerial view of Centre 3, Ardnamurchan Central Complex, from the north-west. Differential weathering of the arcuate gabbroic and dolerite intrusions has resulted in the 'crater'-like topography. The Great Eucrite forms the prominent rim (Photograph: Patricia and Angus Macdonald).*

show significant hydrothermal alteration, with primary pyroxene being chloritised or uralitised. Marginal facies of the Glas Bheinn Granite contain partially assimilated siliceous material, most likely derived from country-rock Triassic sandstones.

Explosion breccias   Several masses of explosion breccia occur along the trace of the ring-fault which is used to define the extent of the Early Caldera. The best examples occur within the south-east sector of the bounding fault, on the eastern side of Sgurr Dearg. The breccias contain subangular to rounded fragments of Paleocene lavas, Mesozoic sedimentary rocks, Moine gneisses and a wide variety of coarse-grained igneous rocks (gabbro, granite, etc.). Moine gneisses are generally absent from the breccias inside the main caldera-bounding fault, indicating that the basement lies at a deeper structural level beneath the caldera and that the explosive brecciation occurred at a fairly shallow level in the crust. Fragmented rhyolitic rocks with flow-banded and perlitic textures also occur in the breccias, which were most likely formed by gas streaming from silicic magmas. Surface volcanic deposits related to this explosive activity are not recognised, due to the level of erosion.

In Coire Mór, on the east side of the central complex, is an outcrop of generally unstratified volcaniclastic breccia containing subangular to rounded blocks of various Paleocene igneous rocks and Pre-Paleocene sedimentary rocks. Also present are large masses of flow-banded rhyolite. Similar material occurs at Barachandroman at the south side of Loch Spelve. The Coire Mór rocks were interpreted by Bailey et al. (1924) as surface accumulations and contemporaneous rhyolite lava flows, but Richey (1961) preferred a model of subsurface gas brecciation, akin to the explosion breccias of Sgurr Dearg.

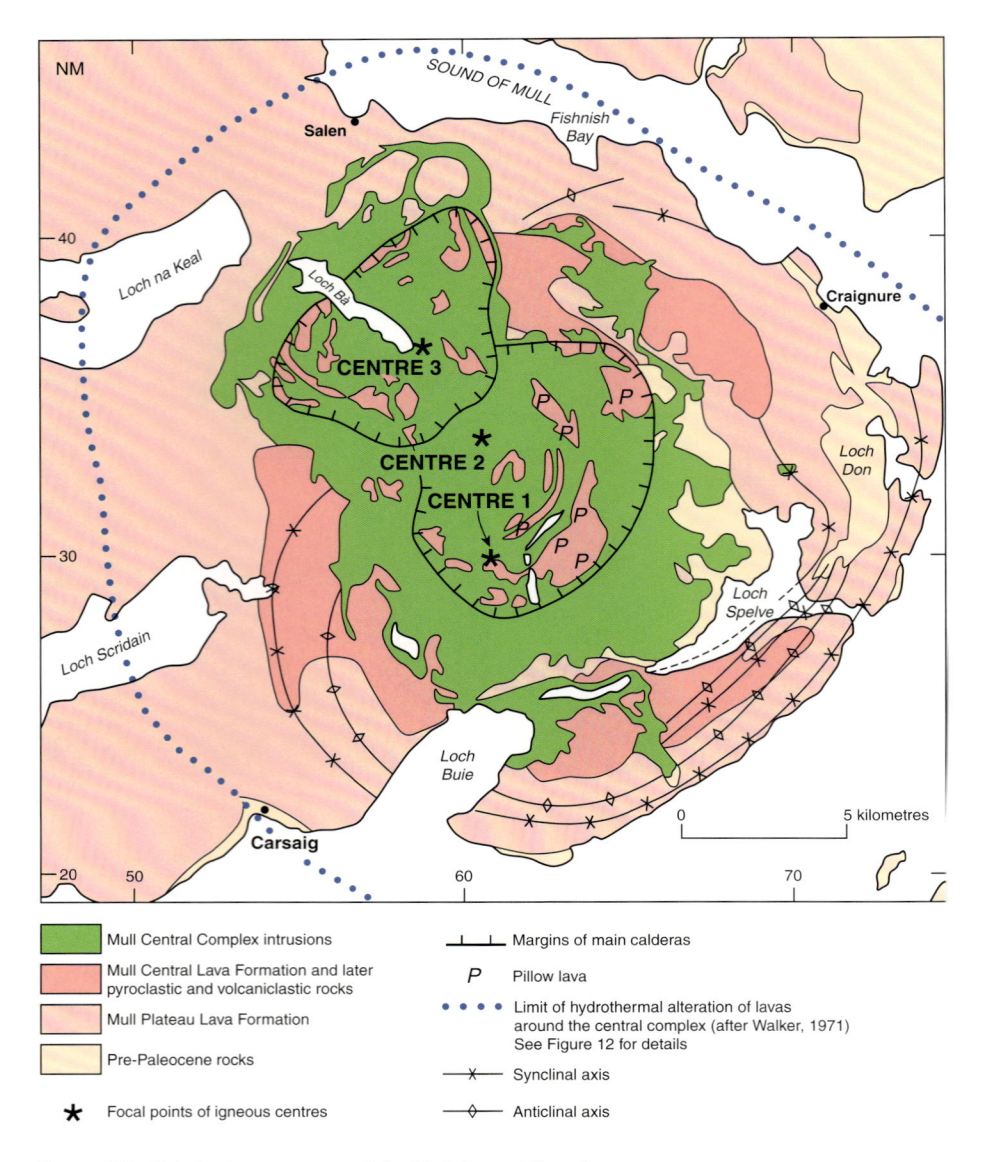

Figure 27 *Principal components of the Mull Central Complex.*

Early felsites The flow-banded Beinn Mheadhon, Torness and Creag na h-Iolaire felsites are approximately contemporaneous, and predate the emplacement of the explosion breccias. The Beinn Mheadhon Felsite is located outside the caldera-bounding fault, but the other two are inside (Figure 28). The felsites are cut by younger basic intrusions (mainly cone-sheets) which obscure the original geometry.

Early cone-sheets A set of early cone-sheets was emplaced into the Glas Bheinn and Derrynaculean granites, the explosion breccias and the early felsites. These cone-sheets are predominantly basic, although a small proportion of intermediate and silicic intrusions is also recognised. They dip inwards at approximately 45° towards a focal point below Beinn Chaisgidle and approach an aggregate thickness of 1000 m, with individual sheets up to 10 m thick. Consequently, significant central uplift will have occurred as a result of their emplacement. The main outcrop can be traced in an arcuate belt that runs from Glen Forsa in the north, close to Loch Spelve, and thence across Glen More to Derrynaculean (Figure 28).

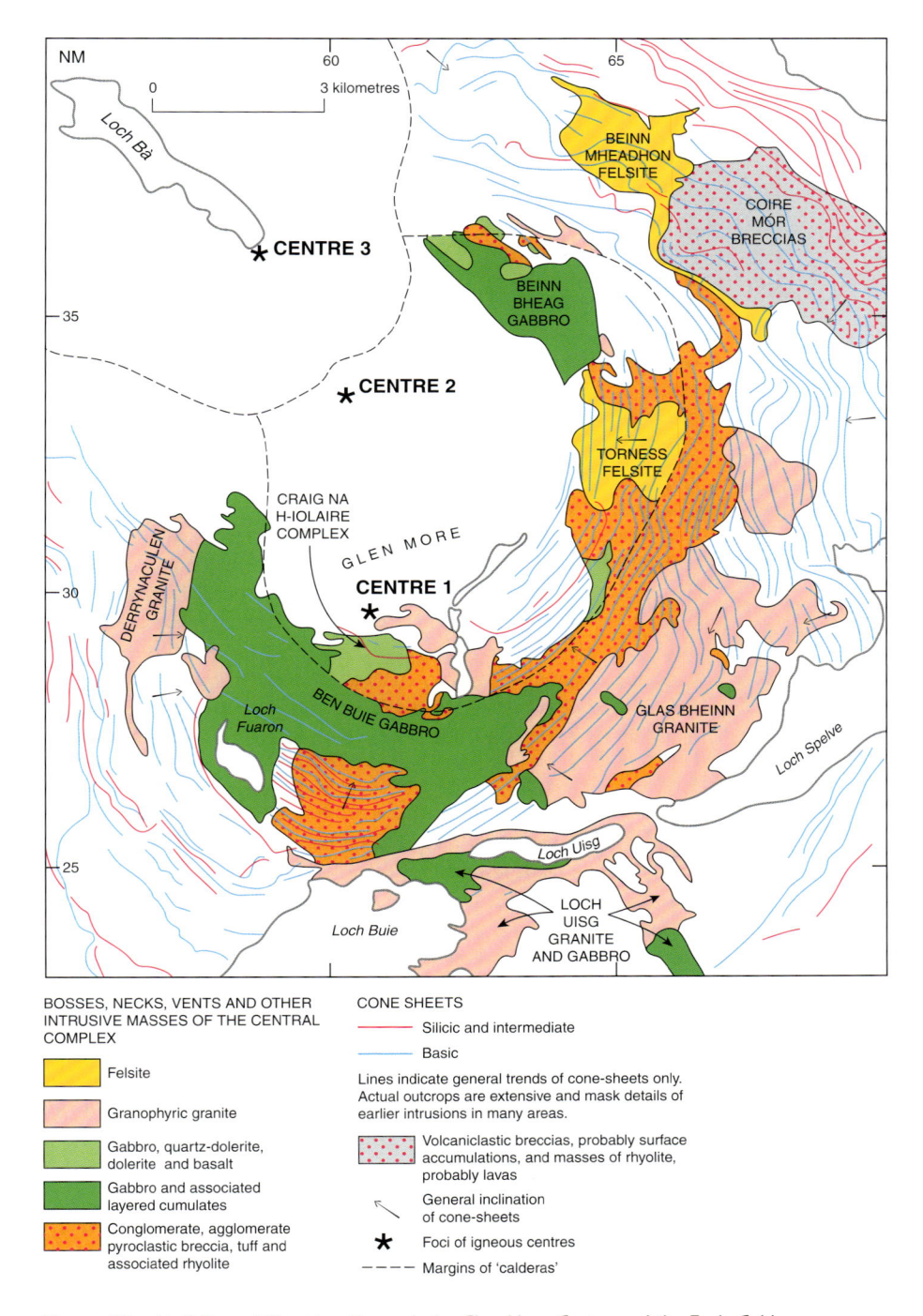

Figure 28 *Mull Central Complex: Centre 1, the Glen More Centre and the Early Caldera.*

The precise timing of the emplacement of the intermediate and silicic intrusions relative to the dominant basic intrusions is unclear, although it is evident that they did overlap. This is confirmed by the presence of a number of composite (basic–silicic) cone-sheets. Movement of the fault defining the Early Caldera had ceased by the time the early cone-sheets were emplaced.

Gabbros

Subsequent to cone-sheet emplacement, two large gabbroic bodies were intruded into the central complex: the Ben Buie Gabbro in the south-west, outside the main ring-fault, and the Beinn Bheag Gabbro inside the fault in the north-east quadrant of the centre. Emplacement of the magmas involved in the formation of the Bein Buie intrusion may have utilised the main ring-fault. However, Skelhorn and Longland (1969) suggested that the Ben Buie mass and possibly the Corra-bheinn Gabbro of Centre 2 were originally circular in plan, and that central subsidence has removed much of the intrusion(s) to a deeper structural level. Furthermore, the inward dips of the mineral layering in the Ben Buie Gabbro increase from about 15° to 20° near the intrusive contacts with earlier rocks, to angles in excess of 35° close to the main ring-fault on the east and north-east sides of the intrusion. The increased dips may have resulted from movement on the fault. The outer, south-eastern, margin of the Ben Buie Gabbro dips outwards at a shallow angle; however, the inward dipping nature of the modal layering of the intrusion may indicate that the base might be at no great depth. The chilled margin of the Ben Buie Gabbro is of tholeiitic basalt composition, akin to that of the Mull cone-sheets, further suggesting a genetic link (Skelhorn et al., 1979).

The cumulate nature of the gabbros results in layers ranging in composition from olivine-dominated assemblages (peridotite), through typical olivine-gabbro assemblages, to plagioclase-dominated assemblages (troctolite and bytownite troctolite). Layers rich in chrome-spinel are common in the more ultrabasic lithologies (Henderson and Wood, 1981). Bailey et al. (1924) defined various facies based upon grain-size and mineral proportions within the Beinn Bheag Gabbro, together with a marginal facies veined with silicic material and a brecciated facies. Xenoliths are common throughout the intrusions, both cognate (peridotite, gabbro, troctolite, etc, essentially unaltered and not recrystallised) and granular-textured rocks, interpreted to be the products of thorough recrystallisation of earlier formed parts of the intrusion, or of country rock basaltic lavas. The two gabbro intrusions were subsequently invaded by various cone-sheets, basic through to silicic, which belong to Centre 2 (see below).

Loch Uisg Granite–Gabbro Intrusion

This intrusion consists of two discrete lithologies. The gabbroic component varies between an olivine-gabbro and an olivine-dolerite, whereas the granite has well-developed granophyric texture and is quite severely hydrothermally altered. The junction between the two comprises a zone of hybrid material formed by magma mixing. This asymmetical, composite intrusion appears to have the geometry of a flat-lying sheet emplaced into lavas, although at the western end of Loch Uisg the upper contact of the granite dips steeply to the north. At the eastern end of the intrusion, part of the roof is formed by volcaniclastic breccias, exposed at Barachandroman (see above), within which the more muddy rocks are thermally altered and thoroughly recrystallised. Emplacement of the Loch Uisg intrusion postdates the formation of the annular folds and the development of the explosion breccias of the Glen More Centre. The relationship with the early basic cone-sheets is less clear since the intrusion is cut by some sheets but in turn cuts others. Overall, the intrusion would seem to be a relatively late component of the Glen More Centre.

Centre 2, the Beinn Chaisgidle Centre

After the development of the Glen More Centre, the focus of igneous activity shifted several kilometres towards the north-west, to the area around Beinn Chaisgidle (Figure 29). Centre 2, also known as the Beinn Chaisgidle Centre, is dominated by thin, steeply inclined, outward-dipping ring-dyke intrusions varying in composition from basalt through to rhyolite, and inwardly inclined basalt and dolerite cone-sheets.

Corra-beinn Gabbro

The Corra-beinn Gabbro is the most westerly of the large gabbro masses in Mull, and is mainly, but not wholly, outside the main bounding fault of the Early Caldera. It contains layered structures that dip to the north-east at 25° to 80°. The gabbro may be a late member of Centre 1; however, since it truncates early basic cone-sheets that in turn intrude the Ben Buie Gabbro it is tentatively assigned to Centre 2.

Ring-dykes The ring-dykes are typically of silicic composition, with steeply inclined margins. Thicknesses vary between 50 and 500 m. They range from relatively coarse-grained rocks such as granite, through to microgranite and rhyolite. Basic ring-dykes are much less common, and vary from gabbro through to dolerite. Some of the ring-dykes are composite, with a range in composition from silicic to basic, but without obvious internal contacts.

Glen More Ring-dyke The Glen More Ring-dyke is probably the best known example of a steeply inclined, compositionally variable hybrid intrusion in the Hebridean Igneous Province. The ring-dyke crops out from the river in Glen More northwards to the summit of Cruach Choireadail, over a vertical distance of almost 500 m. It grades upwards in composition from olivine-gabbro through dioritic rocks to a somewhat

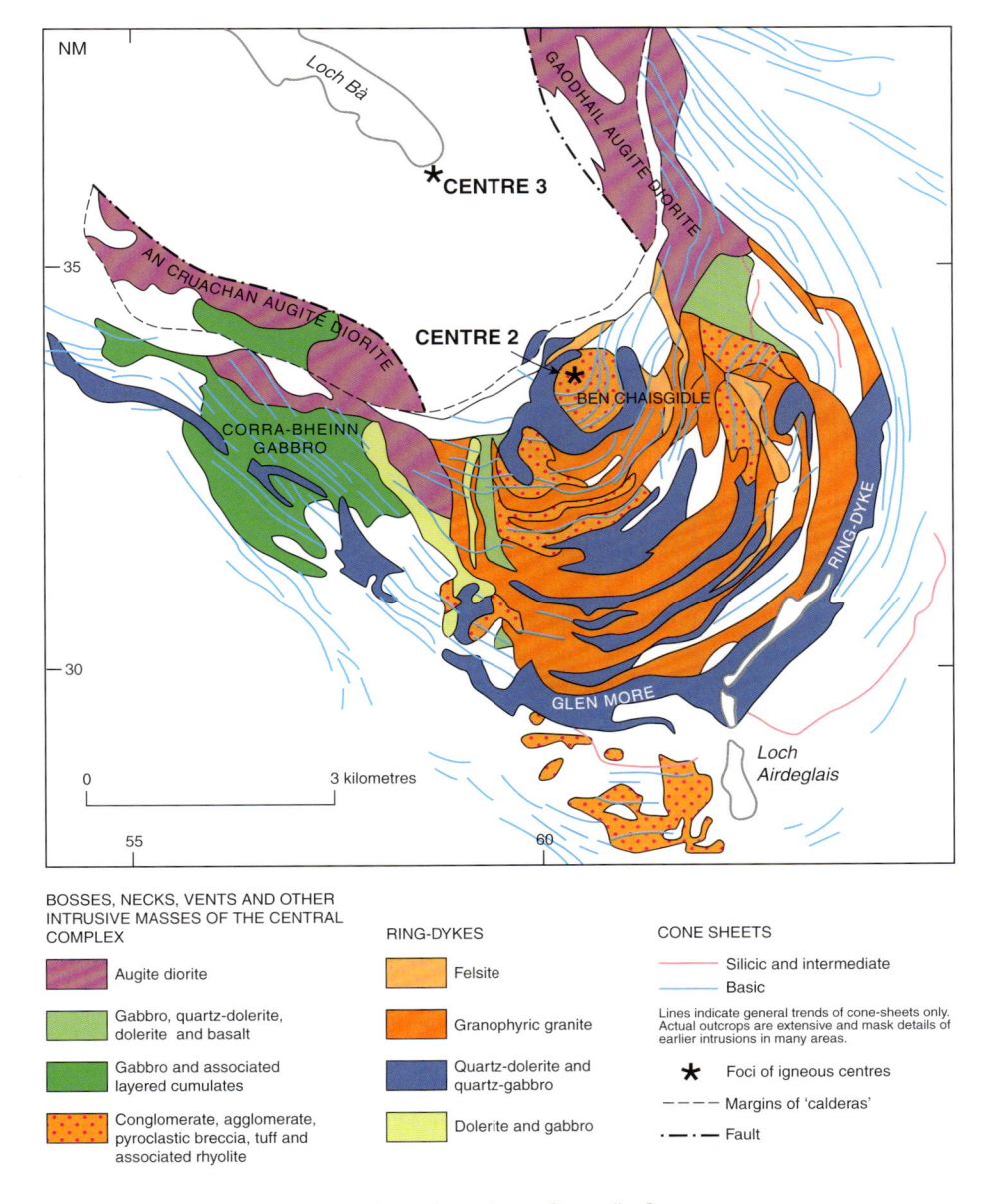

Figure 29 *Mull Central Complex: Centre 2, the Beinn Chaisgidle Centre.*

melanocratic microgranite. The primary mineralogy has largely been replaced by secondary, hydrothermal minerals. It is perhaps the most useful of the Centre 2 intrusions to study in order to observe the processes of differentiation and ring-dyke formation. Bailey et al. (1924) and Koomans and Kuenen (1938) interpreted the vertical variation in composition as the product of in-situ differentiation by liquid–crystal fractionation, whereas others (Holmes, 1936; Fenner, 1937) concluded that the dioritic rocks resulted from the mixing of silicic and basic magmas (see Chapter 10).

Cone-sheets

Most commonly, the basic intrusions are inwardly inclined cone-sheets of basalt and dolerite, in some instances veined by remelted parts of the silicic ring-dykes that they intrude. The cone-sheets are usually less than 10 m thick and dip inwards, generally at 20° to 50°, towards a focal point beneath Beinn Chaisgidle. Thus, complicated relationships between typically silicic ring-dykes and the typically basic cone-sheets are found throughout Centre 2; these relationships are well developed in the Allt Molach stream section in Glen More.

The final intrusive phase unambiguously associated with Centre 2 was the emplacement of the quartz-dolerites that make up the Late Basic Cone-sheets. Emplacement of these cone-sheets continued as the focus of intrusion migrated north-west towards Loch Bà and Centre 3 became established. Conseqently, their emplacement also constitutes the earliest phase of intrusive activity associated with the youngest centre.

Centre 3, the Loch Bà Centre

Centre 3, also known as the Loch Bà Centre, was associated with the development of the Late Caldera (Figure 30).

Late Basic Cone-sheets

These cone-sheets were clearly emplaced during the latter stages of the development of Centre 2 and the earlier part of Centre 3, since plutonic intrusions belonging to both the centres truncate, and in turn are intruded by, cone-sheets belonging to this set. Those that are clearly associated with Centre 3 are symmetrically disposed about an axis trending north-west, parallel to the length of Loch Bà (Figure 30). In places, the density of cone-sheet emplacement is very high, with very little country rock preserved. Central uplift must have been significant.

Glen Cannel Granite

This granite was the first major silicic intrusion to be emplaced within Centre 3. The mildly alkaline granite forms an oval, dome-shaped mass with a north-west-trending long axis. The intrusion contains abundant gas cavities (druses) and is preserved predominantly within the subsided block inside the late-stage Loch Bà Ring-dyke. The granite cuts numerous Late Basic Cone-sheets with n the central subsided block, but outwith the block, to the south-east, it is cut by similar cone-sheets. Thus, it appears that there was an overlap of the intrusive events, or that different intrusions make up the granite, or that more than one set of cone-sheets exists. The granite is partially roofed by volcaniclastic rocks, masses of quartz-dolerite and intrusive felsites. From the disposition of the felsites along the edges of the granite (Figure 30), they might be regarded as a chilled marginal facies were it not for exposures on the east side of Bìth Bheinn and Creag Dubh, south of Loch Bà, which show that the granite is in sharp intrusive contact with the felsites.

Beinn a' Ghraig and Knock granites

The Beinn à Ghraig Granite is located outside the Loch Bà Ring-dyke along its north-west margin. It is of similar petrographic type to that of Glen Cannel, but is considered to be younger as it cuts Late Basic Cone-sheets on Beinn a' Ghraig, but is itself cut by only one or two cone-sheets. The Knock Granite is of similar age, taking the form of a steep-sided, elongate mass, 50 to 300 m wide, separated from the north-west margin of the Beinn a' Ghraig intrusion by a screen of hornfelsed basaltic lavas. Similar, most likely related, granitic, dioritic and hybrid ring-intrusions occur to the north-east of Loch Bà, in the vicinity of Toll Doire, Maol Buidhe and Killbeg. Although the country-rock lavas have been

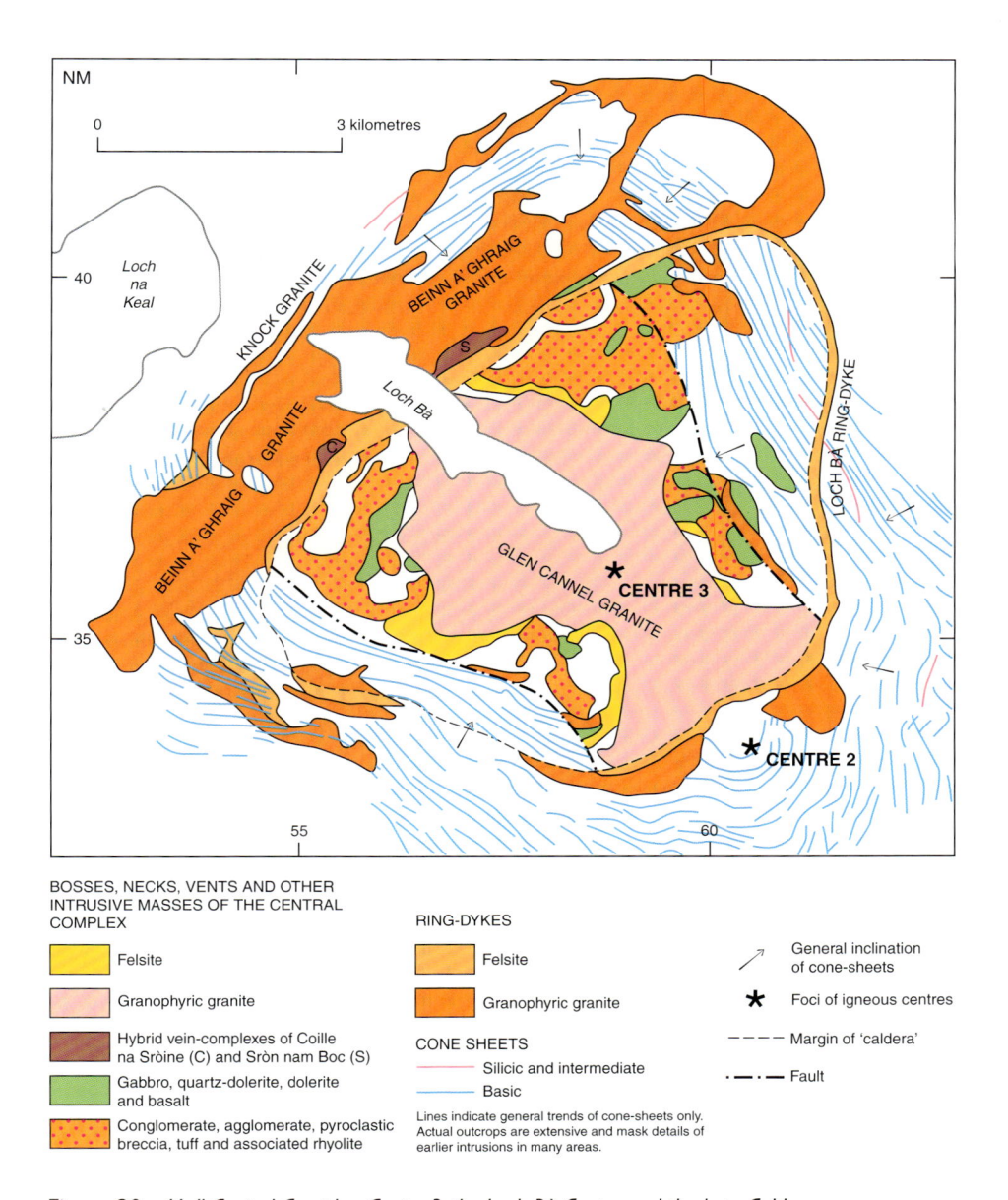

Figure 30 *Mull Central Complex: Centre 3, the Loch Bà Centre and the Late Caldera.*

invaded by the Late Basic Cone-sheets, and are hornfelsed, they do not appear to have been significantly folded or faulted. This suggests the relatively passive emplacement of the granitic magmas, probably by a combination of subsidence and stoping.

Loch Bà Ring-dyke This ring-dyke is the final major silicic intrusion of the Loch Bà Centre. It has an external diameter of about 8 km and a width varying from 400 m down to zero in those areas where the trace of the ring-fault is marked only by brecciation of the country rocks. In general, the ring-dyke walls are close to vertical, although steep outward dips occur along the north-west portion. The Loch Bà Ring-dyke is cut by late members of the north-west-trending regional dyke swarm, but is unique amongst the major intrusions of the Mull Central Complex in being entirely free of cone-sheets.

The intrusion was first described by Bailey et al. (1924) and its petrology and origin have subsequently been investigated by Walker and Skelhorn (1966) and by Sparks (1988). The later studies recognised the hybrid nature of the intrusion, involving dominant silicic rock (rhyolite with sparse phenocrysts of sodic plagioclase, sanidine, hedenbergite, fayalite, magnetite, ilmenite and zircon) containing inclusions (typically less than 10 cm long) of phenocryst-poor basic material ranging in composition from ferrobasalt through to dacite. The inclusions constitute less than 10 per cent of the ring-dyke, are commonly glassy, and range in shape from rounded to lenticular, the latter with distinctive ragged ends. The rhyolite is partially devitrified with an obvious flow banding and the preservation of fiamme (eutaxitic texture) is indicative of a pyroclastic origin. Given the glassy, hybrid nature of the intrusion, even where it is 400 m wide, and the development of textures typical of welded tuff, it is evident that its emplacement involved mixing of magmas during the eruption of pyroclastic material (Chapter 10). Space for this intrusion was most likely created by the combined action of gas brecciation and central subsidence, by a relatively small distance, of the pre-existing block inside the ring-dyke.

BLACKSTONES BANK

The submarine Blackstones Bank Central Complex lies within the Hebridean area (Figures I, VI; 5), about 30 km south-south-west of Tiree, and was first located by seismic, magnetic and gravity surveys (McQuillin et al., 1975; Fyfe et al., 1993). It has a rugged topography, and the shallow depths of some parts have allowed quite extensive sampling by scuba diving (Durant et al., 1976, 1982) and the construction of a geological sketch map (Fyfe et al., 1993). Most of the accessible parts are of gabbro, commonly exhibiting cumulate textures. Granophyric granite, microgranite and tuffisite samples have also been recovered together with calc-silicate hornfels. The bank is the site of both a large positive Bouguer gravity anomaly and a major magnetic anomaly. It is therefore likely that basic and ultrabasic rocks make up the bulk of the central complex, with the calc-silicate hornfels formed from the thermal alteration of the Mesozoic sedimentary rocks which it intrudes. From a geochemical study of the centre, it is inferred that the magmas interacted with crustal rocks similar to those of Proterozoic age exposed in western Islay, implying that such crust extends to the west of the Great Glen Fault (Dickin and Durant, 2002). The age of the complex is most likely Paleocene.

ARRAN

There are two major centres of igneous activity of Paleocene age on Arran (Figure 31). These are the North Arran Granite Pluton, whose mountains dominate the north of Arran, and the Central Arran Ring-complex, which forms part of the high ground between Brodick and the west coast. Additionally, outcrops of microgranite and augite diorite at Tighvein, about 5 km south-west of Lamlash, were once regarded as a possible third centre (Tyrrell 1928) but have subsequently been interpreted as mainly sills (Herriot, 1975) (Chapter 8). The igneous geology was fully described by Tyrrell (1928), with further detail on specific sites by Macgregor (1983), McKerrow and Atkins (1985) and Emeleus and Gyopari (1992).

North Arran Granite Pluton

The North Arran Granite Pluton has a near circular outcrop of 10 to 12 km diameter and consists of a coarse-grained Outer Granite and a finer grained Inner Granite (Figures 31, 32) Both granites are biotite bearing and are generally non-porphyritic. They contain drusy cavities lined with quartz (smoky or 'Cairngorm', purplish amethyst, yellowish 'Scotch Topaz'), alkali feldspar and mica, together with rare topaz, blue beryl, allanite and zircon. The rare-earth-bearing minerals fergusonite and gadolinite have also been identified (Hyslop et al., 1999). Fluorite, recovered from stream sediments, probably also originated in drusy cavities. For most of its circumference, the Outer Granite is in sharply defined, intrusive contact with rocks of the Dalradian Supergroup, which are thermally metamorphosed and indurated for a distance of several hundred metres from the granite. Small xenoliths of hornfelsed sedimentary rocks are present close to many of the excellently exposed contacts, as in Allt a' Chapuill where the

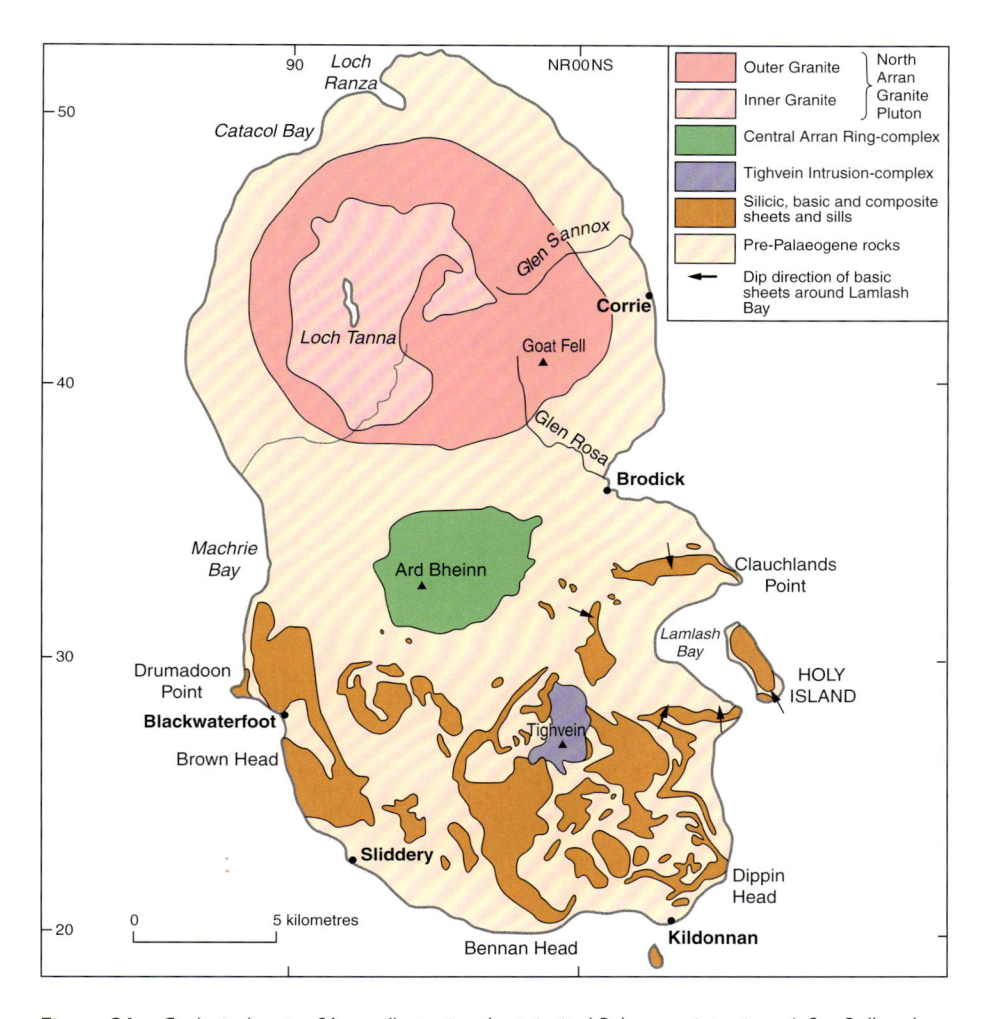

Figure 31 *Geological map of Arran illustrating the principal Paleocene intrusions. (after Bell and Williamson, 2002, fig. 14, 18).*

granite is finer grained and texturally variable and veins the country rocks. In general, however, the granites are remarkably free of inclusions. To the west of Corrie, the granite is in faulted contact with Devonian strata. Emplacement of the granite caused major disturbance of the country rocks (p.151). On the basis of the large-scale deformation, and other evidence, the pluton has long been regarded as an example of diapiric emplacement of granite, an interpretation that has more recently been substantiated and developed (England 1992b; Goulty et al., 2001). However, despite having been forcibly emplaced, there is a complete lack of foliation or other deformation structures in the granite. Within the pluton, the grain-size and textural contrasts between the Inner and Outer intrusions allow the contact to be traced in some detail; well-exposed, sharp intrusive contacts crop out, for example south of Meall nan Damph and towards the northern end of the A' Chir ridge. The younger, Inner Granite is characteristically drusy and finer grained close to the contact and, in stream sections on the west side of Glen Catacol, there is local evidence of multiple, sheeted intrusion in the granite. A well-developed system of joints, more or less parallel to the land-surface, cuts the Outer Granite, giving rise to castellated outcrops as, for example on Goatfell (Plate 38). The joints are considered to have developed as the granite was unloaded during erosion. Unusually amongst the Paleocene rocks of the Hebridean Igneous Province, the intrusions of the North Arran Granite

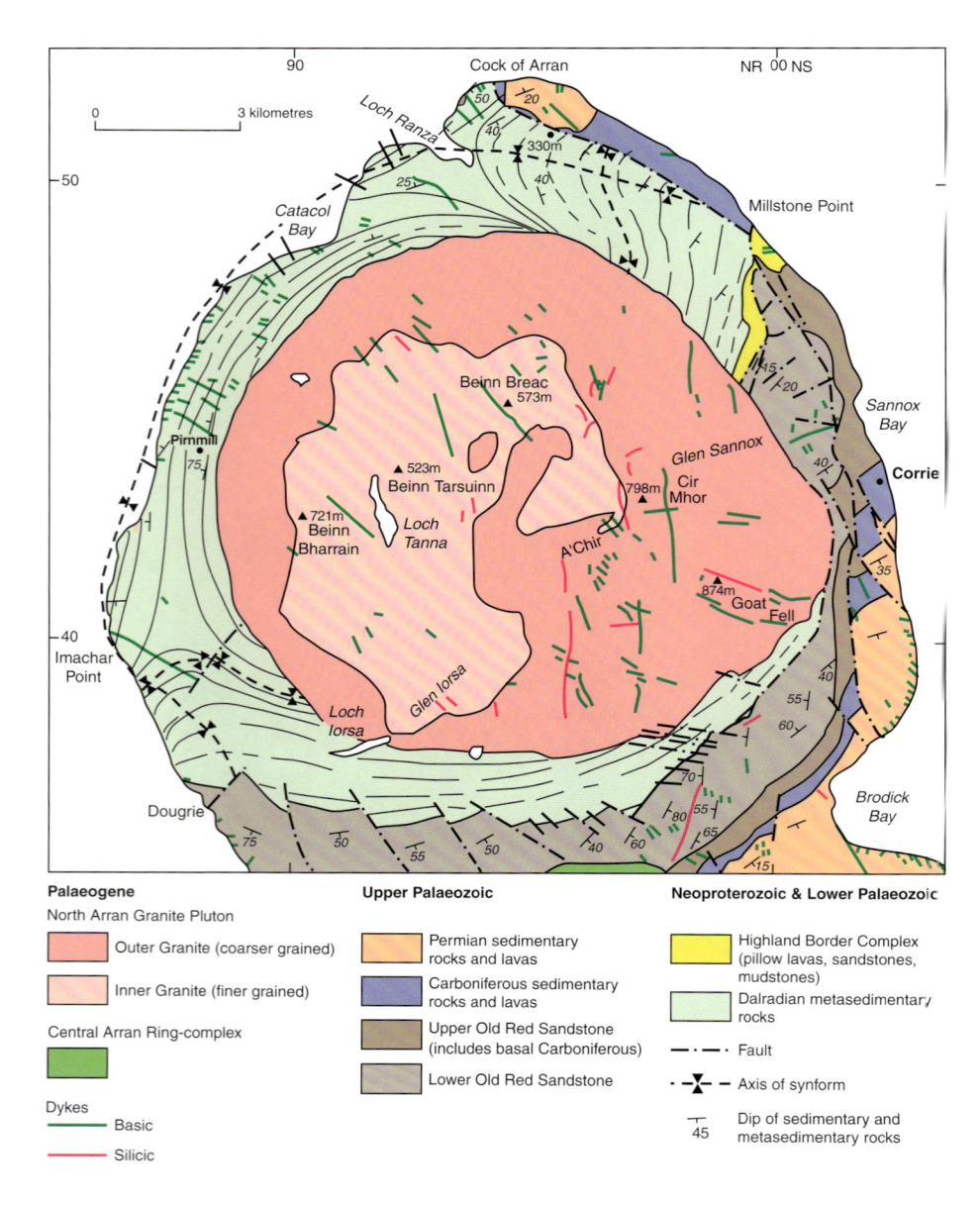

Figure 32 *Map of the North Arran Granite Pluton (after Emeleus, 1991).*

Pluton are normally magnetised, a feature which they share with quartz-porphyry intrusions in the south of Arran and on Bute (Figure 8). This has prompted the suggestion that the granites and quartz-porphyry sills came from a common source (e.g. Mussett et al., 1989).

Central Arran Ring-complex

The rocks of this variably exposed ring-complex (Figure 31) include lavas and minor intrusions of basic, intermediate and silicic composition, tuff, and intrusive masses of granite, gabbro, and dioritic rocks of hybrid aspect. The ring-complex was emplaced into rocks of Devonian to Permian age and, in the south and east, the sedimentary rocks dip steeply away from the complex. Faulted contacts are inferred elsewhere. At its north-east extremity, in and near Glen Ormidale, the ring-complex appears to cut sharply across Devonian, Carboniferous and Permian rocks, which are domed about the North Arran Granite Pluton suggesting that the ring-complex is the younger.

Andesitic and basaltic lavas and volcaniclastic rocks crop out in the Ard Bheinn area (Figures 31; 33) where several centres of eruption have been recognised (King, 1955). Tuffs and volcaniclastic breccias contain fragments of igneous rocks (basalt, andesite, trachyte, felsite) and material of sedimentary origin. The latter includes clasts derived from the Devonian rocks but the most spectacular inclusions are several megablocks hundreds of metres across that contain discrete sequences of fossiliferous Mesozoic strata. They include Cretaceous chalk and sandstone, Lower Jurassic mudstone, and Rhaetian mudstone overlying Triassic sandstone. These megablocks, and others made up of of altered olivine basalt lavas, were probably derived from an original country-rock cover, which was shattered during volcanic explosions and caldera formation. The occurrences of fossiliferous Jurassic and Cretaceous rocks are of particular note since they establish the post-Mesozoic age of this igneous activity, as well as proving the former presence over Arran of a cover of Mesozoic rocks younger than the Triassic.

The intrusions of gabbro, granite and granophyric granite form arcuate masses that are interpreted as partial ring-intrusions. Smaller areas of hybrid dioritic rocks are well exposed in the burn at Glenloig Bridge and in the east of the centre on Tir Dubh (the so-called Hybrid Hill), where complex relationships between silicic and basic rocks suggest the co-existence, and mingling, of basic and silicic melts.

AILSA CRAIG Ailsa Craig is a spectacular, conical island in the Firth of Clyde about 20 km south of Arran (Figure 1). It is formed by a boss of peralkaline microgranite intruded into Triassic rocks. The microgranite is characterised by riebeckitic arfvedsonite and Zr-rich aegirine (Harding, 1983;

Plate 38 Hills composed of the North Arran Granite Pluton. The well-jointed granite (slabs in foreground) forms the glaciated peaks of Cir Mhor (left) and Casteil Abhaill (centre). Deep notches in the ridge mark the sites of easily-weathered basalt dykes (P580487).

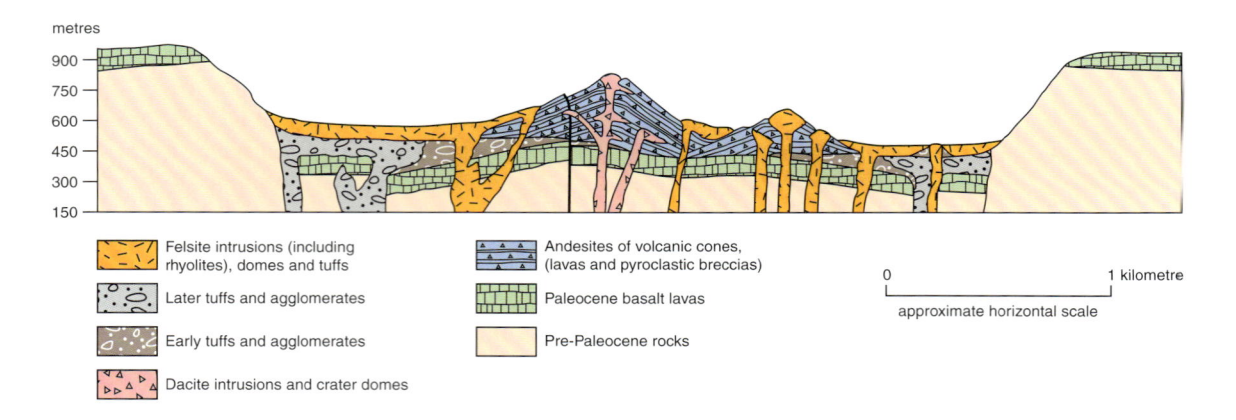

Figure 33 Diagrammatic cross-section showing the inferred structure of the Central Arran Ring-complex, prior to erosion. The section is only approximately to scale but the main cone may be regarded as representing Ard Bheinn (after King, 1955, fig. 1).

Harrison et al., 1987); aenigmatite also occurs (Howie and Walsh, 1981). This distinctive rock-type is a widespread glacial marker southwards on either side of the Irish Sea (p.160). It has traditionally been a favoured lithology for the manufacture of curling stones (p.173).

ROCKALL This body of peralkaline granite forms a precipitous islet, about 20 m high and 30 m across, in the Atlantic Ocean about 250 km west of St Kilda (Harrison, 1975; Figure 2). The nearby micro-gabbro of Helen's Reef is thought to be part of the same intrusive centre, and a variety of igneous rocks have been recorded from the adjacent sea bed (Morton and Taylor, 1991). Gneisses, which have also been recovered from the sea bed, form the country rock and are Early Proterozoic in age. The granite is unusually rich in aegirine and riebeckite (13 to 27 per cent), and there is a particularly high concentration of ferromagnesian minerals (up to 68 per cent) in the variety 'rockallite', which forms vein-like segregations in the granite. Accessory minerals include elpidite, rare monazite and the barium-zirconium silicate bazarite (first described from Rockall). Leucophosphite occurs as a late-stage replacement mineral, formed by reaction between phosphate-rich fluids derived from sea-bird droppings and the potassium feldspars and ferromagnesian minerals of the granite.

ten
Magmas

Much of our present-day understanding of magmas and magmatic processes is the consequence of research undertaken on the Palaeogene rocks of the Hebridean Igneous Province. Following on from pioneering studies on Skye and the Small Isles (Harker, 1904, 1908), the most significant breakthroughs were made by the Geological Survey with their investigations of the volcanic and intrusive rocks of Mull (Bailey et al., 1924), a landmark study that established a framework for most subsequent research. The geochemistry and genesis of the magmatism of the Hebridean Igneous Province have been discussed and summarised in several seminal publications. In particular, Thompson (1982a, b) provided incisive reviews and new ideas on a wide range of topics, and Saunders et al. (1997) discussed the geochemistry of the volcanic rocks of north-west Scotland in the context of the whole North Atlantic Igneous Superprovince. Magmatism in the Hebridean Province has recently been reviewed by Bell and Williamson (2002).

EARLY CONCEPTS

Magma-types and magma-series

The important concepts of magma-types and magma-series were established by Bailey et al. (1924) through their study of the lavas and intrusive rocks of Mull. A dominant *Normal Mull Magma-series* was identified, consisting of: the *Plateau Magma-type* (olivine-rich basalt lavas), the *Non-porphyritic Magma-type* (olivine-poor basalt lavas), the *Intermediate to Sub-acid Magma-type* (various minor intrusions), and the *Acid Magma-type* (granites of the central complex). The Plateau Magma-type was considered to be the parental magma of the Normal Mull Magma-series. The other magma-types were derived from it by magmatic differentiation, and each was regarded as the product of crystallisation of successively more silicic residual liquids. Bowen (1928) concluded that fractionation of olivine was the most likely process involved in the production of the evolved liquids. An alternative model was proposed by Kennedy (1930), who considered that the Non-porphyritic Magma-type was parental to the Plateau Magma-type, but later (Kennedy, 1933) he suggested that both magma-types were parental, renaming them *tholeiitic basalt* (hypersthene-normative) and *olivine basalt* (nepheline-normative, and later renamed alkali olivine basalt by Tilley, 1950). Subsequently, the experimental studies of Yoder and Tilley (1962) confirmed that one type could not yield the other through crystal–liquid fractionation under low pressure conditions.

MAJOR-ELEMENT COMPOSITIONS

The landmark paper by Thompson et al. (1972) on the major-element geochemistry of Skye lavas provided, for the first time, a clear insight into the relationships of parental and derivative magmas. They discovered that, although the Skye lavas are petrographically of alkali olivine basalt type, they vary between nepheline-normative and hypersthene-normative (Figure 34). They assigned the lavas to a Skye Main Lava Series, within which the nepheline-normative flows have a major-element composition typical of alkali olivine basalt, but several of the minor- and trace-element concentrations are typical of tholeiitic basalts (Thompson et al., 1980). On Mull, a similar range in composition occurs within the Plateau Magma-type (Morrison et al., 1980).

Experimental studies led Thompson (1974) to conclude that the most primitive (i.e. magnesium-rich) basalts of the Skye Main Lava Series achieved their final compositions at pressures of up to 17 kbar, well within the upper mantle. These magmas were then erupted relatively fast, with little opportunity for re-equilibration prior to eruption. However, the major-element data for other flows from the Skye Main Lava Series, the Mull Plateau Magma-type and elsewhere in the Hebridean Igneous Province, indicate that subsequent fractionation processes occurred at pressures of around 9 kbar, near to the base of the crust (Thompson and Gibson,

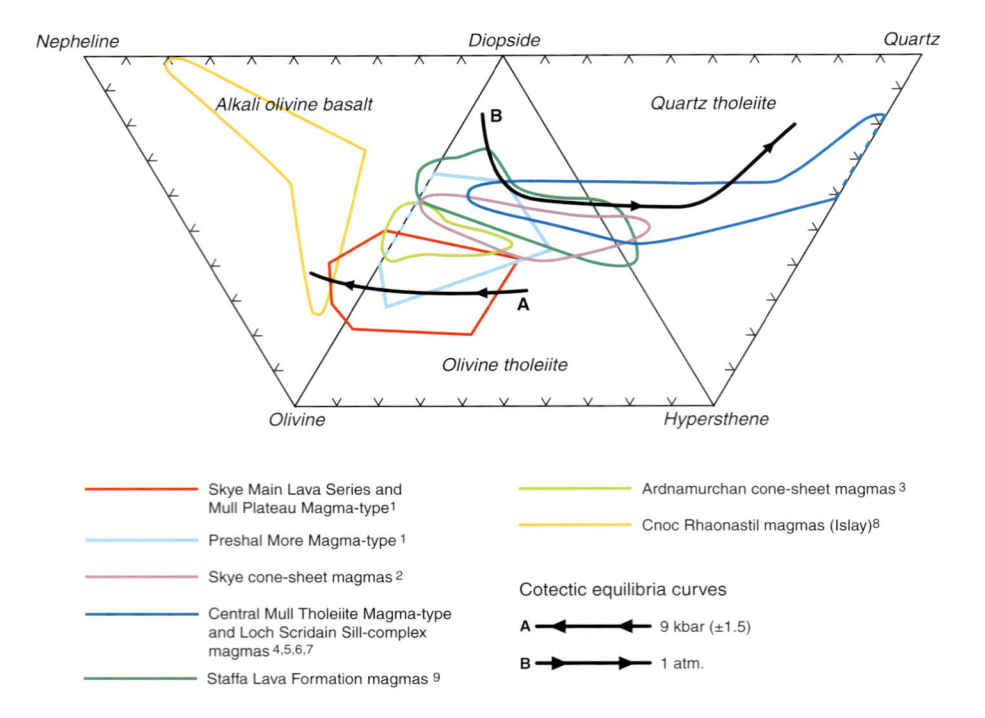

Figure 34 *Normative CIPW (diopside, hypersthene, olivine, nepheline and quartz) plot for the Hebridean Palaeogene magmas showing 1 atmosphere and 9 kbar (anhydrous) cotectic curves for the equilibria olivine + plagioclase + clinopyroxene + natural basic/basaltic liquid (after Thompson, 1982b). See main text for discussion.*
1 Thompson, 1982b; 2 Bell et al., 1994; 3 Geldmacher et al., 1998; 4 Kerr, 1993; 5 Kerr, 1995a; 6 Kerr, 1995b; 7 Preston et al., 1998; 8 Hole and Morrison, 1992; 9 Kerr, 1998

1991). In effect, the continental crust acted as a mechnical filter, causing magmas to pond and fractionate at depth (Saunders et al., 1997). Two evolutionary trends were recognised by Thompson et al. (1972):

- alkali olivine basalt–hawaiite–mugearite–benmoreite
- alkali olivine basalt–Si-rich and Fe-poor intermediate magmas–trachyte (Figure 35)

The Benmoreite Trend evolved under high pressure (deep within the crust) with Fe-enrichment, in contrast to the low pressure (shallow) conditions, with Fe-depletion, of the Trachyte Trend.

A distinctly high concentration of calcium and low potassium, titanium and phosphorous, akin to tholeiitic Mid Ocean Ridge-type basalts has been determined for some of the igneous rocks from Skye (Esson et al., 1975), for example the Preshal More lava (Williamson and Bell, 1994), certain dykes within the axial portion of the Skye Dyke Swarm (Mattey et al., 1977), and the cone-sheets of the Cuillin Centre (Bell et al., 1994). These are referred to as the Preshal More Magma-type and are considered to have evolved under low pressure in the upper crust (Thompson, 1982b; Figure 34). The Non-porphyritic Magma-type of Mull, subsequently renamed the Central Mull Tholeiite Magma-type, probably had a similar low-pressure origin (Kerr, 1995a; Kerr et al., 1999). However, certain tholeiitic basalt lavas belonging to the Staffa Lava Formation on Mull appear to have a different origin, involving fractional crystallisation and contamination of Mull Plateau magmas (Kerr, 1998).

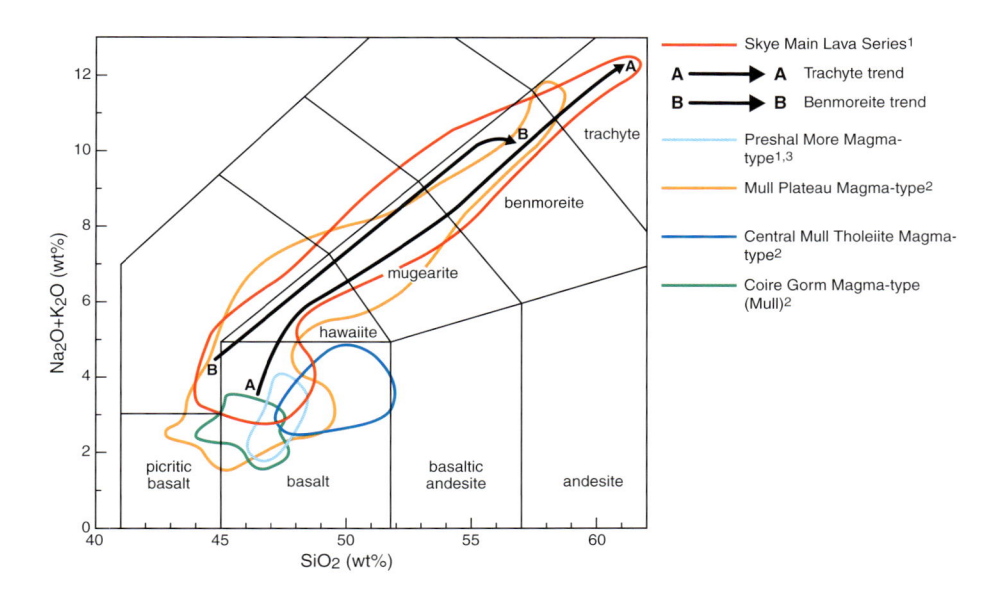

Figure 35 Plot of total alkalis (Na$_2$O + K$_2$O) against silica (SiO$_2$) for the Hebridean Palaeogene magmas. Compositional fields are from LeBas and Streckeisen, 1991.
1 Thompson et al., 1972; 2 Kerr et al., 1999; 3 Esson et al., 1975

CONTAMINATION PROCESSES

Magmas rising through the continental crust beneath the Hebridean Igneous Province were likely to have interacted with and been contaminated by country rock including, granulite-facies (lower crustal) Lewisian gneisses, amphibolite-facies (upper crustal) Lewisian gneisses, Moine pelites and psammites, and Torridonian sandstones and siltstones (Chapter 2). Each of these crustal materials has a distinctive isotopic signature (Figures 36; 37), which would be imparted as a recognisable geochemical imprint if such materials were assimilated by the magmas during ascent. During such contamination processes, the magmas may also have undergone fractional crystallisation, and it is the heat liberated for the crystallisation that provided the thermal energy required by the contamination process. Therefore, it is important to assess whether contamination occurred before, during or after fractional crystallisation. In particular, the combined process of assimilation during fractional crystallization may be recognised from the geochemical signatures of various suites of extrusive and intrusive rocks.

Using the isotopic ratios, and in some instances isotopic abundances, of strontium (Sr), neodymium (Nd) and lead (Pb) in the Skye Main Lava Series of the Skye Lava Field, variable amounts of both lower and upper crustal (Lewisian gneiss) contamination have been detected, with the major contribution coming from granulite-facies lower crust (Carter et al., 1978; Moorbath and Thompson, 1980; Dickin, 1981; Thompson, 1982 a,b; Thompson et al., 1982; Thirlwall and Jones, 1983; Dickin et al., 1987). Similar contamination models have been deduced for the Mull Plateau Magma-type of the Mull Lava Field (Kerr et al., 1999).

The contamination history of the Preshal More Magma-type is completely different from that of the Skye Main Lava Series and the Mull Plateau Magma-type. In the Preshal More magmas, contamination was concurrent with fractionation and the dominant contaminant was upper crustal, amphibolite-facies Lewisian gneisses (Figure 37). The genetically related rocks of the Cuillin Centre of the Skye Central Complex have also had a similar contamination history (Dickin et al., 1984a; Bell et al., 1994). The tholeiitic basalt lavas of the Staffa Lava Formation, close to the base of the Mull Lava Group, are also contaminated, although contrasting models are offered to explain the type of magmas involved, the nature of the contaminants and the timing of contamination (Thompson et al., 1986; Kerr et al., 1999).

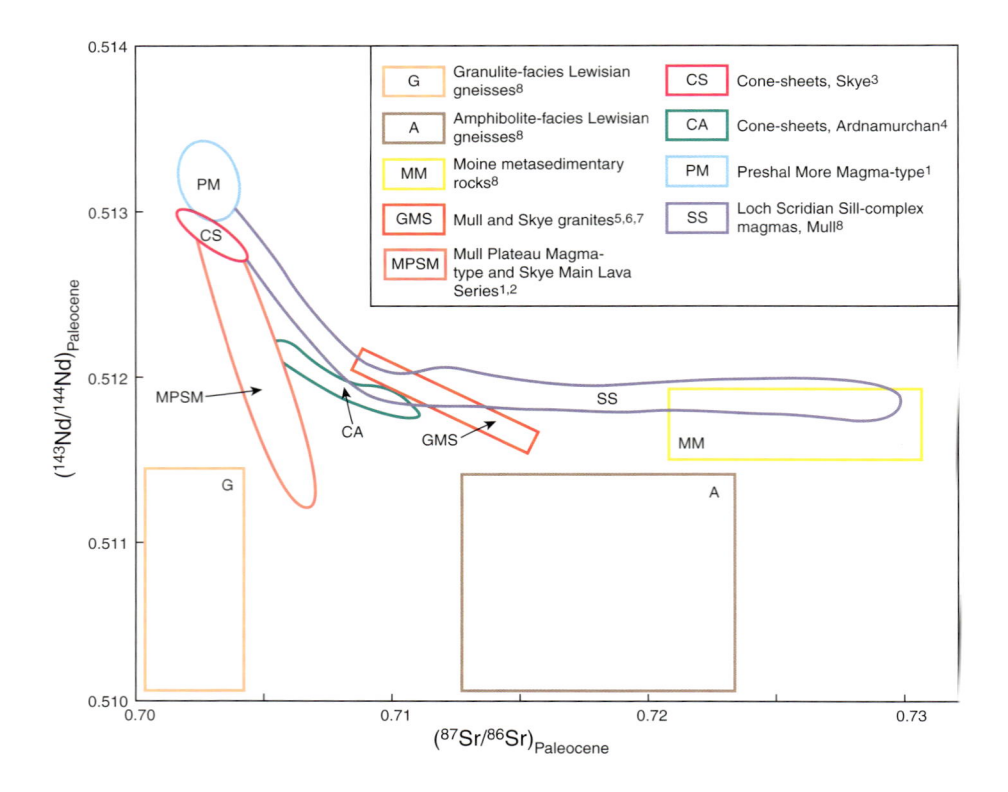

Figure 36 Plot of $(^{87}Sr/^{86}Sr)_{Paleocene}$ against $(^{143}Nd/^{144}Nd)_{Paleocene}$ for basaltic rocks and crustal materials in the Hebridean area. Sr and Nd isotopes corrected for 55 Ma.
1 Thirlwall and Jones, 1983; 2 Kerr et al., 1995a; 3 Bell et al., 1994; 4 Geldmacher et al., 1998; 5 Walsh et al., 1979; 6 Dickin et al., 1984a; 7 Kerr et al., 1999; 8 Preston et al., 1998a

DEPTH OF MAGMA GENERATION

On the basis of experimental studies (Thompson, 1974) and trace-element modelling (Thompson et al., 1980), it may be concluded that the primary melting events that yielded the Skye Main Lava Series magmas occurred in the mantle at a depth of about 60 km and involved partial melting of spinel lherzolite. The Preshal More Magma-type was generated by subsequent larger degrees of melting of the residual mantle material. The remaining unmelted mantle was of harzburgitic composition.

In a re-examination of the Skye lavas, Scarrow and Cox (1995) concluded that the parental magmas were of picritic composition. They developed a model for the production of parental melts involving decompressive melting of abnormally hot mantle, with the final segregation of the melts occurring over the depth range of 60 to 112 km and at a temperature of 1390° to 1510°C. This depth estimate is not at significant variance with that proposed by Thompson et al. (1972) and Thompson (1974). The melts that segregated at the greatest depths were nepheline-normative, whereas hypersthene-normative magmas were produced towards the top of the melting column.

The incompatible trace-element contents of lavas thoughout the Hebridean Igneous Province are low when compared with continental flood lava sequences worldwide. The depletion of these elements from the mantle source probably took place during Permian times when highly alkaline (lamprophyric) magmas were extracted from the mantle below the entire Hebridean area by partial melting (Morrison et al., 1980; Thompson, 1982b). Differences between the Skye Main Lava Series and the Mull Plateau Magma-type could be due to lateral mantle heterogeneity.

ULTRABASIC AND BASIC MAGMAS OF THE CENTRAL COMPLEXES

Interpretation of the geochemical affinities of the coarse-grained ultrabasic and basic units of the central complexes is difficult, as they have all been modified by a variety of fractionation processes. However, it should be possible to deduce the magma-types involved from material forming their chilled margins. In practice, data on these lithologies are extremely sparse and are not always easily interpreted (pp.115–116). A more promising insight may be provided by fine-grained rocks of the abundant minor intrusions, which were intruded contemporaneously with the emplacement of the ultrabasic and basic rocks.

Opinions on the nature of the parental magmas vary. The abundant cone-sheets which focus on the central complexes of Skye (Bell et al., 1994), Mull (Kerr et al., 1999) and Ardnamurchan (Holland and Brown, 1972; Gribble, 1974; Thompson, 1982a, Geldmacher et al., 1998) are of either the Preshal More Magma-type, or the Central Mull Tholeiite Magma-type and its differentiated equivalents. These tholeiitic magma-types, which equilibrated at relatively low pressures (about 3 kbar), are regarded as likely parent magmas, and this conclusion is supported by the similar composition of the chilled margin of the Ben Buie Gabbro on Mull (Skelhorn et al., 1979; Kerr et al., 1999; Chapter 9). Somewhat different parental magma compositions have been proposed for the ultrabasic and basic rocks of Rum. There, direct evidence for the involvement of more-primitive picritic magmas comes from a chilled margin to the Eastern Layered Intrusion, with 15 to 20 per cent MgO (Volker, 1983; Greenwood et al., 1990) and from an aphyric dyke with 13.5 per cent MgO (McClurg, 1982; Upton et al., 2002).

MINOR INTRUSIONS

The dyke swarms, sill-complexes and volcanic plugs or bosses provide a wealth of information about the temporal and areal occurrence of most of the magma-types in the Hebridean

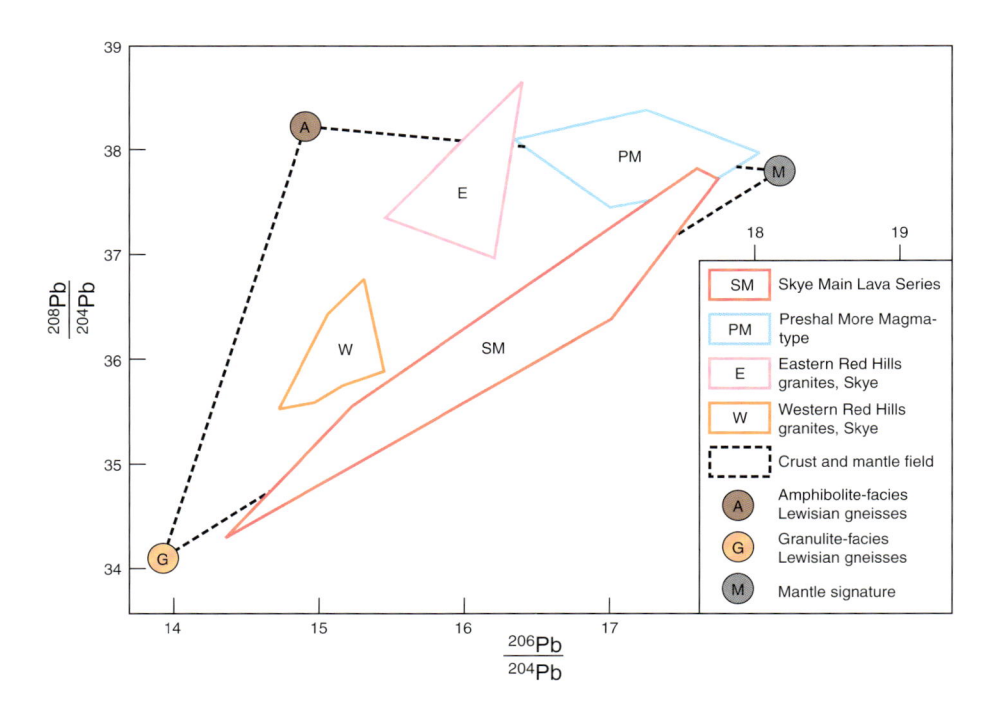

Figure 37 Plot of Pb isotope data for Skye lavas (Skye Main Lava Series and Preshal More Magma-type) and granites (Western and Eastern Red Hills centres) compared with compositions of mixing components involved in their formation (from Dickin et al., 1987).

M is estimated average composition of mantle-derived magmas for Skye; G and A are 'average' granulite- and amphibolite-facies Lewisian gneisses, respectively.

Igneous Province. They may also provide in situ evidence of some of the high crustal level (low pressure) processes considered to have affected the lavas. Thus, crustal xenoliths are found in various stages of incorporation in certain sills, varied lithologies corresponding to near-complete magma-series occur within individual intrusions, dyke swarms show changes from alkali olivine basalt members to tholeiitic basalt members with time, and in composite intrusions completely contrasting rock compositions (commonly silicic and basic) occur together, and are observed in various stages of mixing and mingling.

Dyke swarms

The geochemistry of the Skye Dyke Swarm has been studied in detail by Mattey et al. (1977). Approximately 70 per cent of the dykes are of Preshal More Magma-type and are particularly abundant within the axial region of the swarm, close to the Cuillin Centre, although the type persists throughout the length of the swarm forming, for example, almost all of the Skye dykes on Harris and at Arisaig.

The dominant compositions of the Mull Dyke Swarm are of the Mull Plateau and Central Mull Tholeiite magma-types (Kerr et al., 1999). The tholeiitic dykes of north-east England (Macdonald et al., 1988) are the most south-easterly representatives of the Mull Dyke Swarm. They belong to the Central Mull Tholeiite Magma-type and have elevated initial $^{87}Sr/^{86}Sr$ values (Moorbath and Thompson, 1980), implying substantial contamination by crustal materials, most likely Moine pelite.

The xenolithic monchiquite dyke at Loch Roag, on Harris, is compositionally unique among the Palaeogene dykes of the Hebridean Igneous Province. It is of basic composition (MgO = 9 per cent), rich in K_2O (3.5 per cent) and has elevated concentrations of barium (Ba) and the Light Rare-Earth Elements (Menzies et al., 1987). The composition is similar to minor intrusions of Late Palaeozoic age that are quite common in western Scotland (e.g. Upton et al., 1998).

Volcanic plugs

The Cnoc Rhaonastil Boss on Islay comprises dominant alkali olivine-dolerite together with volumetrically minor pods of coarse-grained nepheline-syenite (Chapter 7; Hole and Morrison, 1992; Preston et al., 1998b, 2000a,b). The major- and trace-element compositions are highly enriched in incompatible elements, which can be adequately explained by the fractionation of olivine and plagioclase from an alkali basalt magma; such enrichment is not reported from elsewhere in the Hebridean Igneous Province. The intrusion provides a unique insight into the composition of liquids produced by extreme fractional crystallisation, most likely in a closed system in a high-level magma chamber. It indicates the possible composition of the end-stage liquids produced by extreme fractionation of the dominant Skye Main Lava Series and Mull Plateau Magma-type that is not recorded in the lava piles, but is approached in the Shiant Isles sills (see below).

Sill-complexes

Three geochemically distinct groups are recognised in the Loch Scridain Sill-complex of Mull (Preston and Bell, 1997; Preston et al., 1998a, 1999).

Group 1 consists of aphyric tholeiitic basalts and basaltic andesites with Preshal More Magma-type affinities
Group 2 comprises plagioclase- and pyroxene-phyric andesites and dacites
Group 3 are rhyolites

The sills show progressive and extreme enrichment in incompatible elements and have elevated initial Sr- and Pb-isotope ratios. These properties are attributed to fractional crystallisation concurrent with assimilation of a partial melt derived from Moine metasedimentary rocks. The Group 3 rhyolites originated as partial melts derived from pelites and the Group 2 sills appear to represent magmas produced by simple bulk mixing of Group 1 and Group 3 magmas. Cognate xenoliths provide samples of early-precipitated cumulate assemblages and quartzite xenoliths from the Moine basement are common. Distinctive mullite-bearing aluminous buchites are considered to represent the end-product of two-stage melting of Moine pelite: initial partial melting resulted in the removal of rhyolitic melts (compare with the Group 3 sills), which was followed by bulk melting of the residual aluminous pelites within the sill conduits. Complex reactions between the aluminous melts and the tholeiitic sill magma produced a hybrid magma that subsequently crystallised to form the plagioclase-spinel-corundum rims that encase the aluminous xenoliths (Dempster et al., 1999).

The sills of the Little Minch Sill-complex on Skye range in composition from picrite through picrodolerite to analcime-bearing olivine-dolerite ('crinanite') and, on the Shiant Isles, to silica-undersaturated alkali syenites (Gibb and Gibson, 1989; Gibson, 1990; Gibb and Henderson, 1996). The sills are of alkaline affinity, with geochemical characteristics comparable with those of the earlier Skye Main Lava Series (see above). Isotopic data indicate contamination with up to 20 per cent of amphibolite-facies Lewisian gneisses, concurrent with fractionation, thus implying that the sill magmas underwent distinctly different evolutionary paths to those of the Skye Main Lava Series lavas as they ascended through, and reacted with, the crust.

SILICIC ROCKS

The silicic rocks include the main granitic lithologies of the central complexes, which range in composition between monzogranite and peralkaline granite, together with rare minor intrusions (mainly dykes), rhyolitic lavas and pyroclastic rocks. The origins of the silicic rocks have been the subject of controversy and the topic is covered in several reviews (J D Bell, 1976; Gass and Thorpe, 1976; Thompson, 1982a, b).

One of the major points of controversy has been the relative importance of fractional crystallisation of basaltic melts compared with the partial fusion of country-rock lithologies in yielding granitic liquids. Fractional crystallisation was preferred in the early models (Harker, 1904, 1908), although this was contested by some (e.g. Reynolds, 1954), and subsequently partial melting models became favoured (e.g Brown, 1963; Wager et al., 1965; Dunham, 1968; Thompson, 1969). Within the environs of the central complexes, where significant masses of ultrabasic and basic magmas were fractionating within crustal reservoirs, there was the potential for *both* processes to have operated.

Isotopic studies have shown that certain of the Skye granites have modified (elevated) radiogenic strontium and lead signatures, because of the presence of crustal material. Consequently, the granites were most likely formed, in part, by partial melting of Lewisian gneisses (Moorbath and Bell, 1965), and Pb isotope data (Moobath and Welke, 1969) clearly point towards a mixed origin, involving both mantle and crustal lead (Figure 37). Some of the granites have isotopic signatures that are very similar to those of the basic rocks of the lava fields and central complexes (Walsh et al., 1979; Moorbath and Thompson, 1980; Dickin, 1981; Dickin et al., 1984a). As with the basic rocks, contributions from various combinations of lower (granulite-facies) and upper (amphibolite-facies) crustal Lewisian gneisses, Moine pelite and psammite and Dalradian pelite are recognised.

Thompson (1982b) identified three compositional groups of granites throughout the British and Irish sectors of the North Atlantic Igneous Superprovince, based on mineralogical (and hence major-element) composition, trace-element composition and radiogenic isotope signatures. He argued that the primitive and peralkaline granites, which are the only types represented in the Hebridean Igneous Province, can be related to basaltic parental magmas through simple crystal fractionation involving plagioclase, alkali feldspar, quartz, iron-titanium oxides and apatite.

The two main basic magma-types that could fractionate to yield granitic liquids are those of the Skye Main Lava Series/Mull Plateau Magma-type and those of the Preshal More Magma-type. Investigations of the rare-earth element geochemistry of the Skye and Mull granites (Thorpe et al., 1977; Thorpe, 1978; Meighan, 1979; Walsh et al., 1979) showed an apparent match between the light-rare-earth-element profiles of the granites and those of the Skye Main Lava Series and Mull Plateau Magma-type, suggesting that they are related through fractional crystallisation. Thompson (1982b), however, advocated an origin through the mixing of 80 per cent of uncontaminated Preshal More basalt with 10 per cent each of granulite-facies and amphibolite-facies Lewisian gneisses. The abundant availability of Preshal More magmas during the growth of the Hebridean central complexes (with the possible exception of Rum), and the intimate association of tholeiitic rocks of this type with silicic rocks in many composite and other intrusions (see below), strongly support this contention, although derivation of the silicic rocks by fractionation of crustally contaminated Skye Main Lava Series/Mull Plateau magmas cannot be ruled out.

Peraluminous silicic liquids have been generated experimentally by melting Lewisian gneisses (granodiorite–tonalite) and Torridonian sedimentary rocks (feldspathic sandstone and siltstone) at 1 kbar water pressure (Thompson, 1981). These rocks are all completely melted by 930°C, a similar temperature to that required for the complete melting of granitic rocks from the Hebridean Igneous Province (Thompson, 1983). Therefore, if the granites are the product of the wholesale melting of the Lewisian and Torridonian lithologies, they should have similar trace-element signatures. However, the trace-element profiles for the granitic components of the Lewisian gneisses are very different from those of the Paleocene granites (Thorpe et al., 1977; Thompson, 1982b). Consequently, the granites cannot represent high-percentage melts of these relatively fusible lithologies.

Therefore, the granites most likely have a dual origin. From a detailed isotope study of the Skye granites, Dickin (1981) concluded that the dominant contaminant is Lewisian amphibolite-facies gneiss, similar to the contaminant introduced into the Preshal More Magma-type (Figure 37). On rising to shallower crustal levels, contaminated Preshal More magmas underwent fractionation in the amphibolite-facies Lewisian gneisses of the upper crust. The heat generated by this process resulted in partial melting of the gneisses, generating silicic melts which then mixed with silicic melts produced by fractional crystallisation of the contaminated Preshal More magmas. It is estimated that the proportion of crustal input was between 5 and 33 per cent of the mass of each intrusion, with a larger crustal fraction occurring in the younger granites, which had undergone a longer period of differentiation in the upper crust.

MAGMA MIXING

Direct evidence for the co-existence of basic and silicic magmas is provided by a wide range of intrusions, including composite dykes and sills with contrasting basaltic and silicic components (Chapters 7; 8), intrusions such as the Loch Bà Ring Dyke of Mull where a suite of intermediate and basic inclusions is scattered through a rhyolitic matrix, and the Marsco Hybrids of Skye in which magma mixing has resulted in compositionally homogeneous hybrids (Chapter 9).

Examination of the composite intrusions commonly reveals that both members also contain evidence of pre-emplacement magma mixing; diffuse clots and small xenoliths of basic material are scattered throughout the silicic rocks, and corroded xenocrysts of quartz, alkali feldspar and sodic plagioclase are common in the basic members. Thus, the composition of the 'end-member' magmas that were originally involved in mixing may be obscure. Where recognised, the basic units are typically somewhat more evolved than basalt, and are usually basaltic andesite or ferroandesite/ferrodiorite, consistent with their generation by about 50 per cent fractional crystallisation of tholeiitic basaltic magma (Bell, 1983; Marshall and Sparks, 1984; Sparks, 1988; Bell and Pankhurst, 1993).

More-complete magma mixing has occurred in certain intrusions in the central complexes. In the Loch Bà Ring Dyke, the dominant rhyolitic material contains typically glassy inclusions that range in composition from ferrobasalt to rhyolite (including welded tuff), with tholeiitic andesites the most abundant (Blake et al., 1965; Walker and Skelhorn, 1966; Sparks, 1988). The compositional range in the Marsco Hybrids is similar to that in the Loch Bà Ring Dyke (Harker, 1904; Wager and Vincent, 1962; Wager et al., 1965; J D Bell, 1966; Thompson, 1969; B R Bell, 1983). The central ferrodiorite contains rare quartz xenocrysts and is thus, in part, a hybrid composition.

The likely spatial relationships and dynamics of the magmas in the magma chambers have been modelled by several investigators. Wager et al. (1965) envisaged a compositionally stratified magma chamber, with mechanical mixing occurring along the interface between the basic and overlying silicic magmas, each of which was convecting separately. The annular mass of hybrid magma so formed at the interface was intruded as the composite Marsco Hybrids ring-dyke.

The dispersed basic inclusions of the Glamaig Granite were explained by Thompson (1980a) as relicts of rounded blebs of basaltic magma that formed as the consequence of violent disruption in a convecting system initiated when basaltic magma invaded overlying silicic magma (compare with Sparks et al., 1977). From a detailed examination of the Loch Bà Ring-dyke, Sparks (1988) concluded that it was formed from a strongly zoned magma chamber in which there were two separately convecting parts, one vertically zoned in composition from tholeiitic basaltic andesite upwards into dacite, the other an overlying cap of rhyolitic magma. Emplacement occurred when rapid subsidence of the central block within the ring-fracture forced thorough mixing of the lower, hotter magmas with the cooler rhyolitic cap. Superheating of the rhyolitic magma led to exsolution of volatiles and rapid expansion, and as the two magmas mixed the resultant emulsion-like magma was injected up the ring-fracture as a gas-charged pyroclastic fluid.

Consideration of these examples of mixed magma intrusions, and of other bodies including intrusions in Centre 2, Ardnamurchan (Skelhorn and Elwell, 1966), the Mullach Sgar intrusions of St Kilda (e.g. Harding, 1966), the Glen More Ring-dyke, Mull (Bailey et al., 1924; Holmes, 1936; Fenner, 1937; Kerr et al., 1999) and composite intrusions on Arran (Kanaris-Sotiriou and Gibb, 1985), highlight not just the consequences of the co-existence and interaction of basic and silicic magmas, but also the dominant, driving force that the basalt magmas represent in almost all aspects of the magmatism of the Hebridean Igneous Province.

eleven

Palaeogene and later structure

The Palaeogene and later structures of the Inner Hebrides, the adjoining mainland, and Arran were strongly influenced by structures inherited from Mesozoic and earlier times (Chapter 4). These earlier structures include major, north-east-trending faults and lineaments dating from the Caledonian Orogeny or earlier, and large, predominantly north–south-orientated Mesozoic sedimentary basins with their bounding faults (Figure 5). There is evidence that at least some of these structures were reactivated during the Paleocene volcanism and it has been postulated that they were major factors in determining the sites of Paleocene central complexes (e.g. Richey, 1961) and the thick lava successions (Walker, 1979). Parts of the lava fields are extensively faulted, and localised faulting and folding occurs within and around the central complexes.

MAJOR FAULTS

The central complexes of Skye and Rum are situated on, or close to, the north–south-trending **Camasunary–Skerryvore Fault**, which defines the western edge of the Inner Hebrides Basin. At Camasunary, on Skye, the fault affects Torridonian and Mesozoic strata but there is little, if any, displacement of the overlying Paleocene lavas, which are assumed to belong to the Skye Lava Group (p.67). Beyond the Blà Bheinn area, the northern continuation of the fault is uncertain, but it has been suggested that the Loch Screapadal Fault on Raasay represents its northern continuation. This is a major, post-Mid Jurassic fault that is offset by a zone of north-west-trending sinistral strike-slip faults between Raasay and the Red Hills (Butler and Hutton, 1994). South of Skye, there is evidence of Paleocene, or post-Paleocene, movement on the continuation of the Camasunary Fault, since gently south-west-dipping flows of the Eigg Lava Formation crop out on Eigg but not on the east coast of Rum (except as faulted slivers in the central complex; Chapter 9). Still farther south, both Coll and Tiree are formed of Lewisian gneisses although down-faulted Paleocene lavas occur beneath the sea a short distance to the east of both islands (Figures 1; 9).

The **Great Glen Fault** cuts across south-east Mull, where it has disturbed Mesozoic and earlier rocks but not the Paleocene lavas (Chapter 4). There is, however, evidence of displacements on this fault system that post-date dykes near Mull (Holgate, 1969). Basaltic dykes, of presumed Palaeogene age, intrude the Caledonian Strontian Granite and Moine rocks on the west side of Loch Linnhe. The probable south-east continuation of this swarm on Lismore appears to have been offset dextrally by the intervening Great Glen Fault, with further dextral offset by the north-east-trending Lynn of Lorn Fault, situated between Lismore and the mainland to the south-east. There has been a total of almost 30 km of post-Paleocene dextral displacement of the dyke swarm, 20 km of which was on the Lynn of Lorn Fault. From an analysis of the Neogene drainage pattern on both sides of the Great Glen, Holgate (1969) suggested that the final dextral movement was considerably later than the emplacement of the Mull Central Complex. As a consequence, the south-east continuation of the Mull Dyke Swarm also should be offset dextrally, and be represented by dykes well to the south and south-west of Oban. However, there is no indication of any offset of the pronounced linear magnetic anomalies associated with the Mull Dyke Swarm to the south-east of Mull, which head directly towards the Mull Central Complex. Furthermore, the dense dyke swarm at Oban and to the south-east includes significant numbers of felsic dykes and a few composite dykes, which are indicators of proximity to an associated central complex. Thus, the available evidence indicates that, although the southern extension of the Skye Dyke Swarm may have been displaced, little movement has occurred on the Great Glen Fault since the emplacement of the Mull Central Complex and the Mull Dyke Swarm.

STRUCTURE OF THE LAVA FIELDS

Skye

The lava field of northern Skye is dissected by numerous faults and has been gently folded. The lavas have been broken into a large number of blocks by fairly major, continuous faults trending between north-west and north–south, and shorter faults trending approximately north-east. The fault pattern is well displayed in west-central Skye where the lava stratigraphy is known in detail (Williamson and Bell, 1994). The broad structure of the lava field is interpreted as an early, shallow north-west-trending syncline, the core of which is occupied by the more evolved lavas. This earlier folding is attributed to loading that resulted from lava accumulation. Later folds developed, along a north–south to north-east trend and were related to Oligocene and younger events; one such syncline is responsible for the preservation of Oligocene strata in the offshore Canna Basin (England, 1994).

Eigg and Canna

The flows of both the Eigg and the Canna lava formations are either flat-lying or dip at low angles. Normal faults are common. On Eigg there is evidence that a north-north-west-trending fault that affected the Mesozoic strata was reactivated after accumulation of the flows of the Eigg Lava Formation (Emeleus, 1997).

Mull

The detailed stratigraphy of the Mull Lava Field has not been determined (Table 10), and therefore only the broad outline of the structure is known. The lavas of north-west Mull, Ulva, the Ardmeanach peninsula and the eastern part of the Ross of Mull, and the thick Ben More succession are generally flat-lying or dip gently, although evolved lavas south-east of Ben More are involved in folding associated with the emplacement of the central complex (see below). Away from the central complex, the base of the lavas crops out in the south at Carsaig Bay, in the west at Gribun and in the north at intervals on the coast between Tobermory and Bloody Bay. This may indicate that the lavas form a broad north-north-west-trending syncline in north-west Mull. Several north-north-west- to north-west-trending faults occur on the island of Ulva, towards the western end of the Ardmeanach peninsula and south of Loch Scridain. On the Ross of Mull, near Ardtun, the Paleocene lava outcrop terminates at the west-north-west-trending Loch Assapol Fault, where they are thrown down against Moine metasedimentary rocks.

Morvern

The lavas on Morvern are cut by north-north-west- to north-trending faults, and are faulted out in the east at the north-trending Inninmore Fault where they are thrown down against Moine psammitic gneisses. The base of the lavas drops from an altitude of about 400 m on Beinn Iadain in the east to sea level near Auliston Point in the west, a distance of about 13 km. Much of the change in altitude is the result of faulting but some is due to postdepositional tilting of the lavas. Near to the Inninmore Fault, the lavas directly overlie Cretaceous, Jurassic, Triassic and Moine rocks, indicating a measure of uplift and erosion before the lavas were erupted. Elsewhere in Morvern, the lavas generally rest on the thin Gribun (Beinn Iadain) Mudstone Formation, which in turn usually rests on the equally thin Upper Cretaceous Clach Alasdair Conglomerate Member (compare with Hancock, 2000; Mortimore et al., 2001). There is thus only slight evidence for erosion between the Late Cretaceous and the Paleocene in this district.

STRUCTURES ASSOCIATED WITH THE CENTRAL COMPLEXES

Country rocks adjacent to the central complexes are commonly folded and faulted as a consequence of the emplacement of the intrusive units. These features are most apparent around centres where granites or other silicic intrusions are present (Chapter 9).

Skye

The lavas and earlier rocks around the Cuillin Centre on Skye are generally flat lying and appear little disturbed, except for gentle folding in Glen Drynoch. However, there is considerable disruption of country rocks adjoining both the Western and Eastern Red Hills centres. Paleocene lavas, Mesozoic sedimentary rocks and Torridonian sandstones on the northern slopes of Glamaig and on Scalpay appear to define part of a dome over the Western Red Hills granites. In the Eastern Red Hills there is a suggestion of doming of Mesozoic and earlier sedimentary rocks over the Beinn na Caillich Granite. The Beinn an Dubhaich Granite occupies the core of the Broadford Anticline, an anticlinal structure in Lower Palaeozoic Durness Group dolostones. However, the granite cuts across, and hence postdates, the folds (Holroyd, 1994). Folding of lavas and Mesozoic rocks in the Coire Uaigneich area, west of Loch Slapin, is attributed to granite emplacement (Butler and Hutton, 1994).

Rum The gentle regional dip of the Torridonian sandstone towards the west-north-west to north-west is greatly disturbed on the northern margin of the Rum Central Complex (Figures 23; 24). Where they are in faulted contact with the Western Granite and the Northern Marginal Zone, the beds dip outwards at over 50° towards the north-west in the west, changing progressively to a north-east direction in the east. There is little folding or tilting of the strata next to the Eastern Layered Intrusion, but large masses of Torridonian sandstone at Mullach Ard, at Welshman's Rock and in parts of southern Rum appear to have slid away from the area of the central complex, with minor folding in places (p.110). These structures formed during the early stages in the growth of the central complex, when central uplift led to doming, and major peripheral faulting initiated the Main Ring Fault. The central complex is cut by the north–south-trending Long Loch Fault. There is evidence of movement on this fault both before and after the emplacement of the central complex, and that it acted as a feeder zone during emplacement of the layered intrusions.

Ardnamurchan On the eastern margin of the Ardnamurchan Central Complex, the north-north-west-trending Loch Mudle Fault throws down Paleocene lavas and Mesozoic rocks to the east (Figure 26a). Apart from this and some other minor faulting, there is little sign of disturbance of the lavas and earlier rocks. At the western end of the peninsula, however, the Mesozoic sedimentary rocks generally dip away from Centre 2 and the northern edge of Centre 3, suggesting a dome-like structure. In contrast to the doming on Skye and Rum, this appears to be associated with predominantly gabbroic intrusions; however, there must have been a substantial contribution to uplift by the intrusion of the multitude of cone-sheets (e.g. LeBas, 1971).

Mull One of the most notable structural features of the Hebridean Igneous Province is the set of concentric folds that almost encircle the Glen More and Beinn Chaisgidle centres on Mull (Bailey et al., 1924). The folds are developed in the surrounding older lava sequence and the various subjacent pre-Paleocene rocks. They include the Loch Spelve and Loch Don anticlines and the Duart Bay and Coire Mòr synclines (Figure 27). South of Loch Don, Dalradian metalimestones and phyllites form the core of a north-trending anticline, flanked successively by late-Silurian andesitic lavas, Lower and Middle Jurassic sandstones and mudstones, and Paleocene lavas. These major structures continue to the north and north-west, where Jurassic rocks form the core of the Craignure Anticline as far as Craignure Bay, with Moine rocks exposed in the core between Craignure Bay and Scallastle Bay. Elsewhere, the folding generally involves Paleocene basaltic lavas and, on the eastern flanks of Ben More, mugearites near the top of the lava succession. It is suggested that the folds formed in response to the early intrusive events in the central complex, but their age is not well understood. Furthermore, the area enclosed by the folds is domed and the folds may owe their origin to gravity-driven movement as the dome developed. From evidence in the Loch Don area, it has been suggested that doming and folding were initiated prior to eruption of the Paleocene lavas (Cheeney, 1962; Walker, 1975a), but Bailey (1962) considered the evidence to be inconclusive. Near Sgurr Dearg, the folds are cross-cut by, and thus predate, volcaniclastic breccias of the central complex.

Arran There is spectacular deformation of the country rocks surrounding the North Arran Granite Pluton. Throughout the south-west Highlands, the Dalradian rocks strike fairly consistently north-east–south-west, but on Arran they strike approximately parallel to the near-circular outline of the Outer Granite over about 270° of arc (Figure 32). The structure is well illustrated by the outcrops of distinctive black, slaty and phyllitic metamudstones on the west and north-west margin of the pluton. To the north of the granite, between Loch Ranza and Catacol, the Dalradian rocks form a concentric synform, termed the Catacol Synform. Offshore, between Catacol and Imichar Point, Permo-Triassic sandstones occur in the core of a concentric synclinal structure affecting Dalradian and Carboniferous rocks. On the eastern side of the granite, the country rocks responded to granite emplacement by a combination of folding and faulting, with reactivation of pre-existing faults in the structurally complex area near Corrie. Locally, near Pirnmill on the west coast, there are small reverse faults inclined towards the granite, some of which cut Paleocene dykes. South and south-east of the granite, doming affects sedimentary rocks of Devonian, Carboniferous and Permian age, for example in Glen Cloy, west of Brodick. This deformation is attributed to the diapiric rise of the Outer Granite (England, 1992b). The later Central Arran Ring-complex cuts across the domed country rocks of the North Arran Granite Pluton and there is some evidence that the country rocks adjacent to the ring-complex are also domed on its south and south-east side, although indifferent exposure makes verification difficult.

twelve

Late Palaeogene and Neogene

Following the cessation of igneous activity in the early Eocene, the geological evolution of the area was dominated by uplift, weathering and erosion, with the accumulation of clastic debris offshore. Rapid erosion of the Highlands provided detritus which accumulated in the North Sea Basin in considerable thickness during Paleocene to early Eocene times. The major sediment influxes may have been pulsed, in response to the Palaeogene igneous activity and were not equalled later (e.g. Zeigler, 1981; White and Lovell, 1997). In contrast, the late Eocene to Neogene was a period of relatively limited uplift and erosion of the Highlands and western Scotland (Hall, 1991). Terrestrial sedimentary rocks of Oligocene age occur in restricted basins and have been proved in a borehole in the Little Minch, south-east of Harris and in the Canna Basin south-west of Skye where they are about 1000 m thick (Fyfe et al., 1993). Oligocene rocks also occur on the Blackstones Bank where they partly conceal the central complex. The sediments consist of carbonaceous mudstone with plant debris and lignite deposited in a flood-plain or lacustrine environment of swamps and fens.

In common with several areas on the mainland, deep chemical weathering predating the Quaternary glaciations has affected the Inner Hebrides and Arran. Chemical weathering profiles, particularly of basic and ultrabasic rocks, occur on Rum where they are found both beneath glacial deposits and in pockets, up to 21 m deep, on bare rock surfaces (e.g. Ball, 1964). The age of the weathering is uncertain, but much of it may date from the Eocene to the Miocene, when chemical alteration would have been favoured by the prevailing warm, wet conditions. Extensive weathering did occur during the Pleistocene and continues to the present day, with physical rather than chemical processes causing disaggregation.

The present-day landscape has clearly developed over a considerable period of time, but the history of uplift and erosion is poorly understood. The distribution of the Cretaceous rocks (Chapter 3) and the sedimentary lithologies associated with the Eigg Lava Formation (Chapter 6) indicate that when igneous activity began at about 61 Ma, relief over the area was generally subdued. It has been recognised that there was rapid, deep subaerial erosion during the Paleocene (e.g. Hall, 1991). This erosion occurred as the central complexes grew and the lava fields developed; there is evidence that the Rum Central Complex was unroofed in less than 1.6 Ma (Hamilton et al., 1998; Chambers et al., 2005 see also chapters 5, 6 and 9). Evidence from the central complexes indicates that mountain massifs formed by a combination of igneous intrusion and extrusion, faulting and folding. Major differential movements occurred in and around all of the central complexes, and evidence for this is particularly well preserved on Mull and Arran (Figures 27; 32).

The Cuillin hills (about 970 m OD), certain of the Red Hills and the Kyleakin Hills of Skye (about 720 m OD) all show a similar altitude of the main summits and this is also apparent elsewhere. In the past, this similarity in the levels has been attributed to marine planation during a relative rise in sea level, most likely in the Neogene. This interpretation is almost certainly wrong because most uplift occurred during the Paleocene and sea level dropped during the Neogene. Tilted planation surfaces of subaerial origin have been recognised, for example on Skye outwith the Cuillin (at 500 to 700 m, 400 to 450 m. and 250 to 300 m OD) and a lower surface is identified at 80 to 120 m OD at many places in western Scotland, including Skye, Rum, Ardnamurchan and Mull (Le Coeur, 1988). These planation surfaces are now attributed to uplift that occurred in a number of stages, and it can be inferred that some tectonic tilting and deformation took place subsequent to the Paleogene igneous activity. Evidence for faulting of the lower planation surface during the Neogene is found in the offshore extension of the surface east of Lewis. This is thrown down to the east by about 150 m on the Minch Fault. Flexuring of the surface is seen at a number of places; for example, it is tilted towards the west on either side of the Sound of Raasay (Le Coeur, 1988).

thirteen
Quaternary

Since the start of the Quaternary Period, about 1.8 million years ago, the Highlands and Islands of western Scotland have undergone several periods of glaciation during the Pleistocene Epoch. Initially, only the mountain tops were covered by ice, but evidence from offshore deposits indicates that at times during the last 750 000 years, the middle latitudes of Europe supported large ice sheets, which extended across much of the Scottish mainland (Boulton et al., 2002). The ice sheets grew and decayed in a rhythmic manner: cold periods (glacials) when the ice sheets grew were separated by shorter, warmer intervals (interglacials) when conditions were comparable with those of the present day. There were at least five glacial/interglacial cycles, each of about 100 000 years duration. Within the glacials, colder periods are known as stadials, and short, warmer periods, but probably not as warm as the interglacial periods, are known as interstadials. The most recent intense and widespread glaciation occurred during the Dimlington Stadial (Table 17) when much of the evidence of earlier glaciations, and pre-Quaternary weathering, was removed by erosion. The Quaternary record in the Hebrides is therefore very incomplete prior to about 26 000 years BP (before present). Many of the ages quoted for this period, and indeed up to about 60 000 years BP are in radiocarbon years. It should be noted that these may be significantly less than true ages in calendar years (e.g. Walker and Lowe in Gordon, 1997).

The effects of the Quaternary ice ages are visible throughout the Inner Hebrides, the western Scottish seaboard and on Arran. Morainic deposits are widespread (Plate 39). Valleys, hills and rock surfaces commonly exhibit signs of glacial erosion and moulding (Frontispiece, Plate 40), and on the coastline numerous raised beaches and marine-cut benches provide evidence of changes in sea level (Figure 41; Plate 42). The large erosional features such as corries began to form during earlier glaciations, but the glacial deposits preserved on and around the mountain massifs of Skye, Rum, Mull and Arran belong largely to the Main Late Devensian glaciation, and were deposited during the Dimlington Stadial and in a subsequent, less extensive glaciation during the Loch Lomond Stadial (Table 17). The mountains attracted high precipitation, and under the prevailing cold conditions this led to the formation of glaciers and local ice caps. During the Main Late Devensian glaciation these glaciers and ice caps coalesced into ice domes, which interacted with ice from much larger sheets on the Scottish mainland (Figure 38).

By about 10 000 years BP, the climate had become more temperate and the local glaciers and ice caps had melted. This change defines the start of the Holocene Epoch, formerly known as the Postglacial Period, and probably another (as yet incomplete) interglacial. Shortly after this time, Man reached the area. Some of the earliest records of human activity in Scotland, of Mesolithic age, come from Staffin, Skye, and from Rum where the local bloodstone (Chapter 6) was fashioned into artefacts at Kinloch. Radiocarbon determinations obtained from charcoal and from hazelnut shells from this site range from 8590 ± 95 years BP to 7570 ± 50 years BP (Wickham-Jones and Woodman, 1998).

The Quaternary geology of the Inner Hebrides and Arran has been intensively studied over many years. The excellent field guides published by the Quaternary Research Association detail these researches for specific areas and contain useful discussions of Pleistocene processes and descriptions of numerous key localities (Mull: Walker et al., 1985; Skye: Ballantyne et al., 1991; south-west Highlands including Mull: Walker et al., 1992).

PRE-LATE DEVENSIAN GLACIATIONS

An indication of some of the glaciations that occurred prior to the Main Late Devensian Glaciation is provided by high-level, marine-cut platforms and notches in clifflines covered by glacial deposits of supposed Late Devensian age, for example on Skye and Rum (see below). These features must have formed while the land was substantially depressed relative to sea level, suggesting that an extensive ice sheet existed prior to the Late Devensian (Sutherland

Table 17 Late-Pleistocene and Holocene events in Britain (after Ballantyne and Harris, 1994; raised shorelines after Gray, 1978, and Gordon and Sutherland, 1993).

Epoch	Age		Stadials/ interstadials	Glaciations and *raised shorelines*	Boundary age BP
Holocene	Flandrian (interglacial)			*Late-Postglacial shorelines, 3000 BP* *Main Postglacial Shoreline, 7000 BP* *Buried beaches, 9000 BP*	10 000* (11 500)
Pleistocene	Devensian (glacial)	Late Devensian	Loch Lomond Stadial	*Loch Lomond (re)advance*	11 000* (13 000)
			Windermere (Late-glacial) Interstadial	*Main Late-glacial shoreline* *Early Late-glacial shorelines*	13 000* (14 700)
			Dimlington Stadial	*Main Late Devensian Glaciation*	26 000* (30 000)
		Mid Devensian	Upton Warren Interstadial		59 000
		Early Devensian	Brimpton Interstadial	*High rock platforms†* *(Early Devensian glaciation?)*	
			Chelford Interstadial		116 000
	Ipswichian (interglacial)				128 000

† or during the Dimlington Stadial (Sissons, 1982)

* These ages are in radiocarbon (C^{14}) years before present day (BP). Due to difficulties in calibration, they are significantly younger than ages based on other methods, such as counting seasonal layers in ice or sediment. In the Late Devensian and Flandrian, radiocarbon dating is more widely used than any other method and radiocarbon ages are usually quoted for consistency. Approximate equivalent ages in calendar years BP are given in brackets.

and Gordon, 1993). Support for this suggestion comes from St Kilda, where there is evidence of four cold periods (Sutherland et al., 1982). The oldest till on St Kilda is the well-consolidated Ruaival Drift which is up to about 14 m thick. This is overlain by the Abhain Ruavail sandy deposits, which are less than 55 cm thick but contain organic remains that show interstadial characteristics. A minimum radiocarbon date obtained from these deposits indicates that they predate the Late Devensian glacial maximum. Exotic erratics in the till on St Kilda are limited to rare, small fragments of red feldspathic sandstone. However, rounded iron-stained quartz grains and assemblages of minerals that are characteristic of contact and regionally metamorphosed rocks have been recovered from stream sediments (Harding et al., 1984). These exotic erratics, minerals and quartz grains most likely came from Mesozoic or older (possibly Torridonian) sedimentary rocks; some possibly came from rocks contact-metamorphosed by the central complex and now concealed beneath the sea east of St Kilda. The erratics were incorporated in a till that formed during an earlier glaciation, pre-Ruaival Drift, when mainland ice covered the islands. After reworking during two later valley glaciations, these materials were finally released into stream sediments during the Holocene.

The scale of some of the erosional landforms that are attributed to glaciation also provide compelling evidence for a pre-Late Devensian glaciation. The likely rates of erosion beneath an ice sheet are estimated at about 1 mm/year. Thus, it is clear that some large corries could not have formed entirely during the Main Late Devensian and Loch Lomond Stadial glaciations, because the volume of moraines associated with the later glaciation are considerably less than the volume of the corries from which the debris was derived.

DIMLINGTON STADIAL

Main Late Devensian Glaciation

The extent of former glaciers has been deciphered by establishing the source of glacial erratics in the tills and moraines, and by recognising glacial trimlines. Trimlines define the highest levels to which glacier ice has actively eroded, or trimmed, frost-shattered bedrock and debris on protruding peaks (nunataks). Above the trimline, frost-shattered bedrock will persist but on the lower slopes, below the trimline, the bedrock will be glacially moulded and striated (Plate 40). Glacially transported boulders will be present below the trimline but otherwise most of the loose material that characterises nunataks will have been removed. The mineralogy of weathered materials may differ; gibbsite is common in the clay fraction of debris on former nunataks but it is rare, or absent, below the trimlines. Subsequent events may obscure or eliminate the evidence of trimlines, for example the downslope movement of scree debris (Ballantyne, 1997, 1999).

Plate 39 Hummocky morainic drift, Glen More, Mull (P580488).

Plate 40 *Glacially smoothed and polished gabbro surface, Coire Lagan, the Cuillin, Skye (Photograph: D Stephenson; P580489).*

The Inner Hebrides and Arran were almost entirely covered by a thick regional ice sheet during the early part of the Late Devensian (Figure 38). The ice sheet extended west of the Inner Hebrides, but probably not as far as St Kilda, which supported local glaciers. Later, as the ice sheet diminished, independent ice domes formed over the Outer Hebrides, and the islands of Skye, Mull, Arran and probably Rum.

Skye was an important centre for ice sheet glaciation and the Cuillin is an outstanding area for the study of glacial geomorphology (e.g. Gordon and Sutherland, 1993). The distribution of glacial striae, ice-moulded rocks and erratics all indicate that the Cuillin and the Red Hills were not over-ridden by ice from the Scottish mainland during the Main Late Devensian glaciation. Instead, independent ice caps developed and these deflected the ice that flowed westward from the mainland. The distribution of glacial erratics from the mainland over a range of altitudes on Raasay and Scalpay to the north, and on Soay to the south, and their absence from upland areas of central Skye, defines the area unaffected by mainland ice (Harker, 1904, but see also Ballantyne, 1990).

The former height of the Main Late Devensian ice cap has been established by the mapping of glacial trimlines around mountains that stood above the ice as nunataks in central and northern Skye (Ballantyne, 1990). High-level, Main Late Devensian trimlines have been recognised on the higher peaks of the Cuillin and Blà Bheinn, on certain of the high peaks in the Red Hills and or Healabhal Bheag in western Skye (Dahl et al., 1996). On the Trotternish peninsula, the Main Late Devensian trimline descends from 580 to 610 m OD in the south to 440 to 470 m OD in the north, within a distance of about 24 km. Frost-weathered bedrock (regolith) is common above the trimlines in the summit regions of the Cuillin and the Red Hills. In these areas relict solifluction sheets and lobes of Late Devensian (Main Late Devensian and/or Loch Lomond Stadial) age are also preserved, for example on Druim na Ruaige, on the west side of Beinn Dearg in the western Red Hills. Talus slopes of similar age are preserved on the Strathaird peninsula, on Ben Méabost and An Càrnach. These features are identified by their relatively complete vegetation cover when compared with features forming at the present day. Some of the extensive landslips on the Trotternish peninsula also predate the Holocene (see below). In the later stages of the Main Late Devensian Glaciation, as the pressure of the mainland ice decreased, the ice cap over

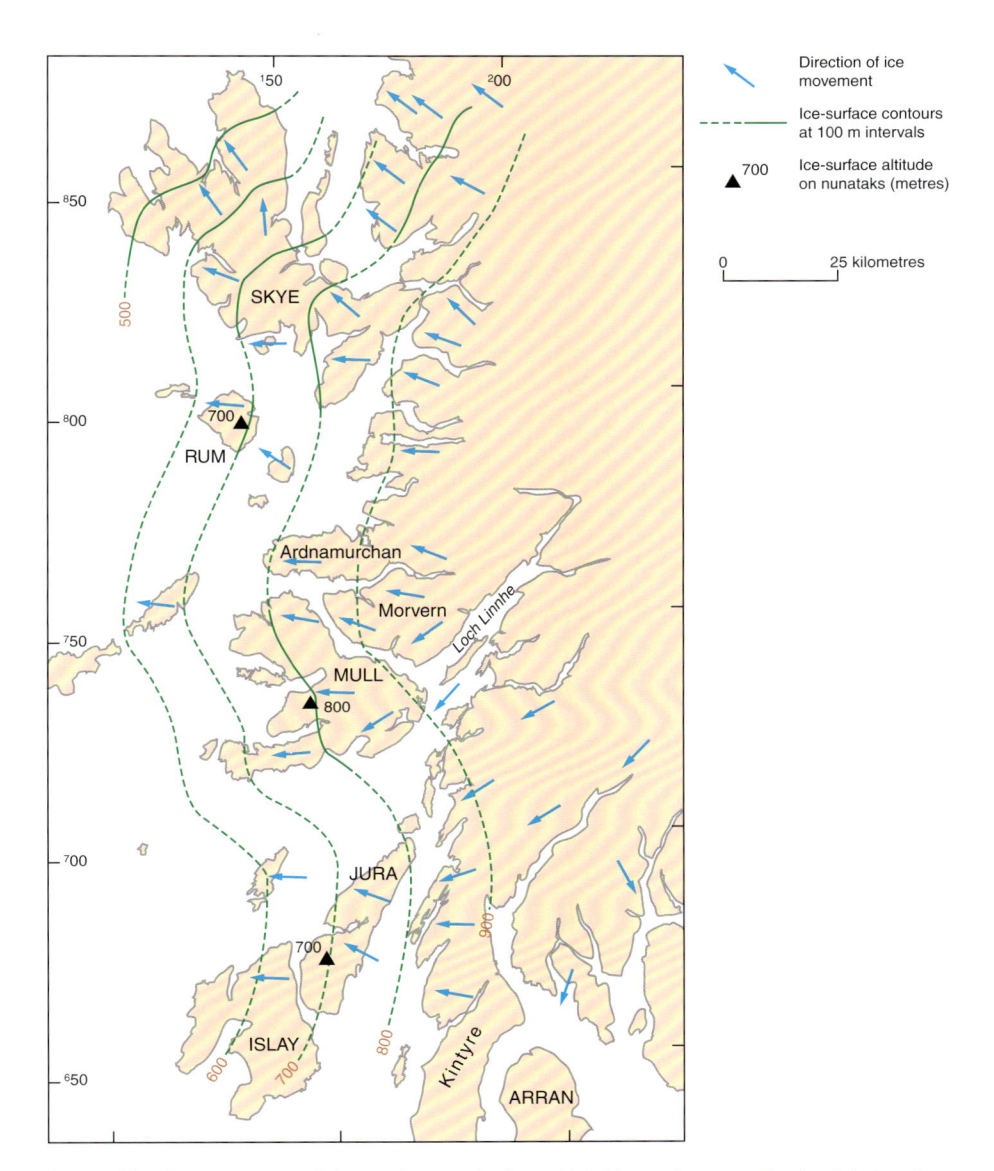

Figure 38 *Reconstruction of the ice sheet in the Inner Hebrides and western Scotland during the maximum of the Main Late Devensian Glaciation, based on an assumed maximum ice surface altitude of c. 800 m on Ben More, Mull (after Ballantyne, 1999, fig. 7).*

the Cuillin and the Red Hills was able to spread outwards. Supporting evidence comes from the Allt Beinn Deirge area, south-east of Beinn na Caillich, where unconsolidated deposits outside the limit of the later, Loch Lomond Stadial glaciation have been deformed by ice moving in an easterly direction (Le Coeur and Kuzucuoglu, 1992).

The mountains of south-central Rum may have supported an independent ice dome during the Main Late Devensian Glaciation. However, the distribution of striae and erratics on the lower ground, and on all the adjacent islands, show that the other Small Isles were over-ridden by mainland ice which was diverted around the mountains of Rum. On Eigg, striae indicate westward movement of ice, with possible diversion around An Sgùrr. There are no convincing striae on Muck, but on Canna and Sanday striae indicate ice flow towards the north-west and west, respectively. Amongst the distinctive erratics found in the Small Isles are:

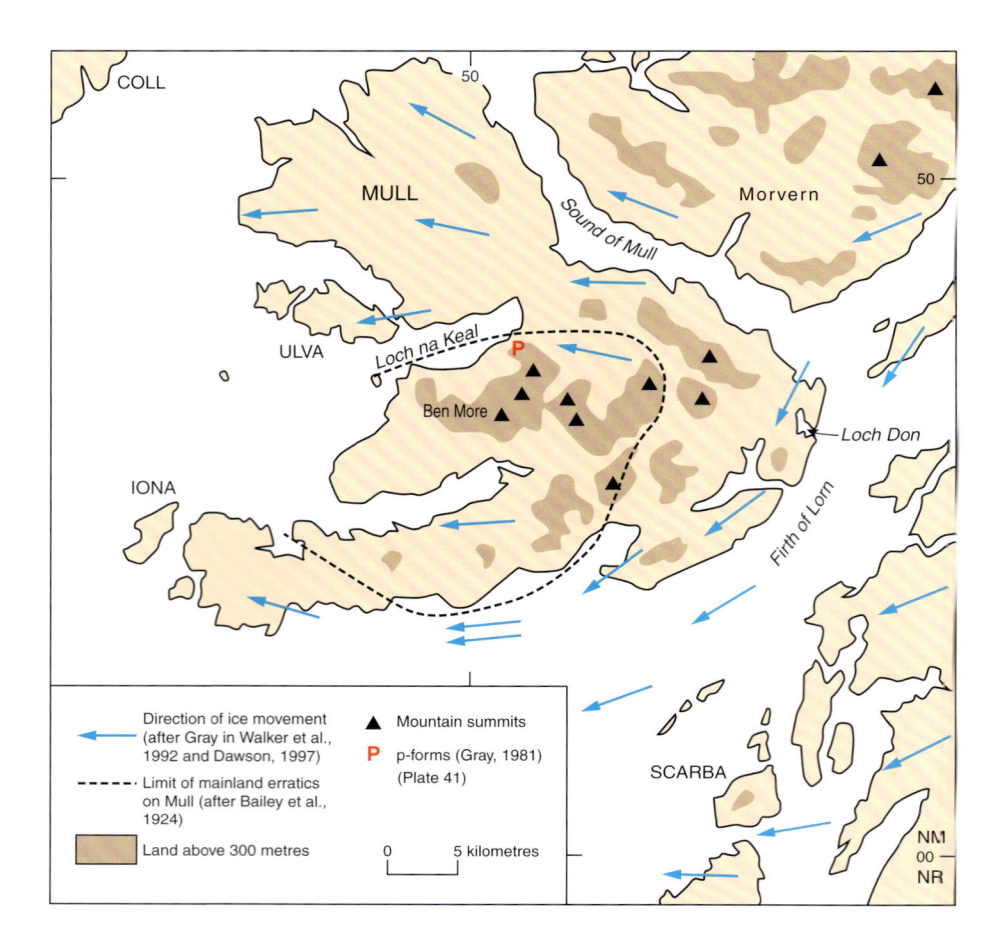

Figure 39 *Direction of ice movement around Mull during the Main Late Devensian Glaciation. A local ice dome over the mountains of southern Mull deflected the ice that flowed from the mainland and hence mainland erratics are absent inside the dashed line (after Ballantyne, 1999, fig. 1).*

- Moine gneiss on Rum
- gneiss, limestone of both Cambro-Ordovician and Jurassic age, and Torridon Group sandstone on Canna and Sanday
- gabbro, basalt, gneiss, mica-schist, quartzite, Torridonian and Mesozoic sandstone on Oigh-sgeir (suggestive of derivation from the mainland, Rum and the surrounding sea bed)
- Moine gneiss on Eigg and Muck

Mainland erratics have not, however, been found on the high ground in northern Eigg. Till deposits attributable to the Main Late Devensian Glaciation are common on Rum and small patches of till occur on the other islands, but glaciofluvial deposits are everywhere rare. There are two prominent north-east-trending morainic ridges in north-east Eigg. The eastern ridge is most likely part of a large crag-and-tail structure produced by ice flowing around the north end of Eigg, whereas the western ridge is a lateral moraine formed as the same ice mass decayed. Good examples of kettleholes occur in Cleadale, on Eigg.

The glaciation of Mull during the Loch Lomond Stadial (see below) removed much of the evidence for earlier glaciations from the central parts of the island. A local ice dome was established over the hills of central Mull during the Main Late Devensian Glaciation, diverting the mainland ice down Glen Forsa and south-west along Glen More and Loch na Keal (Bailey et al., 1924; Figure 39). Ben More and other high peaks probably formed nunataks (Ballantyne, 1999).

Plate 41 'P-form channel', scoured in basalt by glacial meltwater, south shore of Loch na Keal, Scarisdale, Mull. Channel is about 1 m wide (Photograph: D Stephenson; P580490).

On the south side of Loch na Keal, for some distance either side of the outlet of the Scarisdale River, bare basalt slabs on the shoreface have been moulded into a striking assemblage of smoothed, fluted, channel-like forms a metre of so in depth and a few metres in length (Plate 41). These are termed 'p-forms', and are excellent examples of structures that probably originated by a combination of erosion by glacial meltwaters and overlying ice (Gray, 1981; Walker et al., 1992). At the northern end of Loch Don, deltaic sands and gravels were deposited where south- and south-east-flowing meltwater streams entered the sea, to form the lower levels of the 'Loch Don Sand-Moraine' (Bailey et al., 1924; Benn and Evans, 1993; see below).

The Ardnamurchan peninsula was completely over-ridden by mainland ice during the Main Late Devensian Glaciation but apparently escaped renewed glaciation during the Loch Lomond Stadial. Evidence for ice movement from the south-east is provided by numerous striae and by the distribution and character of glacial erratics, including many derived from the Strontian Granite Pluton which occur, for example, on Ben Hiant. Much ice sculpting of the bare rocks has occurred and the hills formed by the gabbroic ring-intrusions now provide numerous examples of *roches moutonnées*.

The mainland ice sheet centred over the Highlands extended across Arran for most of the Main Late Devensian Glaciation. However, by about 13 000 years BP, the southern limit of the ice sheet on Arran lay along a line from Imachar to Corrie, with a detached remnant over the high ground to the south. By 12 400 years BP, the island was free of mainland ice, although a local ice cap may have persisted on the northern hills. The south of Arran escaped glaciation during the Loch Lomond Stadial and, consequently, deposits of an earlier glacial episode are widely preserved in that area. Reddish till, coloured by debris from the New Red Sandstone, is commonly exposed in river and stream sections, and in the Sliddery Water and Kilmory Water a clay bed near the base of the till contains a perfectly preserved shelly fauna (Tyrrell, 1928). Enigmatically, this fauna exhibits both Arctic and temperate characteristics (Sutherland, 1981). A large variety of features associated with the decay of the Main Late Devensian ice are preserved in the low ground of south-west Arran (Gemmell, 1972), including a succession of kame terraces north-east of Shiskine and drumlins near Blackwaterfoot. South-west-draining meltwater channels occur on the western slopes of Beinn Tarsuinn (near Blackwaterfoot) and south of Glen Iorsa. Moraines in Benlister Glen are probably of comparable age, recording the retreat of the Main Late Devensian ice. At the head of Glen Cloy, a raised shoreline estimated to date from about 12 000 BP cuts outwash from moraines in the glen. Deposits of gravelly till derived from readily weathered granite are restricted to the northern hills and the mouths of the valleys draining the granites. These deposits, which are up to 15 m thick, probably formed as the local ice decayed, after Arran became isolated from the mainland ice.

Ailsa Craig, 20 km south of Arran, lay in the path of a south-flowing stream of mainland ice that helped mould it into its present steep-sided dome shape. The distinctive blue-grey Ailsa Craig microgranite is widely distributed as glacial erratics and ice-rafted stones on coasts bordering the southern Firth of Clyde, in the north of County Antrim and on both sides of the Irish Sea as far south as Cork and Pembroke (Harrison et al., 1987). The distinctive pebbles and cobbles are also found on beaches in western Arran, where they may have been carried by drifting icebergs during the decay of the Firth of Clyde ice sheet.

On St Kilda, the Village Bay Till postdates the Abhainn Ruaival organic sand, from which it is separated by up to 3 m of detritus of periglacial origin (the Ruavail Head). The till is generally less than 4 m thick and the upper part lacks a true matrix, consisting only of closely packed angular boulders and smaller fragments, all of local origin.

WINDERMERE INTERSTADIAL

Deposits relating to the warm Windermere (Late-glacial) Interstadial (13 000 to 11 000 years BP; Table 17) are of limited occurrence in the district. Sediments at the base of a succession sampled from Loch an t-Suidhe on the Ross of Mull, yielded a ^{14}C age of 13 100 years BP, while clays with a molluscan fauna of interstadial affinities have been found in a few localities, for example on the floor of Loch Arienas north of Loch Aline in Morvern. Sedimentary deposits from this period are also found on the south shore of Loch Spelve, Mull (Walker et al., 1985), and at Kinlochspelve, where shells have been dated at 11 330 ± 170 years BP. The latter deposits were subsequently disturbed by a glacier. In general, these beds may correlate with the Late-glacial Clyde Beds of the Midland Valley of Scotland (e.g. Peacock, 1981).

LOCH LOMOND STADIAL

During the Loch Lomond Stadial (11 000 to 10 000 years BP; Table 17), small ice fields and glaciers became re-established in the mountainous areas of the district, leading to renewed valley and corrie glaciation, and to the modification or erosion of older glacial deposits; a distinctive suite of deposits is associated with this glaciation.

In Skye, glacial activity was centred over the Cuillin and the western Red Hills (Figure 40a). In addition, there were two small glaciers in north-east-facing valleys on the Trotternish escarpment and glaciers in Kylerhea Glen and Glen Arroch in eastern Skye. The extent of the Cuillin ice field has been defined by the distribution and orientation of glacial striae, roches moutonnées and moraines, and by the lithologies and distribution patterns of glacial erratics. Outlet glaciers flowed northwards down Glen Sligachan into glens Drynoch and Varragill, and into Loch Sligachan; others were located along the north sides of the western and eastern Red Hills. To the south, a large glacier flowed into Loch Slapin and outlet glaciers formed in Srath na Creitheach and from Loch Coruisk into Loch Scavaig. Nunataks protruding through the icefield included the main ridge of the Cuillin together with Sgùrr na Stri and Blà Bheinn, and Marsco and Glamaig in the western Red Hills.

Corries such as Coir' a' Ghrunnda, Coire Lagan, An Garbh-choire and Coir'-uisg, with their precipitous sides, rock falls and striated, ice-scoured rock in the corrie floors, provide some of the most spectacular glacial landforms in the district (Frontispiece, Plate 40). Moraines and glacial trimlines generally help to define the extent and thickness of the corrie glaciers, although in the Red Hills this evidence is generally obscured by scree. Tills are common, and typically have a sheet-like geometry within topographical depressions, although more-linear deposits are also present in the form of moraine ridges, as in the area around Sligachan. Excellent examples of moraines formed at glacier margins occur at the mouth of Coir' a' Ghrunnda in the Cuillin and at Coire Fearchair in the eastern Red Hills. Hummocky moraine is common, containing material supplied from both subglacial and supraglacial environments. In Gleann Torra-mhichaig, east of Glamaig, chains of hummocks oblique to the valley floor are attributed to intermittent bulldozing by advancing ice during the 'overall decay' of the valley glacier. In Coire Choinnich at the head of Loch Ainort, a chaotic assemblage of hummocks, non-aligned ridges and fluvial terrace accumulations resulted from the in-situ decay of less-active glacier ice.

Depositional features from the Loch Lomond Stadial are abundant in the mountains in the southern half of Rum, where there is evidence that twelve local glaciers formed. One occupied Coire nan Gruund and, at the lower end of the corrie, the Kinloch–Dibidil Path crosses a well-defined moraine made of ultrabasic blocks derived from the east side of Hallival and Askival. Rock slabs north of the moraine show north-north-east-directed glacial striae from the Loch Lomond Stadial glaciation cutting across north-west-directed striae from the earlier, Main Late Devensian Glaciation. In upper Glen Harris, linear moraine ridges parallel to the direction of ice movement are strikingly developed in Atlantic Corrie.

A wealth of glacial erosional and depositional features is preserved on the Western Granite hills of Rum, including a fine arcuate terminal moraine composed of microgranite blocks north of Sron an t-Saighdeir. There is also a superb blockfield, of supposed Loch Lomond Stadial age on the southern slopes of Orval and Sròn an t-Saighdeir (Ballantyne and Wain-Hobson, 1980). This has been modified by later solifluction processes, with the development of stone stripes on the lower, steeper slopes.

During the Loch Lomond Stadial, Mull supported an ice sheet that extended from Loch Bà to Loch Spelve, with Beinn Talaidh and other high hills in central Mull forming nunataks (Figure 40b). The most striking scenic features of Mull attributable to glacial processes date from this episode. Glaciers from the main area of ice reached the sea at several points, as did local glaciers flowing north-north-east off the hills into the Sound of Mull in the vicinity of Craignure. Outwash deposits formed at the foot of Loch Bà and Glen Forsa. Terminal moraines are present at Kinlochspelve and at the northern end of Loch Don, where Late-glacial deltaic deposits of the Loch Don Sand Moraine (see above) are partly covered by till from the Loch Lomond Stadial glaciation (Benn and Evans, 1993). Inland, there are widespread deposits of hummocky morainic drift, for example east of Craig in Glen More (Plate 39). Linear drift deposits, or fluted moraines, occur at several localities; well-defined examples on the north-west shoulder of Sgurr Dearg were formed by glaciers converging to the north-west.

The northern hills of Arran supported valley glaciers during the Loch Lomond Stadial and the till formed at this stage augmented earlier till (see above). Valley glaciers from a small ice

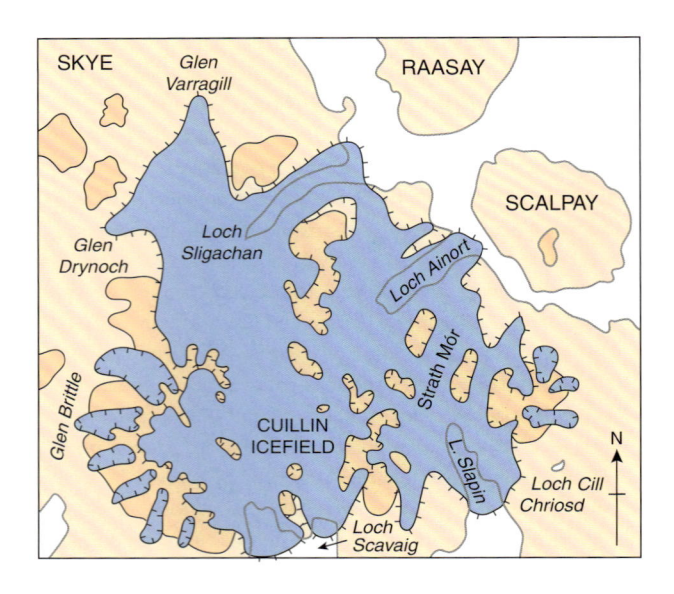

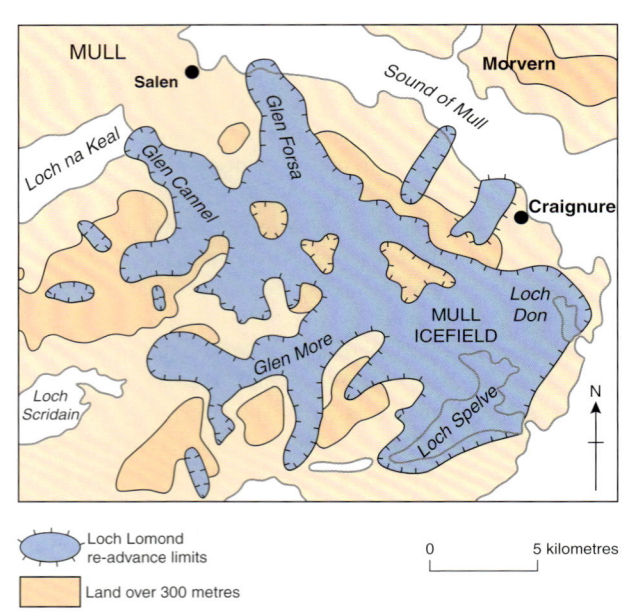

Figure 40 *Limits of local ice fields during the Loch Lomond re-advance.*
a Central Skye (after Benn et al., 1992)
b Mull (after Walker et al., 1992)

field centred on the head of Glen Iorsa extended down glens Catacol, Easan Biorach and Iorsa. Glaciers also occupied North Glen Sannox and the upper parts of glens Rosa and Sannox. Small corrie glaciers formed on the eastern side of the Goat Fell ridge and north-west and north of Beinn Bharrain. There are two generations of moraine in the northern hills. One group of fairly well defined moraines occurs at low levels throughout the area and is covered with grass and heather, for example in lower Glen Rosa. The second group is formed of fresh moraine ridges, studded with boulders and relatively free of vegetation. The latter moraines generally occur above 450 m OD, and are almost entirely restricted to corries in the east of the northern hills, for example at the head of Glen Rosa, on the north side of Casteal Abhail, and to the south-east of Beinn Tarsuinn. Substantial corrie glaciers evidently persisted in the eastern part of the northern hills after most of the ground to the west had become free of ice.

Periglacial deposits are widespread on St Kilda. Hill slopes are mantled by frost-shattered debris, together with inactive solifluction lobes. Two major protalus ramparts north and west of Village Bay overlie the youngest till. On the basis of the relative freshness of the clasts and their stratigraphical relationships, the protalus ramparts are thought to have developed during the Loch Lomond Stadial, the last time that severe cold affected St Kilda.

SEA LEVEL CHANGES

During the Quaternary glaciations, the weight of ice over Scotland caused isostatic depression of the land surface while sea levels were lowered as water became locked up in the major continental ice sheets. When the ice melted in Late-glacial and postglacial times these processes were reversed, but isostatic recovery of the land initially lagged behind the eustatic rise in sea level, with the result that beaches and features of marine erosion were formed well above the present sea level. As isostatic recovery continued (and relative sea level fell), these features were uplifted to become raised beaches and wave-cut platforms. The amount of isostatic recovery was greatest over those areas where the ice had been thickest, with the result that initially horizontal raised marine features have been tilted. Since older shorelines will also have been affected by later events, their gradients tend to be steeper than those of the younger shorelines. Tilting is directed away from the centre of uplift, and ice distribution, which was in the area of Rannoch Moor.

Several groups of features that can be attributed to changes in sea level are recognised in the district, although dating them is difficult, particularly in the Inner Hebrides and Arran where there are very few associated fossiliferous deposits. The amount of tilting provides a rough method of dating, but it does not furnish unequivocal results. For example, correlation of sections of rock platform may be difficult in areas where sedimentary rocks or lavas dip at low angles. Over short distances, platforms may show a gradual rise only to fall and rise again when another resistant layer takes over, a problem that is well illustrated by the lava benches around Treshnish Point, Mull (Walker et al., 1992). Furthermore, it may be difficult to differentiate between benches formed by marine erosion and those picked out by glacial action, as for example in the variably resistant, flat-lying sandstone beds of the Torridon Group at Bagh na h-Uamha, eastern Rum (S B McCann in Peacock, 1969).

High rock platforms

Remnants of high rock platforms are common in the Inner Hebrides on the west coasts of Skye, Mull and Ardnamurchan, and around Rum. Heights vary between 18 and 51 m OD, and tend to be higher towards the east. Some of the high rock platforms are ice-striated and covered by glacial deposits, suggesting a pre-Late Devensian age.

On the eastern coast of the Trotternish peninsula, Skye, there are several high rock platforms at between 17 and 30 m OD. At Staffin Bay, a 500 m-wide platform at 18 m OD is largely cut in till, but a short distance to the north marine gravels rest on a rock platform at the same level (Ballantyne et al., 1991). On Rum there are high coastal platforms between 18 and 38 m OD. The prominent platform eroded from granite on the south-west coast is partly covered by Late-glacial marine gravels and by till, and on the east coast a sandstone platform has smoothed, glacially striated surfaces. On Ardnamurchan, marine notches have been recognised at about 40 m OD west of Kilchoan and on the north coast near Achateny.

The distribution of the high rock platforms in the west has generally been attributed to advancing mainland ice eroding and removing platforms in those areas nearest to the centre of ice dispersal and, especially, those on east-facing coasts. An alternative suggestion by Sissons (1982) was that for much of the Mid and Late Devensian, the approximate western limit of mainland ice was along a north–south line from Skye to Mull. The high rock platforms would then have formed to the west of this line, in seas that were relatively high because of substantial glacio-isostatic depression of the land. Fluctuations in the position of the ice front would explain why some benches are glacially striated while others are not, and why some are cut in till.

Late-glacial shoreline deposits

The higher raised beach deposits in the area were formed during the retreat of the Main Late Devensian ice sheets, when there was much sand and gravel being transported by meltwater. These gravelly deposits occur up to about 30 m OD. Late-glacial raised beaches at lower levels have been identified on Skye where they are generally restricted to sheltered shorelines and the mouths of certain lochs. On the north side of Staffin Bay, near Digg, Late-glacial gravel deposits rest on a high rock platform at 18 m OD, and in southern Skye terraced raised beach deposits are well developed between Broadford and Kyleakin, where they are quarried (p.174). The altitude of the higher raised beach deposits falls westwards from about 30 m OD at Kyleakin to about 15 m OD at Loch Harport, with a gradient of about 0.4 m.km^{-1}. This shoreline, now tilted by variable isostatic recovery, reflects the maximum sea level in Skye during retreat of the Main Late Devensian ice sheet. High raised beaches (over 15 m OD) do not occur within the areas affected by glaciers of the Loch Lomond Stadial; for example, they are absent from Loch Ainort and from Loch Sligachan (Figure 40a). The contrast between Late-glacial and postglacial beach levels may be observed east of Strollamus, Skye. Late-glacial beach sands and gravels occur at over 23 m OD, close to the eastern margin of the former (Loch Lomond Stadial) Strath Beag Glacier, whereas postglacial beach gravels deposited against the moraines from this glacier are at 7 m OD.

Late-glacial raised beach deposits are present on all the Small Isles. Late-glacial storm-deposited shingle ridges rise to about 30 m OD at Harris on Rum. Storm beaches from this period are also present at similar heights above sea level at Tarbert on Canna (21 m), Camas Mór on Muck (30 m) and Camas Sgiotaig on Eigg (25 m).

Raised, marine-cut benches occur at several levels around Mull and on Morvern. On Mull, the highest (at about 30 m OD) and one of intermediate level (at about 20 m above OD) are covered by deposits attributed to glaciers of the Loch Lomond Stadial and hence predate that event. This relationship can also be demonstrated at the lower ends of Glen Cannel (Loch Bà) and of Glen Forsa, where outwash deposits associated with glaciers from the Loch Lomond Stadial are not cut by, and hence postdate, marine notches that are present in adjacent coastal exposures.

Raised beaches were formed on Arran when ice from the Highlands covered the north of the island. The highest altitude, at about 33 m OD, occurs at Imichar on the west coast. Outwash deposits associated with a major glacier from this period in Glen Iorsa are cut by beaches at about 20 m OD (see below).

Main Rock Platform

The Main Rock Platform is a particularly striking example of a well-developed marine shoreline eroded in rock. The height of the platform varies from over 10 m OD in the sea lochs north-east of Oban and in the innermost lochs north of the Firth of Clyde, to below present sea level in the west of Mull and south of Kintyre (Gray, 1989; Figure 41). The clearest development of the platform is in the south-west Highlands, including eastern Mull and Arran, though farther north, rock platforms on Skye and in other areas have also been correlated with the Main Rock Platform. The platform is generally broad and in places is over 200 m wide. Raised sea stacks may be present on the platform, which is commonly backed by relict sea cliffs.

The age of the Main Rock Platform has been much debated (Walker et al., 1992). It was originally regarded as Holocene in age (e.g. Bailey et al., 1924), but subsequently it was pointed out that this would not allow enough time for cutting away of so much solid rock, and a preglacial or interglacial age was suggested (McCallien, 1937). Sissons (1974) argued that the platform was eroded mainly during the Loch Lomond Stadial under periglacial conditions. The evidence favouring a Loch Lomond Stadial age includes:

- the unglaciated character of the platform in many areas
- the surface defined by platform remnants is tilted at a gradient between that typical of the early Late-glacial shoreline and that of the early Holocene shoreline

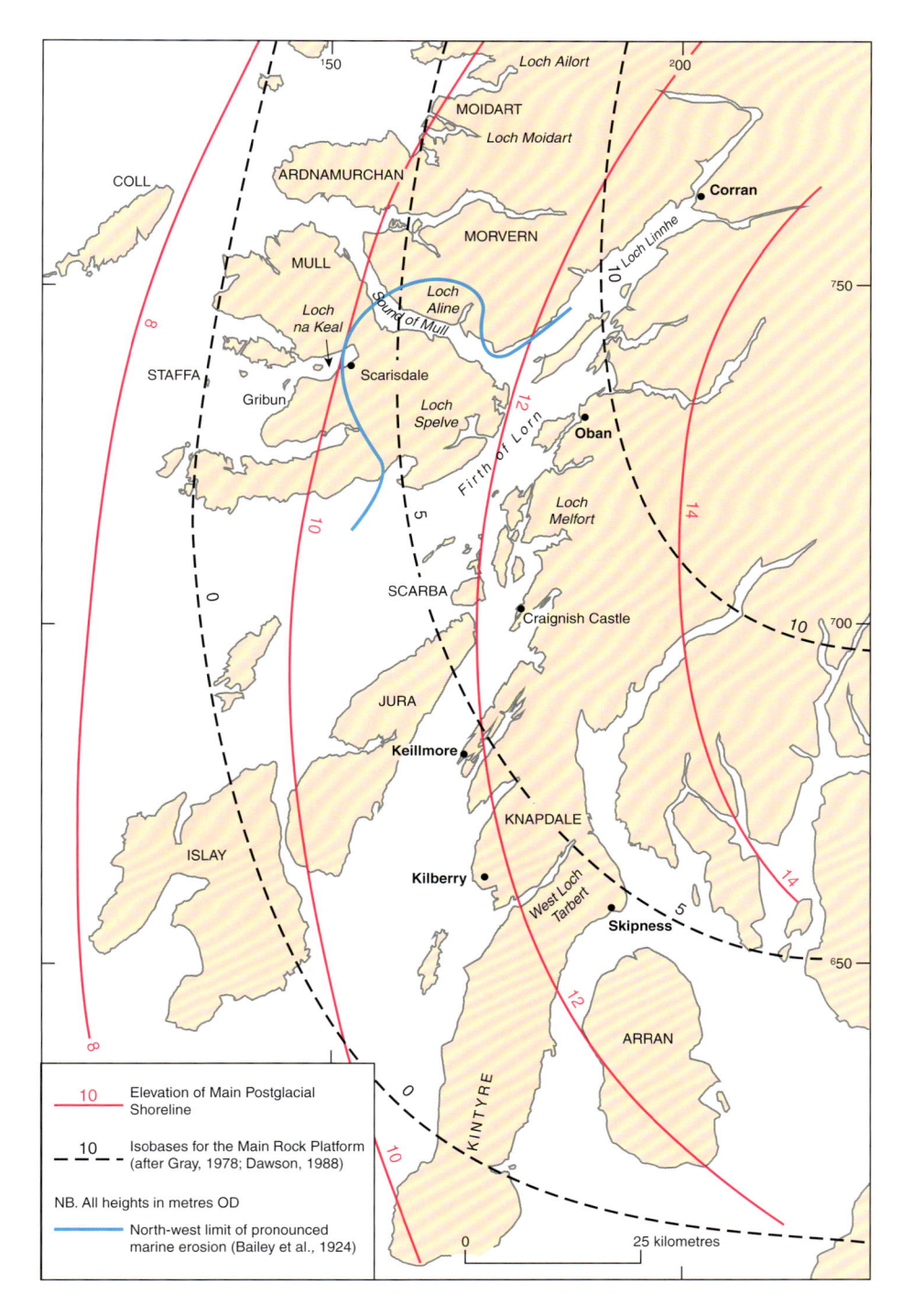

Figure 41 Isobases (in metres OD) for the Main Rock Platform in western Scotland (based on Gray, 1989, fig. 1). The solid blue line represents the north-west limit of pronounced marine erosion, identified by Bailey et al. (1924). See text for details.

- erosion of the platform in some very resistant rocks in sheltered locations where frost riving could have been much more effective than normal coastal erosive processes

Evidence against formation during the Loch Lomond Stadial includes:

- undoubted marine coastal erosional landforms are commonly present, for example sea stacks, caves and undercut cliffs
- the preservation of delicate glacial features, such as striae and p-forms, would hardly be expected to have survived erosion during the Loch Lomond Stadial
- the lack of angular debris in the offshore sediments, which would be expected to be present if there had been rapid erosion to form the platform during the Loch Lomond Stadial
- a few pre-Holocene dates have been obtained that apparently constrain the age of the platform

Remnants of the Main Rock Platform are strikingly preserved on the coast around the Firth of Lorn (Walker et al., 1985). In eastern Mull, Duart Castle is built above the raised platform, at about 8 m OD, and both the platform and raised cliff, which is over 10 m in height in places, are present for several kilometres to the south. South of Loch Don, both the platform and the backing cliffs form prominent features, for example at Port Donain. Remnants of the Main Rock Platform also occur in Ardnamurchan, and in Moidart they extend eastwards up to the limits of the Loch Lomond Readvance (Dawson, 1988). The level of the platform in Ardnamurchan falls from about 2 m OD near Ockle on the north coast, to just above sea level at Ardnamurchan Point, a distance of about 15 km. There are several good examples of the wide (up to 500 m), ice-sculpted platforms backed by raised cliffs, for example south of Rubha Carrach on the north coast where raised sea caves, the 'Glendrian Caves', can also be seen. The Main Rock Platform is extensively developed on Arran but tends to merge with postglacial shoreline features to form a single low raised shoreline around most of the island (Gemmell, 1972, fig. 1; Plate 42), inaccurately termed the '25 foot raised beach' by earlier workers (e.g. Tyrrell, 1928). Much of the coastal road has been built on the platform, for example north of Machrie, between Corrie and Brodick Bay, and west of Loch Ranza. North of Corrie, very large glacial erratics of granite derived from the North Arran Granite Pluton rest on the platform. In north-east Arran, the platform is at about 4.5 m OD, falling to about 2.1 m OD to the south-east of Brodick and at Tormore on the west coast; it drops to 1.4 m near Kildonnan in the south of the island.

Rock platforms below the level of the High Rock Platform are present in eastern Rum and possibly on Canna, but details are sparse. However, on Skye there are well-defined low rock platforms formed by marine erosion, for example for some distance north of Portree Harbour where the platform is at 2 to 7 m OD. The clearest examples, with geos, caves and stacks, occur on Sleat and on the Strathaird peninsula, where the low rock platforms are at 0 to 5 m OD. These occurrences may be the equivalents of the Main Rock Platform, which might be expected to be present at these levels in southern Skye (Ballantyne et al., 1991). The platforms appear to be absent from areas previously occupied by glaciers of the Loch Lomond Stadial, for example Loch Sligachan and Loch Ainort.

Postglacial shorelines

In the early Holocene (Table 17), isostatic rebound of the land initially outpaced the rising sea level, and in general the evidence from Scotland is of falling relative sea level. However, between about 8500 and 6500 years BP, a major rise in sea level occurred which is attributed to the final melting of the ice sheet that covered much of North America (Walker et al., 1992). The widespread Main Postglacial Shoreline dates from this transgression (Figure 41).

Postglacial beaches on Skye are associated with a variety of features, including fossil shingle ridges and raised tombolos. The Main Postglacial Shoreline on Skye is tilted gently towards the north-west (at about 0.07 m.km^{-1}), falling from about 10 m OD at Kyleakin in the south-east to

Plate 42 *Raised shoreline south of Dougarie, Isle of Arran, possibly a combination of the Main Rock Platform and a postglacial raised shoreline (Photograph: Patricia and Angus Macdonald).*

about 6 m OD on the Duirinish peninsula in the north-west. At Staffin Bay, the Main Postglacial Shoreline is marked by well-defined, vegetated shingle ridges at about 6 m OD; to the north, segments of this shoreline are found cut into landslipped material. At Braes, about 12 km south of Portree, a tombolo rising to about 7 m OD extends out to the granite islet of An Aird. At the head of Loch Brittle, there is a rare example of sand dunes that most likely formed during and subsequent to the Main Postglacial Transgression.

Pockets of postglacial raised beach gravels, up to about 6 m OD, are present on most of the Small Isles. On Rum, a thin covering of raised beach sand and gravel at the head of Loch Scresort underlies the made ground at Kinloch Castle. Postglacial raised beaches also occur, for example at the foot of the Late-glacial raised beach deposits at Harris, Guirdil and Kilmory. On northern Eigg, near Talm, a postglacial raised beach is cut into landslip deposits, and is partially covered by later landslip debris, and to the south at the Bay of Laig, Holocene blown sand partly covers a postglacial raised beach.

Remnants of postglacial beaches are common on Mull where they occur as gravel and sand beach terraces, sloping banks and spits, or as notches in morainic deposits, for example in the 'Loch Don Sand Moraine' west of Gorten. At the seaward end of Glen Forsa, an extensive spread of postglacial raised beach deposits at about 9.5 m OD underlies the airstrip and raised shingle ridges, and remnants of terraces up to 11.2 m OD occur around the head of Loch na Keal. More than one beach level may be present, as at Carsaig Bay. At Fishnish, a beach at about 12 m OD, marking the Main Postglacial Shoreline, is accompanied by less well defined beaches at about 7 and 3.3 m OD. Raised beaches are sparsely represented on Ardnamurchan, but are present at Kilchoan Bay where some are regarded as postglacial in age.

Remnants of postglacial raised beach deposits occur at various localities around the coast of Arran (Gemmell, 1972). They are especially well developed where abundant detritus was available from the erosion of moraines; good examples occur at Brodick Bay and north of the mouth of the Sannox River. Postglacial shoreline deposits also occur on the Main Rock Platform (see above; Plate 42). The beach deposits contain shell fragments similar to those of present-day beaches.

OTHER LATE-GLACIAL AND POSTGLACIAL FEATURES

Landslips

As climatic conditions gradually changed, rocks that had been shattered by permafrost conditions over thousands of years thawed out, releasing fragments and providing much surface detritus. Collapse of over-steepened glaciated valley walls was commonplace, and widespread landslipping and rock falls took place, both of which continue to the present day. Throughout the Inner Hebrides, landslips are especially common where thick successions of Paleocene lavas overlie incompetent Mesozoic sedimentary rocks. On Skye, landslips occur at intervals all along the east-facing lava scarp from near the mouth of Loch Sligachan to the northern end of the Trotternish peninsula. In the spectacular examples at Quiraing (Plate 43), many of the slipped masses at lower elevations are ice-sculpted and may predate the Main Late Devensian Glaciation. However, the angular, unmodified outlines of those landslips closest to the present-day scarp, with their steep spires, cliffs and rock pinnacles, could not have survived ice action and must postdate the Main Late Devensian Glaciation. Some of the landslips remain active, as at Flodigarry. In central Skye, there are a few smaller, less spectacular landslips, for example on the north-east side of Glas Bheinn Mhór, where the landslip is superimposed on deposits that might be attributed to the Loch Lomond Stadial. Large landslips, and rock falls occur in eastern Raasay, where a major fall was reported at Hallaig in 1934. Deep fissuring of the thick sandstone at the top of the Middle Jurassic, Bearerraig Sandstone Formation on the summit of Beinn na' Leac can probably be attributed to, or has been accentuated by, landslipping.

Small landslips are present along the granite cliffs of south-west Rum, but the most spectacular examples in the Small Isles occur around the edges of the lava escarpments in northern Eigg. There, a major postglacial landslip on the coast at Talm is likely to be still active, but west of the Bay of Laig, postglacial beach deposits are banked against an older rotational slip. On the north-east coast of Eigg, both ancient scree-covered landslips and more recent landslips can be seen, and there is abundant evidence of recent rockfalls, usually involving the Middle Jurassic Valtos Sandstone Formation and the overlying Paleocene lavas. In the south of the island, at Grulin Cottage, numerous scattered large blocks of columnar pitchstone are part of a rockfall from the south face of An Sgùrr. On Mull, landslips of both pre-Late Devensian and postglacial

Plate 43 *Landslips at Quiraing, Trotternish peninsula, northern Skye. Since the Pleistocene glaciations, large masses of basalt lavas resting upon Jurassic sedimentary rocks have detached from the basalt scarp and slipped towards the sea. Movement of some blocks continues to the present day (P580491).*

or Late-glacial age are recognised at Gribun, and modern examples occur in The Wilderness on the Ardmeanach peninsula. Landslips are prominent south of Loch Teacuis and south of Loch Arienas in Morvern, where Paleocene lavas overlie a thin Mesozoic succession. The well-known Fallen Rocks on Arran are the site of landslips involving Old Red Sandstone conglomerate and farther north, at An Scriadan east of the Cock of Arran, large blocks of Permian sandstone collapsed less than 300 years ago, with a noise that was heard over a wide area.

Talus and other upland features

The present-day environment on the mountain tops of the Inner Hebrides and Arran is dominated by copious rain and high winds. The meagre vegetation is removed by the wind and by grazing animals, laying bare to erosion the silt and fine-sand in the poorly developed soil. This leaves relatively coarse lag deposits and leads to the formation of low, turf-banked terracettes which are very noticeable on the higher hills. Talus (scree) deposits, comprising unmodified rockfall accumulations, are well represented in the mountainous areas and along sea cliffs. Fine examples are found within the Cuillin of Skye, where the coarse, mobile debris forms 'stone chutes', the best known of which is the Great Stone Chute at the head of Coire Lagan. Elsewhere on Skye, the light coloration of the Red Hills (Plate 26) arises in large part from their mantle of granite debris and scree, of which the coarse, bouldery scree on Glamaig and Beinn na Caillich are of particular note. Scree accumulations are common in the mountains and on the coasts of Rum, Mull and Arran. It is not unusual to be able to distinguish older scree, stabilised by soil and vegetation, from younger, mobile scree, for example in the scree banked against the microgranite cliffs of south-west Rum.

Coastal deposits

Boulders and shingle make up a large proportion of the present-day coastal beach deposits, for example on the shoreface at the base of the lava cliffs of Skye and Mull. Sandy beaches are commonly made of material washed out of glacial deposits, as at Brodick Bay, Arran, and sandy detritus that has been derived from the weathering of nearby bedrock comprises other beaches, for example Torridonian sandstone at Kilmory on Rum, New Red Sandstone at Kildonnan and Blackwaterfoot on Arran and Middle Jurassic Valtos Sandstone at the Bay of Laig, Eigg. At Camas Sgiotaig, in the Bay of Laig, the 'Singing Sands' emit shrill squeaks when the dry sand is crushed underfoot. Beaches formed largely of shelly debris also occur, for example at Calgary Bay and Port Langamull in north-west Mull, near Canna Harbour and at Gaillinach on Muck. At Duntulm, in northern Skye, olivine from the weathering of nearby dolerite sills is a prominent constituent of the beach sand. At a number of localities there are gleaming white beaches composed of fragments of the calcareous algae *Lithothamnion calcareum*. On Skye, examples of these so-called 'coral sands' are found north of Rubha na Gairbe on the east shore of Loch Dunvegan and north of Ord in Sleat. Coastal sand dunes back a number of the bays, for example at Drumadoon Point on Arran, Kilmory on Rum, Laig on Eigg and at the east end of Staffin Bay on Skye, where the adjoining dolerite cliffs are polished by sand blasting.

Peat

High atmospheric moisture levels during the fifth millenium BP led to the decline of woodlands, as sites became waterlogged, and promoted the growth of blanket peat, which covers many hillsides to the present day. Extensive spreads of peat occur on the lava plateaux of Skye and Mull, and in broad, ill-drained glens, for example Kinloch Glen on Rum. Peat was formerly an important source of fuel (p.175). The pollen stratigraphy of peat provides a valuable record of climate change during the Quaternary (e.g. Lowe and Walker, 1986, 1991).

Alluvium

Most streams in the mountainous areas have gravelly beds, and deposits of fine-grained alluvium are not extensive, partly due to the absence of major rivers. Small, elongate patches of alluvium occur next to many of the streams and rivers, and some of the broader valleys have more extensive deposits. Relatively large areas of alluvium are gradually encroaching on certain lochs, for example Loch Bà on Mull, and at Loch Cill Chroisd in Strath, Skye, where alluvium, lacustrine mud and vegetation are gradually filling the loch. Also on Skye, there is an extensive tract of alluvium on the site of a former shallow loch at Eilean Chaluim Chille, north of Uig, which was drained in 1947 to give a large area of new pasture. Terraces are cut into the alluvium at more than one level, for example at the bottom of Glen Iorsa, Arran; farther up the glen, alluvium covers the valley floor for a distance of over 5 km above Loch Iorsa. Small placer deposits of heavy minerals occur in some rivers, with chromite-rich examples found on Rum in the Abhainn Rangail, Glen Harris.

Lacustrine deposits of diatomaceous earth are found in many places on the Trotternish peninsula, northern Skye, where they were worked in the past (p.176). The diatoms are the skeletal remains of unicellular plants that were laid down in freshwater lakes, probably under warmer climatic conditions than at present. Extensive deposits of diatoms mixed with clay and peaty material occur at Loch Cuithir, about 6 km north-north-west of The Storr, where they rest on moraine, and hence postdate the last glaciation, but are overlain by peat (Anderson and Dunham, 1966). Diatomaceous earth is also known from near Knock on Mull and has been recorded on Eigg (Haldane et al., 1940).

fourteen
Economic geology

The exploitation pattern of the limited economic resources of the district has undergone notable changes since the first edition of this book. In place of a multiplicity of small-scale operations, the requirements for building materials and aggregate are now met by a few large concerns, and much of what was once obtained locally is now brought into the district, sometimes from considerable distances. Fuel requirements are met by imported coal, oil and (bottled) gas. Electricity is supplied to Mull, Skye and Arran by cable links to the mainland and the only local contribution of any significance comes from the hydroelectric scheme at Loch Fada on Skye.

Iron

Carbonate-rich ironstone was obtained from the Lower Jurassic Raasay Ironstone Formation on Raasay, where the reserves were estimated at about 10 million tonnes. The chamosite-oolite ore occurs in beds up to 2.5 m thick between Churchtown Bay and Hallaig (Lee, 1920). Both underground and opencast workings were employed and over 125 000 tonnes of ore were obtained during the active life of the operations, between 1916 and 1919 (Draper and Draper, 1990). Remains of the workings are visible at several places in the south of Raasay. The Raasay Ironstone Formation also occurs on Skye, Ardnamurchan, Mull and elsewhere in the Hebrides but the beds are thin, generally of low quality and have not been exploited.

Clay-ironstone of Carboniferous age was obtained from sedimentary rocks near Corrie, Arran, and is reputed to have been exported to the mainland. Elsewhere on Arran, there are signs of workings and smelting involving bog iron-ore. The magnetite-bearing skarns close to the margin of the Beinn an Dubhaich Granite on Skye were examined during the Second World War (1939–45) but the deposits, which are associated with small amounts of copper mineralisation, were not exploited.

Chromium, magnesium and precious metals

Substantial amounts of chromite and forsteritic olivine are present in shallow water marine sands off southern Rum (Figure 42) and in Loch Scavaig, Skye (Basham et al., 1989). Although the chromite tends to be iron-rich compared with ores in current commercial use, the abundant, associated magnesian olivine may be of value as an abrasive. The deposits in Loch Scavaig are mixed with glacial clays, which would make exploitation difficult, but those off Rum could easily be recovered. The sources of the chromite and olivine are the layered, olivine-rich ultrabasic rocks of the Rum and Skye central complexes, where chromite occurs in seams and layers generally less than 1 cm thick (Plate 29). It is commonly associated with sulphide minerals and the ultrabasic rocks on Rum, Skye and Mull show generally elevated levels of the platinum-group elements (Butcher et al., 1999; Pirrie et al., 2000). On Rum, electrum (an Au-Ag alloy; Dunham and Wilkinson, 1985), and platinum-group minerals are associated with the chromite seams (Butcher et al., 1999). The platinum-group minerals are also found as a minor constituent of small, chromite-rich placer deposits in streams on Rum. These primary and alluvial deposits are too small to be of direct commercial interest and, furthermore, they occur in environmentally sensitive localities; nonetheless, their paragenesis has proved to be of considerable interest (e.g. Power et al., 2003).

Vein deposits

Good quality baryte has been obtained from a system of north-trending veins in Glen Sannox on Arran. The veins are up to 3 m wide and are hosted by Lower Old Red Sandstone strata near the margin of the Paleocene North Arran Granite Pluton, although it is unlikely that there is any genetic connection with the granite. Imprecise K-Ar dates suggest a Triassic age (Moore, 1979). Several hundred tonnes of ore were raised each year for over 20 years up to 1862 when working ceased. Extraction resumed in 1918, with annual production sometimes exceeding 8000 tonnes of barytes before the mine was abandoned in 1944.

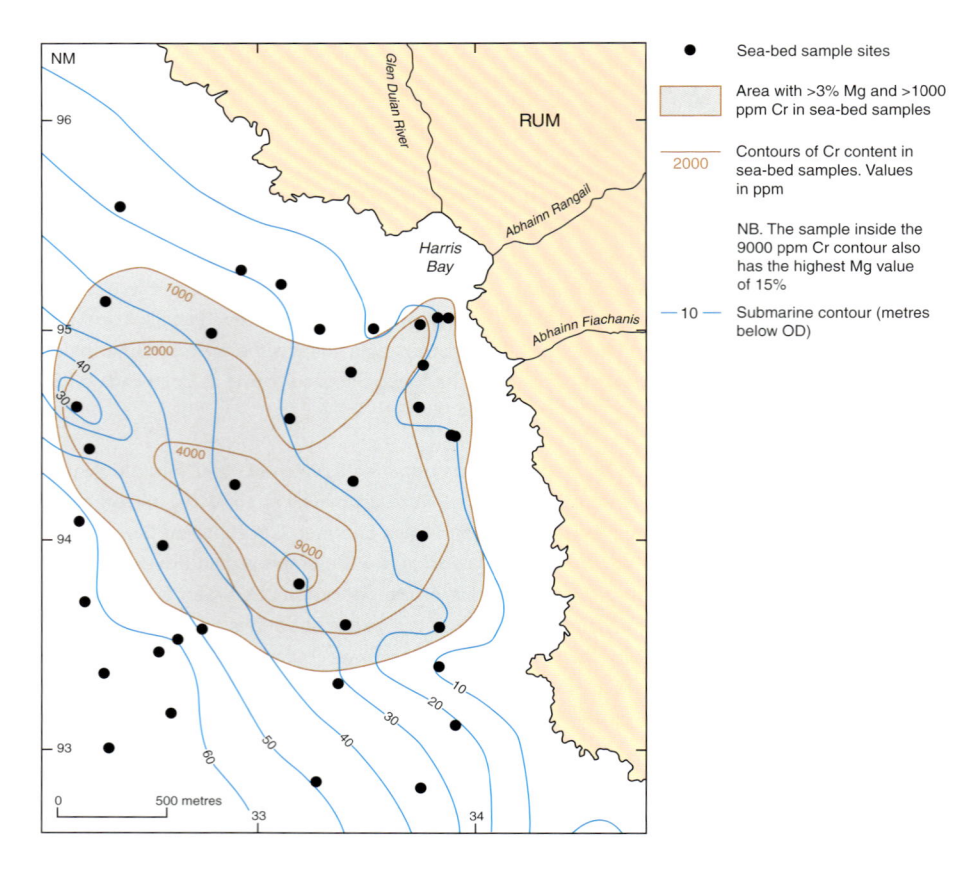

Figure 42 *Chromium and magnesium distribution in the surficial marine sediments off Rum (based on Emeleus, 1997, fig. 64).*

Analyses of stream sediment samples close to the southern margin of the North Arran Granite Pluton show enhanced levels of tin, tungsten, uranium and thorium, and cassiterite has been recovered from sediments in the Glen Sannox River suggesting the presence of mineral veins in the granite or its host rock.

Gemstones Small, impure sapphires have been obtained from hornfelses adjoining a major ring-intrusion at Glebe Hill, Kilchoan, Ardnamurchan (Richey and Thomas, 1930; p.122). Sapphire has also been recovered from tholeiitic basalt sills at Loch Scridain and Carsaig, Mull, where the aluminous, sapphire-bearing accidental inclusions in the sills were probably derived from Moine pelitic rocks (Dempster et al., 1999; Chapter 8). However, the most significant find was made in 1992 at Loch Roag, Isle of Lewis, where large sapphires, some of gem quality and accompanied by corundum and zircon, were recovered from a lamprophye dyke (p.83). An uncut, blue sapphire from this source was sold for several thousand pounds, the highest price paid for a gemstone originating in the British Isles. Bloodstone from lavas in north-west Rum has been used for ornamental purposes (and for weapons by early Man) and small, banded agates have been obtained from the same source.

Building stone Up to the beginning of the 20th century, local stone was employed extensively in houses and public buildings. Many examples remain, commonly roofed with Dalradian slate from Ballachulish and Easdale in Argyll. With improved transport, brick and stone from the Scottish mainland and slate from Wales came to dominate the market, giving way in recent years to concrete blocks, which are brought in from as far afield as Antrim (where the aggregate used is local basalt), and to ceramic tiles for roofing.

The only major granite quarries are those in the attractive pink porphyritic Caledonian granite of the Ross of Mull (Faithfull, 1995). The granite yields large blocks and takes a fine polish. It was quarried extensively in the 19th century, and sporadically since; quarried stone was generally sent for polishing to Aberdeen, Glasgow or Shap in Cumbria. It has been used locally in Iona Cathedral, together with flagstones of Moine strata from the Ross of Mull, and in many buildings on the Ross of Mull. Elsewhere, it has been used widely, for example in the construction of the Liverpool docks, in several Scottish lighthouses including Skerryvore and Ardnamurchan Point, and in London it may be found in Blackfriars and Westminster bridges and the foundations of the Albert Memorial; it was also exported to New York and other places in North America. Production largely ceased before the First World War (1914–18) but, since 1985, previously quarried blocks have been removed from the Tormore quarries and some new stone has also been won.

Despite an abundance of Paleocene granites, the only granitic rock of this age to have been quarried extensively is the distinctive blue-grey riebeckite-bearing microgranite of Ailsa Craig. This was used for paving sets, in Glasgow and elsewhere, and as an ornamental stone, as in the floor of the Scottish National War Memorial, Edinburgh Castle. However, it is best-known for the curling stones that were manufactured in their thousands towards the end of the 19th century. The roughed-out stones were finished on the mainland, with many exported to Canada. The running surfaces for curling stones currently used in all World Championship and Olympic events are from Ailsa Craig, and the polished microgranite continues to find ornamental uses (Nichol, 2001).

Very large amounts of excellent building stone have been obtained from the Carboniferous and New Red Sandstone strata of Arran. White freestone from the Lower Carboniferous at Corrie was used extensively in the late 18th and 19th centuries, for example in the construction of the Crinan Canal, but the most extensive exploitation was of the red-brown rocks of the New Red Sandstone. In addition to local use, the Lower Permian Corrie Sandstone was employed in many works outside Arran, for example at Troon Harbour. In southern Arran, the red sandstones on the south side of Lamlash Bay, at Cordon and Monamore Mill, were extensively quarried. Triassic sandstones in Morvern and elsewhere have been quarried for local building use, and around Broadford, Skye, the local Lower Jurassic rocks provided freestone suitable for building. The quartz porphyry and dolerite sills in the south of Arran have also been used locally.

Aggregate

Small roadside quarries are still used as sources of aggregate on several of the islands, particularly in the Small Isles where shipment facilities were poor until recently. Basalt rubble is applied to the tracks on Muck, Eigg and Canna, although in the past marble chips, probably obtained from Skye, were used to surface the tarred roads on Eigg and Muck. On Skye, the large, well-equipped quarry in Torridonian sandstone at Peinchorran, near Sconser, provides road aggregate for the island, although hornfelsed lava from a quarry near Sligachan and dolerite from a Paleocene sill at Invertote have in the recent past provided material for major road rebuilding programmes. On Mull, altered basalt from within the limit of pneumatolysis around the central complex is used as roadstone and is obtained from a quarry near Salen. Aggregate has also been obtained from a basalt quarry east of Bunessan and from a quarry in a granite ring-dyke south of Salen. On Arran, roadstone has been imported since the 1920s (Tyrrell, 1928). However, a quarry in felsite at Bennecarrigan in south-west Arran produces aggregate for roads and also some building stone. At Dereneneach, on the western margin of the Central Arran Ring-complex, a small quarry in granite supplies aggregate for local use. No large coastal quarries are projected for the district, although the granite of western Rum was identified on geological grounds as a possible site. However, the exposed coast and the National Nature Reserve status of the island make its exploitation improbable.

Sand and gravel

Large amounts of sand and gravel are extracted from quarries in thick (over 10 m) gravel deposits on marine raised beaches at Allt Anavig and Lusa, west of Kyleakin, Skye, and local use has been made of the gravel deposits at the mouth of the River Brittle. On Mull, the extensive moraines north of Loch Don are a source of sand and gravel. Silica-rich beach sands are used locally in Ardnamurchan, Skye and Mull as sharp sand in cement and concrete; on Arran, sand from glacial outwash deposits near Brodick Castle was exported for building purposes.

Silica sandstone and silica sand

At Loch Aline, on Morvern, a deposit of white sandstone of exceptionally high purity (99.7 per cent quartz and generally 0.01 per cent Fe_2O_3 or less) is worked as a source of sand for glass manufacture (Plate 44). The sand is mined from a bed of Cretaceous sandstone that averages 12 m in thickness (of which approximately 5 m are worked), underlying the Paleocene lavas. Although the deposit has been known since the end of the 19th century, the mine was only opened in the 1940s when other European sources of pure sand were cut off during the war. Currently, production is about 120 000 tonnes per year. The deposit is the only United Kingdom source of sand suitable for high-grade domestic, scientific and commercial glass. The sand is crushed at the mine and all is shipped out. About 50 per cent of the production is currently exported, principally to the Republic of Ireland and to Scandinavia where cheap hydroelectricity makes it economic to mix the sand with graphite in the manufacture of 'carborundum' (silicon carbide) abrasive.

Limestone and dolostone

The principal carbonate rocks of the district are in the Cambro-Ordovician Durness Group on Skye, the Carboniferous succession of Arran, and the Mesozoic sequences throughout the Inner Hebrides. Durness Group dolostones are recrystallised to marble adjacent to the Paleocene granites of the Eastern Red Hills Centre and the Broadford Gabbro. The marble, which includes brucite- and serpentine-rich varieties, was formerly quarried extensively in the roof zone of the

Plate 44 *Quartz sand stockpiled at the Loch Aline glass-sand mine, Morvern (P580492).*

Beinn an Dubhaich Granite for ornamental stone slabs and as chips for *terrazzo* flooring. A tramway was constructed to take products from Strath to Broadford, from where they were shipped. The quarries are long abandoned but nearby, at Torrin, the Skye Marble Company produces up to 20 000 tonnes per year of crushed chips, used principally for harling and for roadstone.

The Lower Carboniferous Corrie Limestone in eastern Arran was extensively quarried and mined for lime. Calcining for agricultural lime has been the principal use for the Mesozoic limestone and the ruins of small lime kilns occur near many of the main outcrops, for example where Lias limestone crops out at Broadford, Skye and at Mingary and Swordle on Ardnamurchan. The Middle Jurassic limestones of the Duntulm and Kilmaluag formations of Skye and Eigg were calcined locally. Marble was obtained from a lens within the Lewisian of Iona (see Johnstone and Mykura, 1989). Shelly beach sands and the so-called 'coral sands' with a high calcium carbonate content occur in many places on the coast, but have been little used apart from local exploitation for agricultural lime.

Coal

In the 18th century, there was a small working for coal in the Carboniferous rocks south of the Cock of Arran, where seams about 1 m thick were reported (Tyrrell, 1928). Coal Measures crop out at Inninmore, Morvern, but do not contain any workable coal seams in the exposed 60 m of section. In northern Skye, lignitic coals occur in thin seams in the Great Estuarine Group, where they pass laterally into oil shale; locally, this poor quality coal was used as fuel. Thin seams and streaks of poor-quality coal between Paleocene lava flows occur sparsely throughout the district, for example close to the base of the lava sequence near Portree, Skye, in tuffs to the east of Fionchra, Rum, and at Carsaig on the south coast of the Ross of Mull. Lignites associated with the lavas of Ardtun, Mull, were investigated for the Duke of Argyll, but were found to have high ash and moisture contents. Similar lignitic coal was reported in a seam about 60 cm thick at Ardslignish, Ardnamurchan.

Graphite

An unworked seam of graphite has been reported from Mesozoic strata on the north side of Loch Sligachan, Skye (Peach et al., 1910). Graphite also occurs in small blocks in the Loch Scridain sills, Mull, where attempts were made to extract it in the 19th century. Jet has been reported from the Jurassic sedimentary rocks of Holm, north Skye and from sedimentary rocks underlying basalt lavas on Ben Hiant, Ardnamurchan.

Oil

Although there is considerable interest in the oil-bearing potential of the Mesozoic and Paleocene sedimentary sequences off the west and north-west coast of Scotland, onshore possibilities are limited. However, there has been some exploratory drilling on Skye. In the north of Skye and the south of Raasay, oil-shale up to 3 m thick occurs in the Great Estuarine Group. The oil potential of these rocks has been compared favourably with that of the Carboniferous oil-shales in the Lothians, with up to 54 litres of crude oil per tonne of shale obtained in tests (Anderson and Dunham, 1966).

Peat

This was the principal fuel throughout the Highlands and Islands until well into the 20th century. It was gradually replaced by coal from the Midland Valley of Scotland, and more recently by oil and bottled gas. Small workings continue to the present, notably in northern Skye where consideable areas of blanket bog peat exist, up to 3 m thick in places. Nowhere has machinery been employed to win peat commercially.

Water power

A small hydroelectric station on the Bearreraig River, Skye, makes use of the fall at sea cliffs near the outlet from Loch Leathan, north of Portree. The installation contributes power to the

National Grid. There are local hydroelectric schemes, for example on Rum, which has supplied Kinloch Castle since the early 20th century, and on Eigg.

Water supply

The small communities scattered throughout the region rely extensively on surface water for supplies, which renders them vulnerable to drought and, less commonly, to frost. Water obtained from burns and wells is commonly piped short distances to isolated houses. The larger settlements are generally supplied from small reservoirs and from lochs in nearby hills. The largest groundwater system is on Arran where several boreholes into the Permian sandstones in the Brodick–Lamlash Bay area and in the Shiskine valley serve the southern half of the island.

Diatomite

Deposits of diatomite are common in the small lochs and hollows amongst the landslipped lavas resting on sills and Mesozoic strata in northern Trotternish, Skye (Haldane et al., 1940; Anderson and Dunham, 1966). Many of the occurrences have been worked on a small scale, but the Loch Cuithir deposit, at over 6 m, was the thickest, and was worked up to 1914; the diatomite was transported to the coast at Invertote by a 5 km-long tramway. It was formerly used in the manufacture of dynamite, but later the principal uses were as a filter and as an insulator. Thin diatomite deposits near Loch Bà, Mull have been used locally for whitewash.

Geological hazards

The principal geological hazards in the district are rockfalls and landslips along the edges of the Paleocene lava outcrops. Landslipping has long been a problem in parts of northern Skye, where the roads are affected and movement of buildings has occurred as, for example at Flodigarry. These movements are but the latest manifestation of landslipping which has occurred in this area, on Raasay, and elsewhere, since the Late Devensian glaciation. Rockfalls are a particular problem along the edges of the basalt scarps, especially after frost and rain. On Eigg, a house was demolished at Cleadale and there are extensive, recent falls on the coast north of Kildonan.

References

British Geological Survey holds most of the references listed below, and copies may be obtained via the library service subject to copyright legislation (contact libuser@bgs.ac.uk for details). The library catalogue is available at: http://geolib.bgs.ac.uk

N.B. For a list of additional earlier references the reader is referred to Richey (1961).

ALMOND, D C. 1964. Metamorphism of Tertiary lavas in Strathaird, Skye. *Transactions of the Royal Society of Edinburgh*, Vol. 65, 413–434.

ALLWRIGHT, E A. 1980. The structure and petrology of the volcanic rocks of Eigg, Muck and Canna, NW Scotland. Unpublished MSc thesis, University of Durham (2 volumes).

AMIRI-GARROUSSI, K. 1982. Age of the Camas Malag Formation, Skye. *Scottish Journal of Geology*, Vol. 18, 247–249.

ANDERSON, F W, and DUNHAM, K C. 1966. The geology of northern Skye. *Memoir of the Geological Survey of Great Britain*, Sheet 80 and parts of 81, 90 & 91 (Scotland).

ANDREWS, J E. 1985. The sedimentary facies of a late Bathonian regressive episode: the Kilmaluag and Skudiburgh Formations of the Great Estuarine Group, Inner Hebrides, Scotland. *Journal of the Geological Society of London*, Vol. 142, 1119–1137.

ANDREWS, J E, and HUDSON, J D. 1984. The first Jurassic dinosaur footprint from Scotland. *Scottish Journal of Geology*, Vol. 20, 129–134.

ANDREWS, J E, and WALTON, W. 1990. Depositional environments within Middle Jurassic oyster-dominated lagoons: an integrated litho-, bio-, and palynofacies study of the Duntulm Formation (Great Estuarine Group), Inner Hebrides. *Transactions of the Royal Society of Edinburgh: Earth Sciences*, Vol. 81, 1–22.

BAILEY, E B. 1914. The Sgurr of Eigg. *Geological Magazine*, Vol. 51, 296–305.

BAILEY, E B. 1924. The desert shores of the Chalk Seas. *Geological Magazine*, Vol. 61, 102–116.

BAILEY, E B. 1939. Caledonian tectonics and Moine metamorphism in Skye. *Bulletin of the Geological Survey of Great Britain*, No. 2, 46–62.

BAILEY, E B. 1955. Moine tectonics and metamorphism in Skye. *Transactions of the Edinburgh Geological Society*, Vol. 16, 93–166.

BAILEY, E B. 1962. Early Tertiary fold movements in Mull. *Geological Magazine*, Vol. 99, 478–479.

BAILEY, E B, CLOUGH, C T, WRIGHT, W B, RICHEY, J E, and WILSON, G V. 1924. Tertiary and Post-Tertiary geology of Mull, Loch Aline, and Oban. *Memoir of the Geological Survey of Great Britain*, Sheet 44 (Scotland).

BALL, D F. 1964. Deep weathering profile on the Isle of Rhum, Inverness-shire. *Scottish Geographical Magazine*, Vol. 80, 22–27.

BALLANTYNE, C K. 1990. The Late Quaternary glacial history of the Trotternish Escarpment, Isle of Skye, and its implications for ice-sheet reconstruction. *Proceedings of the Geologists' Association*, Vol. 101, 171–186.

BALLANTYNE, C K. 1997. Periglacial trimlines in the Scottish Highlands. *Quaternary International*, Vols. 38/39, 119–136.

BALLANTYNE, C K. 1999. Maximum altitude of Late Devensian glaciation on the Isle of Mull and Isle of Jura. *Scottish Journal of Geology*, Vol. 35, 97–106.

BALLANTYNE, C K, and HARRIS, C. 1994. *The periglaciation of Great Britain*. (Cambridge: Cambridge University Press.)

BALLANTYNE, C K, and WAIN-HOBSON, T. 1980. The Loch Lomond Advance on the island of Rhum. *Scottish Journal of Geology*, Vol. 16, 1–10.

BALLANTYNE, C K, BENN, D I, LOWE, J J, and WALKER, M J C. 1991. *The Quaternary of the Isle of Skye: Field Guide*. (Cambridge: Quaternary Research Association.)

BASHAM, I R, BEDDOE-STEPHENS, B, and MACDONALD, A. 1989. Mineralogical assessment of submarine heavy mineral sands, southern Rhum. *British Geological Survey Technical Report,* WG/89/26.

BELL, B R. 1983. Significance of ferrodioritic liquids in magma mixing processes. *Nature*, Vol. 306, 323–327.

BELL, B R. 1985. The pyroclastic rocks and the rhyolitic lavas of the Eastern Red Hills district, Isle of Skye. *Scottish Journal of Geology*, Vol. 21, 57–70.

BELL, B R, and CLAYDON, R V. 1992. The cumulus and post-cumulus evolution of chrome-spinels in ultrabasic layered intrusions: evidence from the Cuillin Igneous Complex, Isle of Skye, Scotland. *Contributions to Mineralogy and Petrology*, Vol. 112, 242–253.

BELL, B R, and HARRIS, J W. 1986. *An excursion guide to the geology of the Isle of Skye*. (Glasgow: Geological Society of Glasgow.)

BELL, B R, and JOLLEY, D W. 1997. Application of palynological data to the chronology of the Palaeogene lava fields of the British Province: implications for magmatic stratigraphy. *Journal of the Geological Society*, Vol. 154, 700–708.

BELL, B R, and PANKHURST, R J. 1993. Sr-isotope variations in a composite sill: crystal-liquid processes and the origin of the Skye granites. *Journal of the Geological Society of London,* Vol. 150, 121–124.

BELL, B R, and WILLIAMSON, I T. 1994. Picritic basalts from the Palaeocene lava field of west-central Skye: evidence for parental magma compositions. *Mineralogical Magazine*, Vol. 58, 347–356.

BELL, B R, and WILLIAMSON, I T. 2002. Tertiary igneous activity. 371–407 in *The Geology of Scotland*. Fourth edition. TREWIN, N H (editor). (London: The Geological Society)

BELL, B R, CLAYDON, R V, and ROGERS, G. 1994. The petrology and geochemistry of cone-sheets from the Cuillins Igneous Complex, Isle of Skye: evidence for combined assimilation and

BELL, B R, WILLIAMSON, I T, HEAD, F E, and JOLLEY, D W. 1996. On the origin of a reddened interflow bed within the Palaeocene lava field of north Skye. *Scottish Journal of Geology*, Vol. 32, 117–126.

BELL, J D. 1966. Granites and associated rocks of the eastern part of the Western Redhills Complex, Isle of Skye. *Transactions of the Royal Society of Edinburgh*, Vol. 66, 307–343.

BELL, J D. 1976. The Tertiary intrusive complex on the Isle of Skye. *Proceedings of the Geologists' Association*, Vol. 87, 247–271.

BENN, D I, and EVANS, D J A. 1993. Glaciomarine deltaic deposition and ice marginal tectonics: the 'Loch Don Sand Moraine', Isle of Mull, Scotland. *Journal of Quaternary Science*, Vol. 8, 279–291.

BENN, D I, LOWE, J J, and WALKER, M J C. 1992. Glacier response to climatic change during the

Loch Lomond Stadial and early Flandrian: geomorphological and palynological evidence from the Isle of Skye, Scotland. *Journal of Quaternary Science*, Vol. 7, 125–144.

BENTON, M J, MARTILL, D M, and TAYLOR, M A. 1995. The first Lower Jurassic dinosaur from Scotland: limb bone of a ceratosaur theropod from Skye. *Scottish Journal of Geology*, Vol. 95, 177–182.

BERGGREN, W A, KENT, D V, SWISHER, C C, and AUBREY, M P. 1995. A revised Cenozoic geochronology and chronostratigraphy. 129–212 in *Geochronology, time scales and global stratigraphic correlations*. BERGGREN, W A, KENT D V, AUBRY, M P, and HARDENBOL, J (editors). *Society of Economic Palaeontologists and Mineralogists Special Publication*, No. 54.

BEVAN, J C, and HUTCHISON, R. 1984. Layering in the Gars-bheinn ultrabasic sill, Isle of Skye: a new interpretation, and its implications. *Scottish Journal of Geology*, Vol. 20, 329–342.

BINNS, P E, MCQUILLIN, R, and KENOLTY, N. 1974. The geology of the Sea of the Hebrides. *Report of the Institute of Geological Sciences*, No. 73/14.

BISHOP, A N, and ABBOTT, G D. 1993. The interrelationship of biological marker maturity parameters and molecular yields during contact metamorphism. *Geochimica et Cosmochimica Acta*, Vol. 57, 3661–3668.

BLAKE, D H, ELWELL, R W D, GIBSON, I L, SKELHORN, R R, and WALKER, G P L. 1965. Some relationships resulting from the intimate association of acid and basic magmas. *Quarterly Journal of the Geological Society of London*, Vol. 121, 31–49.

BOTT, M H P, and TANTRIGODA, D A. 1987. Interpretation of the gravity and magnetic anomalies over the Mull Tertiary intrusive complex, NW Scotland. *Journal of the Geological Society of London*, Vol. 144, 17–28.

BOTT, M H P, and TUSON, J. 1973. Deep structure beneath the Tertiary volcanic regions of Skye, Mull and Ardnamurchan, north-west Scotland. *Nature (Physical Sciences)*, Vol. 242, 114–116.

BOULTER, M C, and KVACEK, Z. 1989. The Palaeocene flora of the Isle of Mull. *Palaeontological Association Special Papers in Palaeontology*, No. 42.

BOULTON, G S, PEACOCK, J D, and SUTHERLAND, D. 2002. Quaternary. 409–430 in *The Geology of Scotland*. Fourth edition. TREWIN, N H (editor). (London: The Geological Society.)

BOWEN, N L. 1928. *The evolution of the igneous rocks* (Princeton: Princeton University Press.)

BRALEY, S. 1990. Sedimentology, palaeontology and stratigraphy of the Cretaceous rocks in north-west Scotland. Unpublished PhD thesis, Polytechnic Southwest Plymouth.

BREARLEY, A J. 1986. An electron optical study of muscovite breakdown in pelitic xenoliths during pyrometamorphism. *Mineralogical Magazine*, Vol. 50, 385–397.

BRITISH GEOLOGICAL SURVEY   see pp.200–201.

BROWN, D J. 2003. The nature and origin of breccias associated with central complexes and lava fields of the British Tertiary Igneous Province. Unpublished PhD thesis, University of Glasgow.

BROWN, G M. 1956. The layered ultrabasic rocks of Rhum, Inner Hebrides. *Philosophical Transactions of the Royal Society of London*, Series B, Vol. 240, 1–53.

BROWN, G M. 1963. Melting relations of Tertiary granitic rocks in Skye and Rhum. *Mineralogical Magazine*, Vol. 33, 533–562.

Brown, G M. 1969. *The Tertiary igneous geology of the Isle of Skye.* (with contributions by Drever, H I, Dunham, K C, Thompson, R N, and Weedon, D S). Geologists' Association Guides, No. 13. (Colchester: Benham & Co.)

Butcher, A R. 1985. Channelled metasomatism in Rhum layered cumulates — evidence from late-stage veins. *Geological Magazine,* Vol. 122, 503–518.

Butcher, A R, Young, I M, and Faithfull, J W. 1985. Finger structures in the Rhum Complex. *Geological Magazine,* Vol. 122, 491–502.

Butcher, A R, Pirrie, D, Prichard, H M, and Fisher, P. 1999. Platinum-group mineralization in the Rum layered intrusion, Scottish Hebrides, UK. *Journal of the Geological Society of London,* Vol. 156, 213–216.

Butler, R W H, and Hutton, D W H. 1994. Basin structure and Tertiary magmatism on Skye. *Journal of the Geological Society of London,* Vol. 151, 931–944.

Cameron, I B, and Stephenson, D. 1985. *British Regional Geology: the Midland Valley of Scotland.* Third edition. (London: HMSO for the British Geological Survey.)

Cann, J R. 1965. The metamorphism of amygdales at 'S Airde Beinn, northern Mull. *Mineralogical Magazine,* Vol. 33, 533–562.

Carter, S R, Evensen, N M, Hamilton, P J, and O'Nions, R K. 1978. Neodymium and strontium isotopic evidence for crustal contamination of continental volcanics. *Science,* Vol. 202, 743–747.

Chambers, L M. 2000. Age and duration of the British Tertiary Igneous Province: implications for the development of the ancestral Iceland plume. Unpublished PhD thesis, University of Edinburgh.

Chambers, L M, and Fitton, J G. 2000. Geochemical transitions in the ancestral Iceland plume: evidence from the Isle of Mull Tertiary volcano, Scotland. *Journal of the Geological Society of London,* Vol. 157, 261–263.

Chambers, L M, and Pringle, M S. 2001. Age and duration of activity at the Isle of Mull Tertiary igneous centre, Scotland, and confirmation of the existence of subchrons during Anomaly 26r. *Earth and Planetary Science Letters,* Vol. 193, 333–345.

Chambers, L M, Pringle, M S, and Parrish, R R. 2005. Rapid formation of the Small Isles Tertiary centre constrained by precise $^{40}AR/^{39}AR$ and U-Pb ages. *Lithos,* Vol. 79, 367–384.

Cheeney, R F. 1962. Early Tertiary fold movements in Mull. *Geological Magazine,* Vol. 99, 227–232.

Clark, N D L. 2001. A thyreophoran dinosaur from the early Bajocian (Middle Jurassic) of the Isle of Skye, Scotland. *Scottish Journal of Geology,* Vol. 37, 19–26.

Clark, N D L, Boyd, J D, Dixon, R J, and Ross, D A. 1995. The first Middle Jurassic dinosaur from Scotland: a cetiosaurid? (Sauropoda) from the Bathonian of Skye. *Scottish Journal of Geology,* Vol. 31, 171–176.

Clark, N D L, Booth, P, Booth, C, and Ross, D A. 2004. Dinosaur footprints from the Duntulm Formation (Bathonian, Jurassic) of the Isle of Skye. *Scottish Journal of Geology,* Vol. 40, 13–21.

Claydon, R V, and Bell, B R. 1992. The structure and petrology of ultrabasic rocks in the southern part of the Cuillin Igneous Complex, Isle of Skye. *Transactions of the Royal Society of Edinburgh: Earth Sciences,* Vol. 83, 635–653.

Cope, J W C. 1995. Introduction to the British Jurassic. 1–7 in *Field Geology of the British Jurassic.* Taylor, P D. (editor). (London: Geological Society of London.)

Cox, K G, Bell, J D, and Pankhurst, R J. 1979. *The interpretation of the igneous rocks.* (London: George Allen & Unwin.)

Dagley, P, Mussett, A E, and Skelhorn, A R. 1987. Polarity stratigraphy and duration of the Mull Tertiary igneous activity. *Journal of the Geological Society of London*, Vol. 144, 966–985.

Dahl, S O, Ballantyne, C K, McCarroll, D, and Nesje, A. 1996. Maximum altitude of Devensian glaciation on the Isle of Skye. *Scottish Journal of Geology*, Vol. 32, 107–116.

Dawson, A G. 1988. The Main Rock Platform (Main Lateglacial Shoreline) in Ardnamurchan and Moidart, western Scotland. *Scottish Journal of Geology*, Vol. 24, 163–174.

Dawson, A G. 1997. Introduction. 1–8 in *Quaternary of Islay and Jura.* Dawson A G, and Dawson, S. (editors) (Cambridge: Quaternary Research Association.)

Deer, W A, Howie, R A, and Zussman, J. 1962. *Rock-forming minerals. Vol. 3 Sheet Silicates* (London: Longmans, Green and Co. Ltd.)

Dempster, T J, Preston, R J, and Bell, B R. 1999. The origin of Proterozoic massif-type anorthosites: evidence from interactions between crustal xenoliths and basaltic magma. *Journal of the Geological Society, London*, Vol. 156, 41–46.

De Souza, H A F. 1979. The geochronology of Scottish Carboniferous volcanism. Unpublished PhD thesis, University of Edinburgh.

Dickin, A P. 1981. Isotope geochemistry of Tertiary igneous rocks from the Isle of Skye, NW Scotland. *Journal of Petrology*, Vol. 22, 155–189.

Dickin, A P, and Durant, G P. 2002. The Blackstones igneous complex: geochemical and crustal context of a submerged Tertiary igneous centre in the Scottish Hebrides. *Geological Magazine,* Vol. 139, 199–207.

Dickin, A P, and Exley, R A. 1981. Isotopic and geochemical evidence for magma mixing in the petrogenesis of the Coire Uaigneich Granophyre, Isle of Skye, NW Scotland. *Contributions to Mineralogy and Petrology*, Vol. 76, 98–108.

Dickin, A P, Brown, J L, Thompson, R N, Halliday, A N, and Morrison, M A. 1984a. Crustal contamination and the granite problem in the British Tertiary Volcanic Province. *Philosophical Transactions of the Royal Society of London*, Vol. 310A, 755–780.

Dickin, A P, Henderson, C M B, and Gibb, F G F. 1984b. Hydrothermal Sr contamination of the Dippin sill, Isle of Arran, western Scotland. *Mineralogical Magazine,* Vol. 48, 311–322.

Dickin, A P, Jones, N W, Thirwall, M F, and Thompson, R N. 1987. A Ce/Nd isotope study of crustal contamination processes affecting Palaeocene magmas in Skye, NW Scotland. *Contributions to Mineralogy and Petrology*, Vol. 96, 455–464.

Donaldson, C H. 1977. Petrology of anorthite-bearing gabbro anorthosite dykes in north-west Skye. *Journal of Petrology*, Vol. 18, 595–620.

Donaldson, C H, Troll, V R, and Emeleus, C H. 2001. Felsites and breccias in the Northern Marginal Zone of the Rum Central Complex: changing views, c.1900–2000. *Proceedings of the Yorkshire Geological Society*, Vol. 53, 167–175.

Draper, L, and Draper, P. 1990. *The Raasay iron mine: where enemies became friends.* (Dingwall, Ross-shire: L and P Draper.)

Drever, H I, and Johnston, R. 1965. New petrographical data on the Shiant Isles picrite. *Mineralogical Magazine*, Vol. 34, 194–203.

DUNHAM, A C. 1968. The felsites, granophyre, explosion breccias and tuffisites of the north-eastern margin of the Tertiary igneous complex of Rhum, Inverness-shire. *Quarterly Journal of the Geological Society of London*, Vol. 123, 327–352.

DUNHAM, A C, and WADSWORTH, W J. 1978. Cryptic variation in the Rhum layered intrusion. *Mineralogical Magazine*, Vol. 42, 347–356.

DUNHAM, A C, and WILKINSON, F C F. 1985. Sulphide droplets and the U11/12 chromite band: a mineralogical study. *Geological Magazine*, Vol. 122, 539–548.

DURANT, G P, DOBSON, M R, KOKELAAR, B P, MACINTYRE, R M, and REA, W J. 1976. Preliminary report on the nature and age of the Blackstones Bank Igneous Centre, western Scotland. *Journal of the Geological Society*, Vol, 132, 319–326.

DURANT, G P, KOKELAAR, B P, and WHITTINGTON, R J. 1982. The Blackstones igneous centre, western Scotland. 279–308 in *Proceedings of the symposium Coned. Mondiale des Activities subaquatique*, September 1980. BLANCHARD, J, MAIR, J, and MORRISON, I. (editors). (Swindon: Natural Environment Council.)

ELLIOTT, D K, and FELDMANN, R M. 2003. Stratigraphic range determination for *Pseudoglyphea foersteri* Feldmann, Crisp and Pirrie 2002, from the Jurassic of Raasay. *Scottish Journal of Geology*, Vol. 39, 185–188.

EMELEUS, C H. 1973. Granophyre pebbles in Tertiary conglomerate on the Isle of Canna, Inverness-shire. *Scottish Journal of Geology*, Vol. 9, 157–159.

EMELEUS, C H. 1985. The Tertiary lavas and sediments of northwest Rhum, Inner Hebrides. *Geological Magazine,* Vol. 122, 419–437.

EMELEUS, C H. 1991. Tertiary igneous activity. 455–502 in *Geology of Scotland*. Third edition. CRAIG, G Y (editor). (London: The Geological Society.)

EMELEUS, C H. 1997. Geology of Rum and the adjacent islands. *Memoir of the British Geological Survey*, Sheet 60 (Scotland).

EMELEUS, C H, and GYOPARI, M C. 1992. *British Tertiary volcanic province*. Geological Conservation Review Series, No. 4 (London: Chapman and Hall.)

EMELEUS, C H, ALLWRIGHT, A E, KERR, A C, and WILLIAMSON, I T. 1996a. Red tuffs in the Palaeocene lava successions of the Inner Hebrides. *Scottish Journal of Geology*, Vol. 32, 83–89.

EMELEUS, C H, CHEADLE, M J, HUNTER, R H, UPTON, B G J, and WADSWORTH, W J. 1996b. The Rum layered suite. 403–439 in Layered Intrusions. CAWTHORN, R G (editor). *Developments in Petrology*, Vol. 15. (Amsterdam: Elsevier.)

ENGLAND, R W. 1988. The early Tertiary stress regime in NW Britain: evidence from the patterns of volcanic activity. 381–389 *in* Early Tertiary volcanism and the opening of the NE Atlantic. Morton, A C, and Parson, L M (editors). *Special Publication of the Geological Society of London*, No. 39.

ENGLAND, R W. 1992a. The role of Palaeocene magmatism in the tectonic evolution of the Sea of the Hebrides Basin: implications for basin evolution on the NW Seabord. 163–174 *in* Basins on the Atlantic seabord: petroleum geology, sedimentology and basin evolution. PARNELL, J (editor). *Special Publication of the Geological Society of London*, No. 62.

ENGLAND, R W. 1992b. The genesis, ascent, and emplacement of the Northern Arran Granite Scotland: implications for granitic diapirism. *Geological Society of America Bulletin*, Vol. 104, 606–614.

ENGLAND, R W. 1994. The structure of the Skye lava field. *Scottish Journal of Geology*, Vol. 30, 33–38.

ESSON, J, DUNHAM, A C, and THOMPSON, R N. 1975. Low alkali, high Ca olivine tholeiite lavas from the Isle of Skye, Scotland. *Journal of Petrology*, Vol. 16, 488–497.

EVANS, S E, and WALDMAN, M. 1996. Small reptiles and amphibians from the Middle Jurassic of Skye, Scotland. 219–226 *in* Continental Jurassic. MORALES, M (editor). *Museum of North Arizona, Bulletin*, No. 60.

FAITHFULL, J. 1995. *The Ross of Mull granite quarries*. (Isle of Iona: New Iona Press.)

FAITHFULL, J W. 1985. The Lower Eastern Layered Series of Rhum. *Geological Magazine*, Vol. 122, 459–468.

FARRIS, M A, OATES, M J, and TORRENS, H S. 1999. New evidence on the origin and Jurassic age of palaeokarst and limestone breccias, Loch Slapin, Isle of Skye. *Scottish Journal of Geology*, Vol. 35, 25–29.

FELDMAN, R M, WIEDER, R W, and ROLFE, W D I. 1994. *Urda mccoyi* (Carter 1889), an isopod crustacean from the Jurassic of Skye. *Scottish Journal of Geology*, Vol. 30, 87–90.

FELDMANN, R M, CRISP, G, and PIRRIE, D. 2002. A new species of glypheoid lobster, *Pseudoglyphea foersteri* (Decapoda: Astacidea: Mecochiridae) from the Lower Jurassic (Pliensbachian) of Raasay, Inner Hebrides, UK. *Palaeontology*, Vol. 45, 23–32.

FENNER, C N. 1937. A view of magmatic differentiation. *Journal of Geology*, Vol. 45, 158–168.

FOLAND, K A, GIBB, F G F, and HENDERSON, C M B. 2000. Patterns of Nd and Sr isotopic ratios produced by magmatic and post-magmatic processes in the Shiant Isles Main Sill, Scotland. *Contributions to Mineralogy and Petrology*, Vol. 139, 655–671.

FORESTER, R W, and TAYLOR, H P. 1976. ^{18}O-depleted igneous rocks from the Tertiary complex of the Isle of Mull, Scotland. *Earth and Planetary Science Letters*, Vol. 32, 11–17.

FORESTER, R W, and TAYLOR, H P. 1977. ^{18}O/^{16}O, D/H and ^{13}C/^{12}C studies on the Tertiary igneous complex of Skye, Scotland. *American Journal of Science*, Vol. 277, 136–177.

FORSTER, R M. 1980. A geochemical and petrological study of the Tertiary minor intrusions of Rhum, northwest Scotland. Unpublished PhD thesis, University of Durham.

FREDERIKSEN, K S, CLEMMENSEN, L B, and LAWAETZ, H S. 1998. Sequential architecture and cyclicity in Permian desert deposits, Brodick Beds, Arran, Scotland. *Journal of the Geological Society of London*, Vol. 155, 677–683.

FRIEND, C R L, and KINNY, P D. 2001. A reappraisal of the Lewisian Gneiss Complex: geochronological evidence for its tectonic assembly from disparate terranes in the Proterozoic. *Contributions to Mineralogy and Petrology*, Vol. 142, 198–218.

FRIEND, P F, HARLAND, W B, and HUDSON, J D. 1963. The Old Red Sandstone and the Highland Boundary in Arran. *Transactions of the Edinburgh Geological Society*, Vol. 19, 363–425.

FYFE, J A, LONG, D, and EVANS, D. 1993. *United Kingdom offshore regional report: the geology of the Malin–Hebrides sea area*. (London: HMSO for the British Geological Survey.)

GASS, I G, and THORPE, R S. 1976. *Igneous case study: the Tertiary igneous rocks of Skye, NW Scotland*. (Milton Keynes: Open University Press.)

GELDMACHER, J, MAASE, K M, DEVEY, C W, and GARBE-SCHONBERG, C D. 1998. The petrogenesis of Tertiary cone-sheets in Ardnamurchan, NW Scotland: petrological and geochemical constraints on crustal contamination and partial melting. *Contributions to Mineralogy and Petrology*, Vol. 131, 196–209.

GEMMELL, A M D. 1972. The deglaciation of the Island of Arran, Scotland. *Transactions of the Institute of British Geographers,* Vol. 59, 25–39.

GIBB, F G F. 1968. Flow differentiation in the xenolithic ultrabasic dykes of the Cuillins and Strathaird Peninsula, Isle of Skye, Scotland. *Journal of Petrology,* Vol. 9, 411–433.

GIBB, F G F. 1969. Cognate xenoliths in the Tertiary ultrabasic dykes of south-west Skye. *Mineralogical Magazine,* Vol. 37, 504–514.

GIBB, F G F, and GIBSON, S A. 1989. The Little Minch sill complex. *Scottish Journal of Geology,* Vol. 25, 367–370.

GIBB, F G F, and HENDERSON, C M B. 1978. The petrology of the Dippin sill, Isle of Arran. *Scottish Journal of Geology,* Vol. 14, 1–27.

GIBB, F G F, and HENDERSON, C M B. 1996. The Shiant Isles Main Sill: structure and mineral fractionation trends. *Mineralogical Magazine,* Vol. 60, 67–98.

GIBSON, S A. 1990. The geochemistry of the Trotternish sills, Isle of Skye: crustal contamination in the British Tertiary Volcanic Province. *Journal of the Geological Society of London,* Vol. 147, 1071–1081.

GIBSON, S A, and JONES, A P. 1991. Igneous stratigraphy and internal structure of the Little Minch Sill Complex, Trotternish Peninsula, northern Skye, Scotland. *Geological Magazine,* Vol. 128, 51–66.

GORDON, J E (editor). 1997. *Reflections on the Ice Age in Scotland.* (Glasgow: Scottish Association of Geography Teachers.)

GORDON, J E, and SUTHERLAND, D G (editors). 1993. *Quaternary of Scotland.* Geological Conservation Review Series, No. 6 (London: Chapman and Hall.)

GOULTY, N R, LEGGETT, M, DOUGLAS, T, and EMELEUS, C H. 1992. Seismic reflection test on the granite of the Skye Tertiary igneous centre. *Geological Magazine,* Vol. 129, 633–636.

GOULTY, N R, DOBSON, A R, JONES, G D, AL-KINDI, S A, and HOLLAND, J G. 2001. Gravity evidence for diapiric ascent of the Northern Arran Granite. *Journal of the Geological Society of London,* Vol. 158, 869–876.

GRAY, G M. 1978. Low-level shore platforms in the south-west Scottish Highlands: altitude, age and correlation. *Transactions of the Institute of British Geographers,* New Series, Vol. 3, 151–164.

GRAY, J M. 1981. p-forms from the Isle of Mull. *Scottish Journal of Geology,* Vol. 17, 39–47.

GRAY, J M. 1989. Distribution and development of the Main Rock Platform, western Scotland: comment. *Scottish Journal of Geology,* Vol. 25, 227–231.

GREENWOOD, R C, DONALDSON, C H, and EMELEUS, C H. 1990. The contact zone of the Rhum ultrabasic intrusion: evidence of peridotite formation from magnesian magmas. *Journal of the Geological Society of London,* Vol. 147, 209–212.

GRIBBLE, C D. 1974. The dolerites of Ardnamurchan. *Scottish Journal of Geology,* Vol. 10, 71–89.

GRIBBLE, C D, DURANCE, E M, and WALSH, J N. 1976. *Ardnamurchan: a guide to geological excursions.* (Edinburgh: Edinburgh Geological Society.)

HALDANE, D, EYLES, V A, and DAVIDSON, C F. 1940. Diatomite. *Geological Survey Wartime Pamphlet* No. 5, 1–13.

HALL, A M. 1991. Pre-Quaternary landscape evolution in the Scottish Highlands. *Transactions of the Royal Society of Edinburgh: Earth Sciences,* Vol. 82, 1–26.

HALLIDAY, A N, AFTALION, M, VAN BREEMEN, O, and JOCELYN, J. 1979. Petrogenetic significance of Rb-Sr and U-Pb isotopic systems in the 400 Ma old British Isles granitoids and their hosts. 653–661 in Caledonides of the British Isles — reviewed. HARRIS, A L, HOLLAND, C H, and LEAKE, B E (editors). Special Publication of the Geological Society of London, No. 8.

HAMILTON, M A, PEARSON, D G, THOMPSON, R N, KELLEY, S P, and EMELEUS, C H. 1998. Rapid eruption of Skye lavas inferred from precise U-Pb and Ar-Ar dating of the Rum and Cuillin plutonic complexes. Nature, Vol. 394, 260–263.

HANCOCK, J M. 2000. The Gribun Formation: clues to the latest Cretaceous history of western Scotland. Scottish Journal of Geology, Vol. 36, 137–141.

HARDING, R R. 1966. The Mullach Sgar Complex, St Kilda, Outer Hebrides. Scottish Journal of Geology, Vol. 2, 165–178.

HARDING, R R. 1967. The major ultrabasic and basic intrusions of St Kilda, Outer Hebrides. Transactions of the Royal Society of Edinburgh, Vol. 66, 419–444.

HARDING, R R. 1983. Zr-rich pyroxenes and glauconitic minerals in the Tertiary alkali granite of Ailsa Craig. Scottish Journal of Geology, Vol. 19, 219–227.

HARDING, R R, MERRIMAN, R J, and NANCARROW, P H A. 1984. St Kilda: an illustrated account of the geology. Report of the British Geological Survey, Vol. 16, No. 7.

HARKER, A. 1904. The Tertiary Igneous Rocks of Skye. Memoir of the Geological Survey of Great Britain. Sheets 70 and 71 (Scotland).

HARKER, A. 1908. The geology of the Small Isles of Inverness-shire. Memoir of the Geological Survey of Great Britain, Sheet 60 (Scotland).

HARLAND, W B, and HACKER, J L F. 1966. 'Fossil' lightning strikes 250 million yeas ago. The Advancement of Science, Vol. 22, 633–671.

HARRIS, A L, and JOHNSTONE, M R. 1991. Moine. 87–123 in Geology of Scotland. Third edition. CRAIG, G Y (editor). (London: The Geological Society.)

HARRIS, J P. 1989. The sedimentology of a Middle Jurassic lagoonal delta system: Elgol Formation (Great Estuarine Group), NW Scotland. 147–166 in Deltas: sites and traps for fossil fuels. WHATLEY, M K G, and PICKERING, K T (editors). Special Publication of the Geological Society of London, No. 41.

HARRIS, J P. 1992. Mid-Jurassic lagoonal delta systems in the Hebridean Basins: thickness and facies distribution patterns of potential reservoir sandbodies. 111–144 in Basins on the Atlantic seaboard: petroleum geology, sedimentology and basin evolution. PARNELL, J (editor). Special Publication of the Geological Society of London, No. 62.

HARRISON, R K. 1975. Expeditions to Rockall. 1971–1972. Report of the Institute of Geological Sciences. No. 75/1.

HARRISON, R K, STONE, P, CAMERON, I B, ELLIOT, R W, and HARDING, R R. 1987. Geology, petrology and geochemistry of Ailsa Craig, Ayrshire. Report of the British Geological Survey, Vol. 16 (No. 9).

HENDERSON, C M B, GIBB, F G F, and FOLAND, K A. 2000. Mineral fractionation and pre- and post-emplacement processes in the uppermost part of the Shiant Isles Main Sill. Mineralogical Magazine, Vol. 64, 779–790.

HENDERSON, P, and WOOD, R J. 1981. Reaction relationships of chrome-spinels in igneous rocks — further evidence from the layered intrusions of Rhum and Mull, Inner Hebrides, Scotland. Contributions to Mineralogy and Petrology, Vol. 78, 225–229.

HERRIOT, A. 1975. Observations on the Tighvein 'Complex', Arran. *Proceedings of the Geological Society of Glasgow*, for the years 1972–1974, 7–11.

HESSELBO, S P, and COE, A C. 2000. Jurassic sequences in the Hebridean Basin, Isle of Skye, Scotland. 41–58 in *Field Guide Book, International Association of Sedimentologists Meeting, Dublin 2000*. GRAHAM, J R, and RYAN, A. (editors) (Dublin: University of Dublin.)

HESSELBO, S P, OATES, M J, and JENKYNS, H C. 1998. The lower Lias Group of the Hebrides Basin. *Scottish Journal of Geology*, Vol. 34, 23–60.

HOLDSWORTH, R E, STRACHAN, R A, and HARRIS, A L. 1994. Precambrian rocks in northern Scotland east of the Moine Thrust: the Moine Supergroup. 23–32 *in* A revised correlation of the Precambrian rocks in the British Isles. GIBBONS, W, and HARRIS, A L (editors). *Geological Society of London Special Report*, No. 22.

HOLE, M J, and MORRISON, M A. 1992. The differentiated boss, Cnoc Rhaonastil, Islay: a natural experiment in the low pressure differentiation of an alkali olivine-basalt magma. *Scottish Journal of Geology*, Vol. 28, 55–70.

HOLGATE, N. 1969. Palaeozoic and Tertiary transcurrent movements on the Great Glen fault. *Scottish Journal of Geology*, Vol. 5, 97–139.

HOLLAND, J G, and BROWN, G M. 1972. Hebridean tholeiitic magmas: a geochemical study of the Ardnamurchan cone sheets. *Contributions to Mineralogy and Petrology*, Vol. 37, 139–160.

HOLMES, A. 1936. The idea of contrasted differentiation. *Geological Magazine*, Vol. 73, 228–238.

HOLNESS, M B. 1992. Metamorphism and fluid infiltration of the calc-silicate aureole of the Beinn an Dubhaich Granite, Skye. *Journal of Petrology*, Vol. 33, 1261–1293.

HOLNESS, M B. 1999. Contact metamorphism and anatexis of Torridonian arkose by minor intrusions of the Rum Igneous Complex, Inner Hebrides, Scotland. *Geological Magazine*, Vol. 136, 527–542.

HOLROYD, J D. 1994. The structure and stratigraphy of the Suardal area, Isle of Skye, north-west Scotland: an investigation of Tertiary deformation in the Skye Volcanic Complex. Unpublished PhD thesis, University of Manchester.

HOWARTH, M K. 1992. The ammonite family Hildoceratidae in the Lower Jurassic of Britain. *Monographs of the Palaeontographical Society, London*, Vols. 145 and 146.

HOWIE, R A, and WALSH, J N. 1981. Riebeckite, arfvedsonite and aenigmatite from the Ailsa Craig microgranite. *Scottish Journal of Geology*, Vol. 17, 123–128.

HUDSON, J D. 1960. The Laig Gorge Beds, Isle of Eigg. *Geological Magazine*, Vol. 97, 313–325.

HUDSON, J D. 1966. Hugh Miller's reptile bed and the Mytilus Shales, Middle Jurassic, Isle of Eigg, Scotland. *Scottish Journal of Geology*, Vol. 2, 265–281.

HUDSON, J D. 1983. Mesozoic sedimentation and sedimentary rocks in the Inner Hebrides. *Proceedings of the Royal Society of Edinburgh*, Vol. 83B, 47–63.

HUDSON, J D, and ANDREWS, J E. 1987. The diagenesis of the Great Estuarine Group, Middle Jurassic, Inner Hebrides, Scotland. 259–276 *in* Diagenesis of sedimentary sequences. MARSHALL, J D (editor). *Special Publication of the Geological Society of London*, No. 36.

HUDSON, J D, and HARRIS, J P. 1979. Sedimentology of the Great Estuarine Group (Middle Jurassic) of north west Scotland. 1–13 *in* La sedimentation du Jurassique W-Europeen. (No editor named). *Publication Speciale 1, Association Sedimentologistes Francais*.

HUDSON, J D, and WAKEFIELD, M I. 1999. New observations on the Kildonnan Member, Lealt Shale Formation, Middle Jurassic, Isle of Eigg. *Scottish Journal of Geology*, Vol. 35, 63–64.

HUGHES, C J. 1960. An occurrence of tilleyite-bearing limestone in the Isle of Rhum, Inner Hebrides. *Geological Magazine*, Vol. 97, 384–388.

HUNTER, R H. 1996. Texture development in cumulate rocks. 77–101 in Layered intrusions. CAWTHORN, R G (editor). *Developments in Petrology*, No. 15. (Amsterdam: Elsevier.)

HUTCHISON, R. 1966. Intrusive tholeiites of the western Cuillin, Isle of Skye. *Geological Magazine*, Vol. 103, 352–363.

HUTCHISON, R. 1968. Origin of the White Allivalite, Western Cuillin, Isle of Skye. *Geological Magazine*, Vol. 105, 338–347.

HUTCHISON, R, and BEAVAN, J C. 1977. The Cuillin layered igneous complex — evidence for multiple intrusion and former presence of a picritic liquid. *Scottish Journal of Geology*, Vol. 13, 197–210.

HYSLOP, E K, GILLANDERS, R J, HILL, P G, and FAKES, R D. 1999. Rare-earth-bearing minerals fergusonite and gadolinite from the Arran granite. *Scottish Journal of Geology*, Vol. 35, 65–69.

JASSIM, S Z, and GASS, I G. 1970. The Loch na Creitheach volcanic vent, Isle of Skye. *Scottish Journal of Geology*, Vol. 6, 285–294.

JOHNSTON, D R. 1996. A reassessment of the age and form of the Broadford Gabbro, Isle of Skye: new field evidence from Creag Strollamus. *Scottish Journal of Geology*, Vol. 32, 51–58.

JOHNSTONE, G S, and MYKURA, W. 1989. *British Regional Geology: the Northern Highlands of Scotland*. (Fourth edition) (Edinburgh: HMSO for British Geological Survey.)

JOLLEY, D W. 1997. Palaeosurface palynofloras of the Skye lava field and the age of the British Tertiary Volcanic Province. 67–94 in Palaeosurfaces: recognition, reconstruction and palaeoenvironmental interpretation. WIDDOWSON, M (editor). *Geological Society of London Special Publication*, No. 120.

JOLLEY, D W, CLARKE, B, and KELLEY, S. 2002. Palaeogene time scale miscalculation: evidence from the dating of the North Atlantic igneous province. *Geology*, Vol. 30, 7–10.

JOLLY, R J H, and SANDERSON, D J. 1995. Variation in the form and distribution of dykes in the Mull swarm, Scotland. *Journal of Structural Geology*, Vol. 17, 1543–1557.

JUDD, J W. 1878. The secondary rocks of Scotland. Third paper. The strata of the Western Coast and Islands. *Quarterly Journal of the Geological Society of London*, Vol. 34, 660–743.

KANARIS-SOTIRIOU, R, and GIBB, F G F. 1985. Hybridization and the petrogenesis of composite intrusions: the dyke at An Cumhann, Isle of Arran, Scotland. *Geological Magazine*, Vol. 123, 693–697.

KENNEDY, W Q. 1930. The parent magma of the British Tertiary Province. *Geological Survey of Great Britain, Summary of Progress*, No. 11, 61–73.

KENNEDY, W Q. 1933. Trends of differentiation in basaltic magmas. *American Journal of Science*, Vol. 25, 239–256.

KENT, R W, THOMSON, B A, SKELHORN, R R, KERR, A C, NORRY, M J, and WALSH, N J. 1998. Emplacement of Hebridean Tertiary flood basalts: evidence from an inflated pahoehoe lava flow on Mull, Scotland. *Journal of the Geological Society of London*, Vol. 155, 599–607.

KERR, A C. 1993. Elemental evidence for an enriched small-fraction melt input into the Tertiary Mull basalts, Western Scotland. *Journal of the Geological Society of London*, Vol. 150, 763–769.

Kerr, A C. 1995a. The geochemistry of the Mull–Morvern lava succession, NW Scotland: an assessment of mantle sources during plume-related volcanism. *Chemical Geology*, Vol. 122, 43–58.

Kerr, A C. 1995b. The geochemical stratigraphy, field relations and temporal variation of the Mull–Morvern lava succession, NW Scotland. *Transactions of the Royal Society of Edinburgh: Earth Sciences*, Vol. 86, 35–47.

Kerr, A C. 1997. The geochemistry and significance of plugs intruding the Tertiary Mull–Morvern lava succession, western Scotland. *Scottish Journal of Geology*, Vol. 33, 157–167.

Kerr, A C. 1998. Mineral chemistry of the Mull–Morvern Tertiary lava succession, western Scotland. *Mineralogical Magazine*, Vol. 62, 295–312.

Kerr, A C, Kent, R W, Thomson, B A, Seedhouse, J K, and Donaldson, C H. 1999. Geochemical evolution of the Tertiary Mull Volcano, western Scotland. *Journal of Petrology*, Vol. 40, 873–908.

Kille, I C, Thompson, R N, Morrison, M A, and Thompson, R F. 1986. Field evidence for turbulence during flow of basaltic magma through conduits from southwest Mull. *Geological Magazine*, Vol. 123, 693–697.

King, B C. 1955. The Ard Bheinn area of the Central Igneous Complex of Arran. *Quarterly Journal of the Geological Society of London*, Vol. 110, 323–356.

King, B C. 1982. Composite intrusions: associations of acid and basic magmas. 441–447 in *Igneous rocks of the British Isles*. Sutherland, D S (editor). (Chichester: John Wiley & Sons.)

Knox, R W O'B. 1977. Upper Jurassic pyroclastic rocks in Skye, west Scotland. *Nature*, Vol. 265, 323–324.

Knox, R W O'B, and Morton, A C. 1988. The record of early Tertiary North Atlantic volcanism in the sediments of the North Atlantic. 407–409 in *Early Tertiary volcanism and the opening of the NE Atlantic*. Morton, A C, and Parsons, L M (editors). *Special Publication of the Geological Society of London*, No. 39.

Koomans C, and Kuenen, P H. 1938. On the differentiation of the Glen More ring-dyke, Mull. *Geological Magazine*, Vol. 75, 145–160.

LeBas, M J. 1971. Cone-sheets as a mechanism of uplift. *Geological Magazine*, Vol. 108, 373–376.

LeBas, M J, and Streckeisen, A L. 1991. The IUGS systematics of igneous rocks. *Journal of the Geological Society of London*, Vol. 148, 825–833.

Le Coeur, C. 1988. Late Tertiary warping and erosion in western Scotland. *Geografisk Annaler*, Vol. 70A, 361–367.

Le Coeur, C, and Kuzucuoglu, C. 1992. Glaciotectonic structures in altered dolostones in the Isle of Skye, western Scotland. *Scottish Journal of Geology*, Vol. 28, 159–166.

Lee, G W. 1920. The Mesozoic rocks of Applecross, Raasay and north-east Skye. *Memoir of the Geological Survey of Great Britain,* Sheet 81 (Scotland).

Lee, G W, and Bailey, E B. 1925. Pre-Tertiary geology of Mull, Loch Aline and Oban. *Memoir of the Geological Survey of Great Britain.* Parts of sheets 43, 44, 51 and 52 (Scotland).

Lovell, J P B. 1991. Permian and Triassic. 421–438 in *Geology of Scotland*. Third edition. Craig, G Y (editor). (London: The Geological Society).

LOWDEN, B, BRALEY, S, HURST, A, and LEWIS, J. 1992. Sedimentological studies of the Cretaceous Lochaline Sandstone, NW Scotland. 159–162 in Basins on the Atlantic seaboard: petroleum geology, sedimentology and basin evolution, PARNELL, J (editor). *Special Publication of the Geological Society of London*, No. 62.

LOWE, J J, and WALKER, M J C. 1986. Lateglacial and early Flandrian environmental history of the Isle of Mull, Inner Hebrides, Scotland. *Transactions of the Royal Society of Edinburgh: Earth Sciences*, Vol. 77, 1–20.

LOWE, J J, and WALKER, M J C. 1991. Vegetational history of the Isle of Skye: II The Flandrian. 119–142 in *Quaternary of the Isle of Skye. Field Guide*. BALLANTYNE, C K, BENN, D I, LOWE, J J, and WALKER, M J C (editors). (Cambridge: Quaternary Research Association.)

MCCALLIEN, W J. 1937. Late and early post-glacial Scotland. *Proceedings of the Society of Antiquaries of Scotland*, Vol. 71, 174–206.

MCCLURG, J E. 1982. Petrology and evolution of the northern part of the Rhum ultrabasic complex. Unpublished PhD thesis, University of Edinburgh (2 volumes).

MACDONALD, R, WILSON, L, THORPE, R S, and MARTIN, A. 1988. Emplacement of the Cleveland Dyke: evidence from geochemistry, mineralogy and physical modelling. *Journal of Petrology*, Vol. 29, 559–583.

MACGREGOR, M. 1983. *Excursion guide to the geology of Arran*. Third edition, revised by MACDONALD, J G, and HERRIOT, A with contributions by KING, B C. (Glasgow: Geological Society of Glasgow.)

MCKERROW, W S, and ATKINS, F B. 1985. *Isle of Arran: a field guide for students of geology*. (London: Geologists' Association.)

MACKINNON, A. 1974. The Madadh Rocks: a Tertiary olivine dolerite sill in the Outer Hebrides. *Scottish Journal of Geology*, Vol. 10, 67–70.

MCQUILLIN, R, and TUSON, J. 1963. Gravity measurements over the Rhum Tertiary plutonic complex. *Nature*, Vol. 199, 1276–1277.

MCQUILLIN, R, BACON, M, and BINNS, P E. 1975. The Blackstones Tertiary igneous complex. *Scottish Journal of Geology*, Vol. 11, 179–192.

MARSHALL, L A, and SPARKS, R S J. 1984. Origins of some mixed-magma and net-veined ring intrusions. *Journal of the Geological Society*, Vol. 141, 171–182.

MATTEY, D P, GIBSON, I S, MARRINER, G F, and THOMPSON, R N. 1977. The diagnostic geochemistry, relative abundance and spatial distibution of high-calcium, low-alkali olivine tholeiite dykes in the Lower Tertiary regional swarm of the Isle of Skye, NW Scotland. *Mineralogical Magazine*, Vol. 41, 273–285.

MEIGHAN, I G. 1979. The acid igneous rocks of the British Tertiary Province. *Bulletin of the Geological Survey of Great Britain*, No. 70, 10–22.

MEISSNER, R, MATTHEWS, D, and WEVER, TH. 1986. The 'Moho' in and around Great Britain. *Annales Geophysicae*, Vol. 4, B6, 659–664.

MENZIES, M A, HALLIDAY, A N, PALACZ, Z, HUNTER, R H, UPTON, B G J, ASPEN, P, and HAWKSWORTH, C J. 1987. Evidence for mantle xenoliths from an enriched lithospheric keel under the Outer Hebrides. *Nature*, Vol. 325, 44–47.

MITCHELL, W I (editor). 2004. *The geology of Northern Ireland — our natural foundation*. Second edition. (Belfast: Geological Survey of Northern Ireland.)

MOORBATH, S, and BELL, J D. 1965. Strontium isotope abundance studies and rubidium-strontium age determinations on Tertiary igneous rocks from the Isle of Skye, north-west Scotland. *Journal of Petrology*, Vol. 6, 37–66.

MOORBATH, S, and THOMPSON, R N. 1980. Strontium isotope geochemistry and petrogenesis of the Early Tertiary lava pile of the Isle of Skye, and other basic rocks of the British Tertiary Province: an example of magma-crust interaction. *Journal of Petrology*, Vol. 21, 295–321.

MOORBATH, S, and WELKE, H. 1969. Lead isotope studies on igneous rocks from the Isle of Skye, northwest Scotland. *Earth and Planetary Science Letters*, Vol. 5, 217–230.

MOORE, D J. 1979. The barite deposits of central and southern Scotland. Unpublished PhD thesis, University of Leeds.

MORRISON, M A, THOMPSON, R N, GIBSON, I L, and MARRINER, G F. 1980. Lateral chemical heterogeneity in the Palaeocene upper mantle beneath the Scottish Hebrides. *Philosophical Transactions of the Royal Society of London*, Vol. 297A, 229–244.

MORRISON, M A, THOMPSON, R N, and DICKIN, A P. 1985. Geochemical evidence for complex magmatic plumbing during development of a continental volcanic centre. *Geology*, Vol. 13, 581–584.

MORTIMORE, R, WOOD, C, and GALLOIS, R. 2001. *British Upper Cretaceous stratigraphy.* Geological Conservation Review Series, No. 23. (Peterborough: Joint Nature Conservation Committee.)

MORTON, A C, and TAYLOR P N. 1991. Geochemical and isotopic constraints on the nature and age of basement rocks from the Rockall Bank, NE Atlantic. *Journal of the Geological Society of London*, Vol. 148, 631–634.

MORTON, N. 1989. Jurassic sequence stratigraphy in the Hebrides Basin, NW Scotland. *Marine and Petroleum Geology*, Vol. 6, 243–260.

MORTON, N. 1999a. Discussion on 'The Lower Lias Group in the Hebrides Basin' reply by HESSELBO, S P, OATES, M J, and JENKYNS, H C. *Scottish Journal of Geology*, Vol. 35, 85–88.

MORTON, N. 1999b. Middle Hettangian (Lower Jurassic) ammonites from Isle of Raasay, Inner Hebrides, and correlation of the Hettengian–lowermost Sinemurian Breakish Formation in the Skye area, NW Scotland. *Scottish Journal of Geology*, Vol. 35, 119–130.

MORTON, N, and DIETL, G. 1989. Age of the Garantiana Clay (Middle Jurassic) in the Hebrides Basin. *Scottish Journal of Geology*, Vol. 25, 153–159.

MORTON, N, and HUDSON, J D. 1995. Field guide to the Jurassic of the isles of Raasay and Skye, Inner Hebrides, NW Scotland. 209–280 in *Field geology of the British Jurassic.* TAYLOR, P D (editor). (London: Geological Society of London.)

MORTON, N, SMITH, R M, GOLDEN, M, and JAMES, A V. 1987. Comparative stratigraphic study of Triassic–Jurassic sedimentation and basin evolution in the North Sea and north-west of the British Isles. 697–709 in *Petroleum geology of north west Europe.* BROOKS, J, and GLENNIE, K W (editors). (London: Graham and Trotman.)

MUSSETT, A E, DAGLEY, P, and SKELHORN, R R. 1988. Time and duration of activity in the British Tertiary Igneous Province. 337–348 in *Early Tertiary volcanism and the opening of the North Atlantic.* MORTON, A C, and PARSON, L M (editors). *Geological Society of London Special Publication*, No. 39.

MUSSETT, A E, DAGLEY, P, and SKELHORN, R R. 1989. Further evidence for a single polarity and a common source for the quartz-porphyry intrusions of the Arran area. *Scottish Journal of Geology*, Vol. 25, 241–257.

NICHOL, D. 2001. Ornamental granite from Ailsa Craig, Scotland. *Journal of Gemmology*, Vol. 27, 286–290.

NICHOLSON, P G. 1993. A basin reappraisal of the Proterozoic Torridon Group, northwest Scotland. 183–202 in Tectonic controls and signatures in sedimentary successions. FROSTICK, L E, and STEEL, R G (editors). International Association of Sedimentologists Special Publication, No. 20.

NICHOLSON, R. 1978. The Camas Malag Formation: an interbedded rhythmite/conglomerate sequence of probable Triassic age, Loch Slapin, Isle of Skye. Scottish Journal of Geology, Vol. 14, 301–309.

NICHOLSON, R. 1985. The intrusion and deformation of Tertiary minor sheet intrusions, west Suardal, Isle of Skye, Scotland. Geological Journal, Vol. 20, 53–72.

OATES, M J. 1978. A revised stratigraphy of the western Scottish Lower Lias. Proceedings of the Yorkshire Geological Society, Vol. 42, 143–156.

OFOEGBU, C O, and BOTT, M H P. 1985. Interpretation of the Minch linear magnetic anomaly and of a similar feature on the shelf north of Lewis by non-linear optimization. Journal of the Geological Society of London, Vol. 142, 1077–1087.

PANKHURST, R J, WALSH, J N, BECKINSALE, R D, and SKELHORN, R R. 1978. Isotopic and other geochemical evidence for the origin of the Loch Uisg granophyre, Isle of Mull, Scotland. Earth and Planetary Science Letters, Vol. 38, 355–363.

PARK, R G. 1991. The Lewisian Complex. 25–65 in Geology of Scotland. Third edition. CRAIG, G Y (editor). (London: The Geological Society.)

PARK, R G, CLIFF, R A, FETTES, D J, and STEWART, A D. 1994. Precambrian rocks in northwest Scotland west of the Moine Thrust: the Lewisian complex and the Torridonian. 6–22 in A revised correlation of the Precambrian in the British Isles. GIBBONS, W, and HARRIS, A L (editors). Geological Society of London Special Report, No. 22.

PARK, R G, STEWART, A D, and WRIGHT, D T. 2002. The Hebridean terrane. 45–80 in The Geology of Scotland. Fourth edition. TREWIN, N H (editor). (London: The Geological Society.)

PEACH, B N, HORNE, J, GUNN, W, CLOUGH, C T, GEIKIE, A, HINXMAN, L W, and TEALL, J J H. 1907. The geological structure of the north-west Highlands of Scotland. Memoir of the Geological Survey of Great Britain.

PEACH, B N, HORNE, J, WOODWARD, H B, CLOUGH, C T, BARROW, G, FLETT, J S, HARKER, A, KITCHIN, F L, TEAL, J J H, and WEDD, C B. 1910. The geology of Glenelg, Lochalsh and south-east part of Skye. Memoir of the Geological Survey of Great Britain, Sheet 71 (Scotland).

PEACOCK, D. 1969. Coastal features of Rhum. Scottish Journal of Geology, Vol. 5, 301–302.

PEACOCK, J D. 1981. Scottish Late-glacial marine deposits and their environmental significance. 222–236 in The Quaternary in Britain. NEALE, J, and FLENLEY, J (editors). (Oxford: Pergamon Press.)

PEARSON, D G, EMELEUS, C H, and KELLEY, S P. 1996. Precise $^{40}Ar/^{39}Ar$ age for the initiation of Palaeogene volcanism in the Inner Hebrides and its regional significance. Journal of the Geological Society of London, Vol. 153, 815–818.

PIRRIE, D, POWER, M R, ANDERSEN, J C O, and BUTCHER, A R. 2000. Platinum-group mineralisation in the Tertiary Igneous Province: new data from Mull and Skye, Scottish Inner Hebrides, UK. Geological Magazine, Vol. 137, 651–658.

PLATTEN, I M. 2000. Incremental dilation of magma filled fractures: evidence from dykes on the Isle of Skye, Scotland. Journal of Structural Geology, Vol. 22, 1153–1164.

POTTS, G J. 1982. The origin of recumbent fold nappes: the Lochalsh Fold as the main example. Unpublished PhD thesis, University of Leeds.

POTTS, G J, HUNTER, R H, HARRIS, A L, and FRASER, F M. 1995. Late-orogenic extensional tectonics at the NW margin of the Caledonides in Scotland. *Journal of the Geological Society of London*, Vol. 153, 907–910.

POWER, M R, PIRRIE, D, and ANDERSON, J C Ø. 2003. Diversity of platinum-group element mineralisation styles in the North Atlantic Igneous Province: new evidence from Rum, UK. *Geological Magazine*, Vol. 140, 499–512.

PRESTON, J. 1982. Eruptive volcanism. 351–368 in *Igneous rocks of the British Isles*, SUTHERLAND, D S (editor). (Chichester: Wyllie.)

PRESTON, J. 2001. Tertiary igneous activity. 353–373 in *Geology of Ireland*. HOLLAND, C J (editor). (Edinburgh: Dunedin Academic Press.)

PRESTON, R J, and BELL, B R. 1997. Cognate gabbroic xenoliths from a tholeiitic subvolcanic sill complex: implications for fractional crystallization and crustal contamination processes. *Mineralogical Magazine*, Vol. 61, 329–349.

PRESTON, R J, BELL, B R, and ROGERS, G. 1998a. The Loch Scridain sill complex, Isle of Mull, Scotland: fractional crystallization, assimilation, magma-mixing and crustal anatexis in sub-volcanic conduits. *Journal of Petrology*, Vol. 39, 519–550.

PRESTON, R J, HOLE, M, BOUCH, J, and STILL, J. 1998b. The occurrence of zirconian aegirine and calcic catapleite ($CaZrSi_3O_9.2H_2O$) within a nepheline syenite, British Tertiary Igneous Province. *Scottish Journal of Geology*, Vol. 34, 173–80.

PRESTON, R J, DEMPSTER, T J, BELL, B R, and ROGERS, G. 1999. The petrology of mullite-bearing peraluminous xenoliths: implications for contamination processes in basaltic magmas. *Journal of Petrology*, Vol. 40, 549–573.

PRESTON, R J, HOLE, M J, and STILL, J. 2000a. The occurrence of Zr-bearing amphiboles and their relationships with the pyroxenes and biotites in the teschenite and nepheline syenites of a differentiated dolerite boss, Islay, NW Scotland. *Mineralogical Magazine*, Vol. 64, 459–468.

PRESTON, R J, HOLE, M J, and STILL, J. 2000b. Exceptional REE-enrichment in apatite during the low pressure fractional crystalisation of alkali olivine basalt: an example from the British Tertiary Igneous Province. *Transactions of the Royal Society of Edinburgh: Earth Sciences*, Vol. 90, 273–285.

RAINBIRD, R H, HAMILTON, M A, and YOUNG, G M. 2001. Detrital zircon geochronology and provenance of the Torridonian, NW Scotland. *Journal of the Geological Society of London*, Vol. 158, 15–27.

REYNOLDS, D L. 1954. Fluidization as a geological process, and its bearing on the problem of intrusive granites. *American Journal of Science*, Vol. 252, 577–614.

RICHEY, J E. 1932. Tertiary ring structures in Britain. *Transactions of the Geological Society of Glasgow*, Vol. 19, 42–140.

RICHEY, J E. 1961. *British Regional Geology, Scotland: the Tertiary volcanic districts*. Third edition, revised by MACGREGOR, A G, and ANDERSON, F W. (Edinburgh: HMSO for British Geological Survey.)

RICHEY, J E, and THOMAS, H H. 1930. The geology of Ardnamurchan, north-west Mull and Coll. *Memoir of the Geological Survey of Great Britain*, Sheet 51 and part of Sheet 52 (Scotland).

RIDING, J B. 1992. On the age of the Upper Ostrea Member, Staffin Bay Formation (Middle Jurassic) of north-west Skye. *Scottish Journal of Geology*, Vol. 28, 155–158.

RITCHEY, J D, and HITCHEN, K. 1996. Early Palaeogene offshore igneous activity to the northwest of the UK and its relationship to the North Atlantic Igneous Province. 63–78 in Correlation of the early Palaeogene in northwest Europe. KNOX, R W O'B, CORFIELD, R M, and DUNAY, R E (editors). *Geological Society of London Special Publication*, No. 101.

ROBERTS, A M, and HOLDSWORTH, R E. 1999. Linking onshore and offshore structures: Mesozoic extension in the Scottish Highlands. *Journal of the Geological Society*, Vol. 156, 1061–1064.

ROGERS, N W, and GIBSON, I L. 1977. The petrology and geochemistry of the Creag Dubh composite sill, Whiting Bay, Arran, Scotland. *Geological Magazine*, Vol. 114, 1–8.

SAUNDERS, A D, FITTON, J G, KERR, A C, NORRY, M J, and KENT, R W. 1997. The North Atlantic Igneous Province. 45–94 in Large igneous provinces. MAHONEY, J J, and COFFIN, M F (editors). *American Geophysical Union, Geophysical Monograph*, No. 100.

SCARROW, J H, and COX, K G. 1995. Basalt generated by decompressive adiabatic melting of a mantle plume: a case study from the Isle of Skye, NW Scotland. *Journal of Petrology*, Vol. 36, 3–22.

SIMKIN, T. 1967. Flow differentiation in the picritic sills of north Skye. 64–69 in *Ultramafic rocks*. WYLLIE, P E (editor). (New York: John Wyllie.)

SISSONS, J B. 1974. The Quaternary in Scotland: a review. *Scottish Journal of Geology*, Vol. 10, 311–338.

SISSONS, J B. 1982. The so-called high 'interglacial' shoreline of western Scotland. *Transactions of the Institute of British Geographers*, Vol. 7 (New Series), 205–216.

SKELHORN, R R, and ELWELL, R W D. 1966. The structure and form of the granophyric quartz-dolerite intrusion: Centre II Ardnamurchan, Argyllshire. *Transactions of the Royal Society of Edinburgh*, Vol. 66, 285–306.

SKELHORN, R R, and LONGLAND, P J N. 1969. The Tertiary igneous geology of the Isle of Mull. *Geologists' Association Guide*, No. 20. (Colchester: Benham & Co.)

SKELHORN, R R, HENDERSON, P, WALSH, J N, and LONGLAND, P J N. 1979. The chilled margin of the Ben Buie layered gabbro, Isle of Mull. *Scottish Journal of Geology*, Vol. 15, 161–167.

SMITH, D G W. 1969. Pyrometamorphism of phyllites by a dolerite plug. *Journal of Petrology*, Vol. 10, 20–55.

SMITH, N J. 1985. The age and structural setting of limestones and basalts on the Main Ring Fault in southeast Rhum. *Geological Magazine*, Vol. 122, 439–445.

SPARKS, R S J. 1988. Petrology of the Loch Ba ring dyke, Mull (NW Scotland): an example of the extreme differentiation of tholeiitic magmas. *Contributions to Mineralogy and Petrology*, Vol. 100, 446–461.

SPEIGHT, J M, SKELHORN, R R, SLOAN, T, and KNAAP, R J. 1982. The dyke swarms of Scotland. 449–454 in *Igneous rocks of the British Isles*. SUTHERLAND, D S (editor). (Chichester: John Wylie & Sons.)

STEEL, R J. 1974a. New Red Sandstone piedmont and floodplain sedimentation in the Hebridean province, Scotland. *Journal of Sedimentary Petrology*, Vol. 44, 336–357.

STEEL, R J. 1974b. Cornstone (fossil caliche) — its origin, stratigraphy and sedimentological importance in the New Red Sandstone, western Scotland. *Journal of Geology*, Vol. 82, 351–369.

STEEL, R J, NICHOLSON, R, and KALANDER, L. 1975. Triassic sedimentation and palaeogeography in central Skye. *Scottish Journal of Geology*, Vol. 11, 1–13.

STEPHENSON, D, and GOULD, D. 1995. *British Regional Geology: the Grampian Highlands*. Fourth edition. (London: HMSO for the British Geological Survey.)

STEWART, A D. 1991. Geochemistry, provenance and palaeoclimate of the Sleat and Torridon groups in Skye. *Scottish Journal of Geology*, Vol. 27, 81–95.

STEWART, A D. 2002. The later Proterozoic Torridonian rocks of Scotland: their sedimentology, geochemistry and origin. *Geological Society of London Memoir*, No. 24.

STONE, P, and KIMBELL, G S. 1995. Caledonian terrane relationships in Britain: an introduction. *Geological Magazine*, Vol. 132, 461–464.

STOKER, M S, HITCHEN, K, and GRAHAM, C C. 1993. *United Kingdom offshore regional report: the geology of the Hebrides and West Shetland shelves, and adjacent deep-water areas.* (London: HMSO for the British Geological Survey).

STRACHAN, R A, HARRIS, A L, FETTES, D J, and SMITH, M. 2002. The Highland and Grampian terranes. 81–148 in *Geology of Scotland*. Fourth edition. TREWIN, N H (editor). (London: The Geological Society.)

SUTHERLAND, D G. 1981. The high-level marine shell beds of Scotland and the build-up of the last Scottish ice-sheet. *Boreas*, Vol. 10, 247–254.

SUTHERLAND, D G, and GORDON, J E. 1993. The Quaternary in Scotland. 11–47 in *Quaternary of Scotland. Geological Conservation Review Series*, No. 6. GORDON, J E, and SUTHERLAND, D G (editors). (London: Chapman and Hall.)

SUTHERLAND, D G, BALLANTYNE, C K, and WALKER, M J C. 1982. A note on the Quaternary deposits and landforms of St Kilda. *Quaternary Newsletter*, Vol. 37, 5.

SUTHERLAND, D S. 1982. *Igneous rocks of the British Isles.* (Chichester: John Wiley & Sons.)

SYKES, R M. 1975. The stratigraphy of the Callovian and Oxfordian stages (Middle–Upper Jurassic) in Northern Scotland. *Scottish Journal of Geology*, Vol. 11, 51–78.

SYKES, R M, and CALLOMON, J H. 1979. The *Amoeboceras* zonation of the Boreal Upper Oxfordian. *Palaeontology*, Vol. 22, 839–903.

TAIT, S R. 1985. Fluid dynamic and geochemical evolution of cyclic unit 10, Rhum, Eastern Layered Series. *Geological Magazine*, Vol. 122, 469–484.

TAYLOR, H P, and FORESTER, R W. 1971. Low ^{18}O igneous rocks from the intrusive complexes of Skye, Mull and Ardnamurchan, western Scotland. *Journal of Petrology*, Vol. 12, 465–497.

THIRLWALL, M F, and JONES, N W. 1983. Isotope geochemistry and contamination mechanics of Tertiary lavas from Skye, northwest Scotland. 186–208 in *Continental basalts and mantle xenoliths*. HAWKSWORTH, C J, and NORRY, M J (editors). (Norwich: Shiva.)

THOMPSON, R N. 1969. Tertiary granites and associated rocks of the Marsco area, Isle of Skye. *Quarterly Journal of the Geological Society of London*, Vol. 124, 349–385.

THOMPSON, R N. 1974. Primary basalts and magma genesis. *Contributions to Mineralogy and Petrology*, Vol. 45, 317–341.

THOMPSON, R N. 1980a. Askja 1875, Skye 56 Ma: basalt-triggered Plinian, mixed-magma eruptions during the emplacement of the Western Redhills Granites, Isle of Skye, Scotland. *Geologische Rundschau*, Vol. 69, 249–262.

THOMPSON, R N. 1980b. An assessment of the Th-Hf-Ta diagram as a discriminant for tectonomagmatic classification and in the detection of continental contamination of magmas. *Earth and Planetary Science Letters*, Vol. 50, 1–10.

THOMPSON, R N. 1981. Thermal aspects of the origin of Hebridean Tertiary acid magmas. I An experimental study of partial fusion of Lewisian gneisses and Torridonian sediments. *Mineralogical Magazine*, Vol. 44, 161–170.

THOMPSON, R N. 1982a. Magmatism of the British Tertiary Volcanic Province. *Scottish Journal of Geology*, Vol. 18, 49–107.

THOMPSON, R N. 1982b. Geochemistry and magma genesis. 461–477 in *Igneous rocks of the British Isles*. SUTHERLAND, D S (editor). (Chichester: John Wylie & Sons.)

THOMPSON, R N. 1983. Thermal aspects of the origin of Hebridean Tertiary acid magmas. II Experimental melting behaviour of the granites at 1 kbar PH_2O. *Mineralogical Magazine*, Vol. 47, 111–121.

THOMPSON, R N, and GIBSON, S A. 1991. Subcontinental mantle plumes, hotspots and pre-existing thinspots. *Journal of the Geological Society of London*, Vol. 148, 973–978.

THOMPSON, R N, ESSON, J, and DUNHAM, A C. 1972. Major element variation in the Eocene lavas of the Isle of Skye, Scotland. *Journal of Petrology*, Vol. 13, 219–253.

THOMPSON, R N, MORRISON, M A, MATTEY, D P, DICKIN, A P, and MOORBATH, S. 1980. Trace-element evidence of multistage mantle fusion and polybaric fractional crystallisation in the Palaeocene lavas of Skye, NW Scotland. *Journal of Petrology*, Vol. 21, 265–293.

THOMPSON, R N, DICKIN, A P, GIBSON, I L, and MORRISON, M A. 1982. Elemental fingerprints of isotopic contamination of Hebridean Palaeocene mantle-derived magmas by Archaean sial. *Contributions to Mineralogy and Petrology*, Vol. 79, 159–168.

THOMPSON, R N, MORRISON, M A, DICKIN, A P, GIBSON, I L, and HARMON, R S. 1986. Two contrasting styles of interaction between basaltic magma and continental crust in the British Tertiary Volcanic Province. *Journal of Geophysical Research*, Vol. 91 (B6), 5985–5997.

THORPE, R S. 1978. The parental basaltic magma of granites from the Isle of Skye, NW Scotland. *Mineralogical Magazine*, Vol. 42, 157–158.

THORPE, R S, POTTS, P J, and SARRE, M B. 1977. Rare earth evidence concerning the origin of granites on the Isle of Skye, northwest Scotland. *Earth and Planetary Science Letters*, Vol. 36, 111–120.

TILLEY, C E. 1950. Some aspects of magmatic evolution. *Quarterly Journal of the Geological Society of London*, Vol. 106, 37–61.

TILLEY, C E, and MUIR, I D. 1964. Intermediate members of the ocean basalt–trachyte association. *Geologiska Föreningens i Stockholm Förhandlingar*, Vol. 85, 436–444.

TREWIN, N H (editor). 2002. *The geology of Scotland*. Fourth edition. (London: The Geological Society.)

TROLL, V R, EMELEUS, C H, and DONALDSON, C H. 2000. Caldera formation in the Rum Central Igneous Complex, Scotland. *Bulletin of Volcanology*, Vol. 62, 301–317.

TROLL, V R, DONALDSON, C H, and EMELEUS, C H. 2004. Pre-eruptive magma mixing in ash-flow deposits of the Tertiary Rum Igneous Centre, Scotland. *Contributions to Mineralogy and Petrology*, Vol. 147, 722–739.

TURNBULL, M J M, WHITEHOUSE, M J, and MOORBATH, S. 1996. New isotope age determinations for the Torridonian, Scotland. *Journal of the Geological Society of London*, Vol. 153, 955–964.

TYRRELL, G W. 1928. The geology of Arran. *Memoir of the Geological Survey of Great Britain*. Isle of Arran Special Sheet, parts of Sheets 12, 13, 20 and 21 (Scotland).

UPTON, B G J. 1988. History of Tertiary igneous activity in the N Atlantic borderlands. 429–454 *in* Early Tertiary volcanism and the opening of the NE Atlantic. MORTON, A C, and PARSON, L M (editors). *Geological Society of London Special Publication*, No. 39.

UPTON, B G J, ASPEN, P, REX, D C, MELCHER, F, and KINNY, P. 1998. Lower crustal and possible shallow mantle xenoliths from beneath the Hebrides: evidence from a xenolithic dyke at Gribun, western Mull. *Journal of the Geological Society of London*, Vol. 155, 813–828.

UPTON, B G J, MCCLURG, J, SKOVGAARD, A C, KIRSTEIN, L, CHEADLE, M, EMELEUS, C H, WADSWORTH, W J, and FALLICK, A E. 2002. Picritic magmas and the Rum ultramafic complex. *Geological Magazine*, Vol. 139, 437–452.

VIERECK, L G, TAYLOR, P N, PARSON, L M, MORTON, A C, HERTOGEN, J, GIBSON, I L, and the ODP LEG 104 SCIENTIFIC PARTY. 1988. Origin of the Palaeogene Voring Plateau volcanic sequence. 69–84 *in* Early Tertiary volcanism and the opening of the NE Atlantic. MORTON, A C, and PARSON, L M (editors). *Geological Society of London Special Publication*, No. 39.

VOLKER, J A. 1983. The geology of the Trallval area, Rhum, Inner Hebrides. Unpublished PhD thesis, University of Edinburgh.

VOLKER, J A, and UPTON, B G J. 1990. The structure and petrogenesis of the Trallval and Ruinsival areas of the Rhum ultrabasic complex. *Transactions of the Royal Society of Edinburgh: Earth Sciences*, Vol. 81, 69–88.

WADSWORTH, W J. 1961. The layered ultrabasic rocks of south-west Rhum, Inner Hebrides. *Philosophical Transactions of the Royal Society of London*, Vol. 244B, 21–64.

WADSWORTH, W J. 1992. Ultrabasic igneous breccias of the Long Loch area, Isle of Rhum. *Scottish Journal of Geology*, Vol. 28, 103–113.

WADSWORTH, W J. 1994. The peridotite plugs of northern Rum. *Scottish Journal of Geology*, Vol. 30, 167–174.

WAGER, L R, and BROWN, G M. 1968. *Layered igneous rocks*. (Edinburgh: Oliver and Boyd.)

WAGER, L R, and VINCENT, E A. 1962. Ferrodiorite from the Isle of Skye. *Mineralogical Magazine*, Vol. 33, 26–36.

WAGER, L R, BROWN, G M, and WADSWORTH, W J. 1960. Types of igneous cumulates. *Journal of Petrology*, Vol. 1, 73–85.

WAGER, L R, VINCENT, E A, BROWN, G M, and BELL, J D. 1965. Marscoite and related rocks of the Western Red Hills complex, Isle of Skye. *Philosophical Transactions of the Royal Society of London*, Vol. 257A, 273–307.

WALDMAN, M, and EVANS, S E. 1994. Lepidosauromorph reptiles from the Middle Jurassic of Skye. *Zoological Journal of the Linnean Society of London*, Vol. 112, 135–150.

WALDMAN, M, and SAVAGE, R J G. 1972. The first Jurassic mammal from Scotland. *Journal of the Geological Society of London*, Vol. 128, 119–125.

WALKER, F. 1960. The Islay–Jura dyke swarm. *Transaction of the Geological Society of Glasgow*, Vol. 24, 121–137.

WALKER, G P L. 1971. The distribution of amygdale minerals in Mull and Morvern (Western Scotland). 181–194 *in Studies in earth sciences: a volume in honour of William Dixon West*. MURTY, T V V G R K, and RAO, S S (editors). (New Delhi: Today and Tomorrow's Printers and Publishers.)

WALKER, G P L. 1975a. A new concept in the evolution of the British Tertiary intrusive centres. *Journal of the Geological Society of London*, Vol. 131, 121–142.

WALKER, G P L. 1975b. Intrusive sheet swarms and the identity of Crustal Layer 3 in Iceland. *Journal of the Geological Society of London*, Vol. 131, 143–162.

WALKER, G P L. 1979. The environment of Tertiary igneous activity in the British Isles. 5–6 in Mesozoic and Tertiary volcanism in the North Atlantic and neighbouring regions: Proceedings of the Flett Symposium, 3rd November, 1976. *Bulletin of the Geological Survey of Great Britain*, No. 70.

WALKER, G P L. 1993a. Basaltic-volcano systems. 3–38 in Magmatic processes and plate tectonics. PRICHARD, H M, ALABASTER, T, HARRIS, N B, and NEARY, C R (editors). *Geological Society of London Special Publication*, No. 76.

WALKER, G P L. 1993b. Re-evaluation of inclined intrusive sheets and dykes in the Cuillins volcano, Isle of Skye. 489–497 in Magmatic Processes and Plate Tectonics. PRICHARD, H M, ALABASTER, T, HARRIS, N B W, and NEARY, C R (editors). *Geological Society of London Special Publication*, No. 76,

WALKER, G P L, and SKELHORN, R R. 1966. Some associations of acid and basic intrusions. *Earth Science Reviews*, Vol. 2, 93–109.

WALKER, M J C, GRAY, J M, and LOWE, J J. 1985. *Field Guide, Isle of Mull, Inner Hebrides, Scotland*. (Cambridge: Quaternary Research Association.)

WALKER, M J C, GRAY, J M, and LOWE, J J. 1992. *The south-west Scottish Highlands, Field Guide*. (Cambridge: Quaternary Research Association.)

WALSH, J N, BECKINSALE, R D, SKELHORN, R R, and THORPE, R S. 1979. Geochemistry and petrogenesis of Tertiary granitic rocks from the Island of Mull, northwest Scotland. *Contributions to Mineralogy and Petrology*, Vol. 71, 99–116.

WARRINGTON, G. 1973. Microspores of Triassic age and organic-walled microplankton from the Auchenhew Beds, southeast Arran. *Scottish Journal of Geology*, Vol. 9, 109–116.

WARRINGTON, G, AUDLEY-CHARLES, M G, ELLIOTT, R E, EVANS, W B, IVIMEY-COOK, H C, KENT, P E, ROBINSON, P L, SHOTTON, F W, and TAYLOR, F M. 1980. A correlation of Triassic rocks in the British Isles. *Geological Society of London Special Report*, No. 13.

WEEDON, D S. 1960. The Gars Bheinn ultrabasic sill, Isle of Skye. *Quarterly Journal of the Geological Society of London*, Vol. 116, 37–54.

WEEDON, D S. 1961. Basic igneous rocks of the southern Cuillin, Isle of Skye. *Transactions of the Geological Society of Glasgow*, Vol. 24, 190–212.

WEEDON, D S. 1965. The layered ultrabasic rocks of Sgurr Dubh, Isle of Skye. *Scottish Journal of Geology*, Vol. 1, 42–68.

WELLS, M K. 1954. The structure and petrology of the hypersthene-gabbro intrusion, Ardnamurchan, Argyllshire. *Quarterly Journal of the Geological Society of London*, Vol. 109, 367–397.

WHITE, N, and LOVELL, B. 1997. Measuring the pulse of a plume with the sedimentary record. *Nature*, Vol. 387, 888–891.

WHITE, R S. 1988. A hot-spot model for early Tertiary volcanism in the N. Atlantic. 393–414 in Early Tertiary volcanism and the opening of the NE Atlantic. MORTON, A C, and PARSON, L M (editors). *Geological Society of London Special Publication*, No. 39.

WHITE, R S, and MCKENZIE, D P. 1989. Magmatism at rift zones: the generation of volcanic continental margins and flood basalts. *Journal of Geophysical Research*, Vol. 94, 7685–7729.

WICKHAM-JONES, C R, and WOODMAN, P C. 1998. Studies on the early settlement of Scotland and Ireland. *Quaternary International*, Vols. 49/50, 13–20.

WILLIAMSON, I T, and BELL, B R. 1994. The Palaeocene lava field of west-central Skye, Scotland: stratigraphy, palaeogeography and structure. *Transactions of the Royal Society of Edinburgh: Earth Sciences*, Vol. 85, 39–75.

WILKINSON, M. 1992. Concretionary cements in Jurassic sandstones, Isle of Eigg, Inner Hebrides. 145–154 *in* Basins on the Atlantic Seabord: petroleum geology, sedimentology and basin evolution, PARNELL, J (editor). *Geological Society of London Special Publication*, No. 62.

YODER, H S, and TILLEY, C E. 1962. Origin of basalt magmas: an experimental study of natural and synthetic rock systems. *Journal of Petrology*, Vol. 3, 342–532.

ZEIGLER, P A. 1981. Evolution of sedimentary basins in north-west Europe. 3–39 in *Petroleum geology of the continental shelf of north-west Europe*. ILLING, L V, and HOBSON, G D (editors). (London: Heyden.)

British Geological Survey Publications

BGS publishes a range of maps, books and reports, which are listed in our current catalogue. We also maintain a large number of digital databases; information on these is summarised in *Britain beneath our feet* (2004). This book and the BGS catalogue are available on request and can be accessed online. The BGS website at www.bgs.ac.uk gives access to our online shop and enquiry service, as well as some of our databases, including the Lexicon of Named Rock Units, National Archive of Geological Photographs and the BGS library catalogue.

Some maps and books of particular relevance to this area are listed below.

British Regional Geology

Geology of Northern Ireland — our natural foundation	2004
Grampian Highlands	1995
Midland Valley of Scotland	1985
Northern Highlands of Scotland	1989

Offshore reports

Geology of the Malin–Hebrides sea area	1993

A landscape fashioned by geology
(joint publications with Scottish Natural Heritage)

Arran and the Clyde Islands	1997
Mull and Iona	2005
Rum and the Small Isles	2004
Skye	2002

Memoirs

Sheets 13, 21 *Arran*	1928
Sheets 41, 43, 51, 52 *Tertiary and post-Tertiary Mull, Loch Aline and Oban*	1924
Sheet 51 *Ardnamurchan, North-west Mull and Coll*	1930
Sheet 60 *Rum and the adjacent islands*	1997

Geology maps

Small scale maps

Geology of the UK, Ireland and the adjacent continental shelf (North sheet) 1:1 000 000	1991
Solid geology map of the UK (North sheet) 1:625 000	2001
Tectonic map of Britain, Ireland and adjacent areas. Sheet 1. 1:1 500 000	1996
Sub-Pleistocene geology of the British Islands and adjacent continental shelf. 1:2 500 000	1979

1:250 000

57N06W Great Glen. Solid geology	1989
57N08W Little Minch. Solid geology	1988
56N08W Tiree. Solid geology	1986

56N06W Argyll. Solid geology	1987
55N06W Clyde. Solid geology	1986

Also sea-bed sediments and Quaternary maps are available

1:50 000

Abbreviations: S solid;   D drift

Sheet 21, 20 and part of 13, Arran Special sheet	1985
Sheet 43S Ross of Mull (S&D)	1999
Sheet 43N Staffa (S&D)	1996
Sheet 44W Eastern Mull (S)	1992
Sheet 51E Caliach Point (S&D)	1976
Sheet 52W Ardnamurchan	1977
Sheet 52 Tobermory (D)†	1968
Sheet 60 Rum (S&D)	1994
Sheet 61 Arisaig (S&D)†	1971
Sheet 70 Minginish (S&D)	2001
Sheet 71W Broadford (S&D)	2002
Sheet 80W Dunvegan (S&D)	1975
Sheet 80E with part of 81 Portree (S&D)	1975
Sheet 81W Raasay (S&D)	1980
Sheet 90 and part of 91 Staffin (S&D)	1976

†   1:63 360 scale

1:25 000

Skye Central Complex (Bedrock)	2005

Geophysical maps and data

1:1 500 000

Colour shaded relief gravity anomaly map of Britain, Ireland and adjacent areas,	1996
Colour shaded relief magnetic anomaly map of Britain, Ireland and adjacent areas	1996

1:250 000 scale magnetic and Bouguer gravity maps are also available

Regional geophysics of north Scotland — an interactive guide to the subsurface structure based on regional gravity and magnetic data.	in preparation

Hydrogeology

Hydrogeology map Scotland. 1:625 000 scale	1988
Hydrogeology of Scotland. N S Robins	1990

Regional geochemical atlases

Hebrides	1984
Argyll	1990
Southern Scotland	1993

Index

a'a 56, 68
Abhain Ruavail 154
Achosnich 122, 124
agate 54, 68, 172
age 1, 6, 8, 9, 10
age, magnetic 44, 46, 97, 110, 160
age, palynological 44, 45
age, radiocarbon 153–155
age, radiometric 15, 44–48, 59, 60, 62, 91, 126
agglomerate 93, 173
aggregate 173–175
Ailsa Craig 1, 137, 160, 173
Ailsa Craig Central Complex 137–138
Ailsa Craig microgranite 48, 160
Alasdair Conglomerate Member 40
algae 30, 31, 169
allanite 134
allivalite 115
Allt a'Chapuill 134
Allt Beinn Deirge 157
Allt Ceann a'Gharaidh 37, 38
Allt Fearna Granite 108
Allt Fearns 27
Allt Geodh' a'Ghamhna Member 62, 64
Allt Mór 56
Allt Mòr Member 62
Allt nan Dris 19
Allt Slapin 108
Allt Stapaig 37, 38
alluvial fan 64
alluvium 169, 170
Am Màm Breccias 114, 117
amethyst 134
ammonite 23, 26–28, 30, 36, 38
  zonal sequence 25, 35
amygdales 54, 78
An Aird 91, 167
An Càrnach 156
An Cròcan Member 63
An Garbh-choire 161
An Leac Member 62, 63
An Sgùrr 77, 157, 168
An Stac 67
An t-Sron Formation 12, 14
andesite 53, 108–109, 121, 144
annular folds 95, 108, 126, 130
anorthosite 102

Antrim Lava Field 1, 6
Antronestheria [arthropod] 34
Appin Group 12
Applecross Formation 9, 10, 12
Archaean 8
Ard Bheinn 137
Ard Mheall 115
Ardmeanach 56, 60, 72, 73, 74
Ardnamurchan 1, 2, 44, 69, 81, 84, 147, 160
  Mesozoic 22–24, 34, 38
  minerals and aggregate 172, 174, 175
  Moine 5, 11
  sea level change 163, 166, 167
Ardnamurchan Central Complex 45, 47, 71, 92, 120–126, 127, 150,
Ardnish Formation 23, 25-27
Ardtun 56
Ardtun Conglomerate Member 72
Arnaval 57
Arnaval Member 62, 63
Arnioceras [ammonite] 26, 27
Arran 1, 5, 7, 45, 151, 176
  Mesozoic 19, 23, 40
  minerals and aggregate 171–175
  pre-Mesozoic 11, 12, 15, 16, 17, 18
  Quaternary 156, 160, 161, 164–167, 169
  sills 85, 89, 90
Arran Central Ring Complex 2, 48, 92–93, 134–137
Arran Dyke Swarm 78, 81, 83
Arran Northern Granite Pluton 2, 134–136
artefact, human 87, 153
arthropod 22, 31, 34
Ascrib Islands 87
Assapol Fault 71
Asteroceras [ammonite] 27
Asterophyllites [plant] 17
Atlantic Corrie 114, 161
Auchenhew Mudstone Formation 17, 19
Auchnacraig 37, 40
Aultbea Formation 10
aureole 44, 84, 108

Barachandroman 127
baryte 171
Basal Lava Member 67

basalt 53–54, 56, 83, 85, 87, 139
Bay of Laig 32, 34, 80
bazarite 138
beach sands 169, 174
Bearreraig Sandstone Formation 25, 28–30, 33, 34, 86
Beinn à Ghraig Granite 132–133
Beinn Airein 59
Beinn an Dubhaich Granite 47, 108
Beinn Bheag Gabbro 130
Beinn Chaisgidle 128, 130, 132
Beinn Chaisgidle Centre 126, 130–132
Beinn Dearg Mhòr 27, 105
Beinn Dearg Mhor Granite 107
Beinn Edra Formation 66
Beinn Iadain 37, 39, 40
Beinn Iadain Mudstone Formation 36, 39, 40, 71
Beinn Leacach 38
Beinn Mheadhon felsite 128
Beinn na Caillich 84, 106
Beinn na Caillich (or Inner) Granite 108
Beinn na Crò 106, 108
Beinn na'Leac 22, 29
Beinn nan Codhan 122
Beinn nan Dubh 55, 64
Beinn nan Stac 26, 60, 114
Beinn Suardal 41
Beinn Tarsuinn 160, 163
Beinn Totaig Formation 62, 66
belemnite 28, 30
Belig 105
Ben Buie Gabbro 130, 143
Ben Hiant 22, 37, 121, 122
Ben Méabost 156
Ben More 71, 73, 75, 158
Ben More Main Member 71, 73, 75
Ben More Pale Member 71, 73, 75
Ben Scaalan 55
Ben Scaalan Member 63
Ben Suardal Member 14
Ben Suardal Thrust 41
Benlister 160
benmoreite 53, 54
Bennan Head 90
berthierine 28
beryl 134
Bheinn Buidhe 37, 38
Bidean Boideach 76, 77
Biod Mór 55
Biod Mór Member 63
bivalve 12, 19, 21

bivalve *continued*
   Cretaceous 38, 40
   Lower Jurassic 22, 23, 26, 27, 28
   Mid Jurassic 30, 31, 32, 34
Blà Bheinn 105, 106, 156
Blackstones Bank 2
Blackstones Bank Central Complex 48, 134
blanket bog peat 175
Blaven Granite 105
blockfield 161
bloodstone 54, 68, 153, 172
Bloodstone Hill 68
Bloody Bay 72
Blue Lias Formation 23, 24, 25, 26
bog iron-ore 171
Bracadale 67, 83
Bracadale Formation 66, 67
brachiopod 15, 16, 27, 30
Breakish Formation 23, 25, 26, 27
breccia 65, 93, 98, 101, 111, 110
   explosion 105, 114, 127
   fissure 120
   intrusion 99, 102, 116, 121
   sedimentary 15, 17, 22, 26, 36, 39, 60, 62
British Geological Survey publications 200–201
Broadford 22
Broadford Anticline 150
Broadford Gabbro 108
Brodick Bay 167, 169
Brodick Beds Formation 17
Brown Head 90
brucite 15, 174
Bualintur Formation 62–63
buchite 84, 87, 89, 145
building stone 172
Burg 72
burial depth 23, 54
burial metamorphism 95
Bute 85

Calamites [plant] 17
calcareous concretions 33
caldera 75–77, 93, 126, 127, 129, 133
   formation 95, 110–111, 126
Caledonian Orogeny 1–5, 41
   intrusions 5, 15
caliche 33
Camas Malag Formation 26
Camas Mór 33, 34, 59, 60
Camas Mór Breccia 60
Camas na Cairdh 78

Camasunary 42, 99
Camasunary Fault 36, 42, 43, 148
Cambir Dolerite 97
Cambrian 5, 12, 14
Canna 1, 83, 149, 157, 158,
    sea level change 164, 166
Canna Lava Formation 45–49, 60, 68, 70
Canna at Cùil a'Bhainne 54
carbonate concretions 19, 21, 27, 31, 32, 34
    see also palaeosols
Carboniferous 6, 16–17, 173
Cardinia [bivalve] 21, 27, 28
Cardioceras [ammonite] 38
Carn Mor Sandstone Member 86
Carsaig 37, 40
cassiterite 172
Casteil Abhaill 137
Catacol Synform 151
cauldron subsidence 93, 107, 109
cave 56, 73, 166
central complex 2, 6–7, 20, 43, 44, 92–138
    structure 189–151
Central Intrusion, Rum 117, 119, 121
Central Mull Tholeiite Magma-type 140, 143, 144
chalk 36–40
chamosite 26, 27, 28, 29, 38, 171
chevkinite 97, 98
Chlamys [bivalve] 19, 21
chrome-spinel 101, 115, 130
chromite 116, 171–172
Cir Mhor 137
Clach Alasdair 37, 38
Clach Alasdair Conglomerate Member 36, 38, 40
Clach Glas 106
clasts 8, 60, 62, 67, 68, 75
clay with flints 49
Cleadale 158
Cleiteadh nan Sgarbh 91
Cleveland Dyke 81
Cleviceras [ammonite] 28
climate, palaeo- 49, 169
Clyde Plateau Volcanic Formation 17
Clyde Sandstone Formation 17
Cnapan Breaca 110, 113
Cnoc Dubh Heilla 54
Cnoc Rhaonastil Boss 144
Cnoc Rhaonstil 84
Cnoc Scarall 54, 55
Cnoc Scarall Member 62, 63
coal 17, 49, 62, 71, 72, 175

Coal Measures 16
Coir' a'Ghrunnda 161
Coir' an Rathaid 55
Coire Choinnich 161
Coire Dubh 112
Coire Dubh Breccias 111
Coire Fearchair 161
Coire Lagan 156, 161
Coire Mór 127
Coire nan Gruund 161
Coire Riabhach Phospatic Formation 37
Coire Uaigneich Granite 104
Coir'-uisg 161
columnar jointing 56, 80, 91
    in lava 58, 59, 72, 73, 74, 77
    in sills 87, 88, 121
Compass Hill 52, 68
composite dyke 78, 83, 91
composite sill 85, 87, 90, 108
Conachair Granite 98
conchostracan fauna 22, 30, 31, 34
cone sheet 47, 81, 92, 93-95, 114,
    Ardnamurchan 121–124, 126
    Mull 128–132
    Skye 103, 104, 106
conglomerate 49, 52, 59, 62, 64, 69
    fluvialite 68, 71, 77
    Mesozoic 15, 17, 19, 21–22, 26, 36–38, 40
conifer 74, 76
contamination process 141
continental shelf 1
coral 26
coral sand 169, 175
cordierite 67, 89
Corloch Fault 41
cornstone 19, 21, 22
Coroniceras [ammonite] 26
Corra-bheinn Gabbro 47, 130
Corrie Limestone 16, 175
Corrie Sandstone 17
corries 153, 155, 160, 161
corundum 84, 172
cotectic curves 140
crag-and-tail 158
Craignure Anticline 150
Creag Mhór Member 63
Creag na h-Iolaire felsite 128
Creag Strollamus 108
Creagan Dubh 8
Cretaceous 6, 20, 36–40, 42
crinanite 83, 86, 90, 145
crinoid 27, 30

Croggan 71
Cruachan Formation 63
Crucilobiceras [ammonite] 28
crust 18, 43, 92, 141
   dilation 80, 93
crustacean 36
crystal fractionation 132
Cùil a'Bhainne 68
Cuillin 7, 11, 156, 157, 161
Cuillin Centre 67, 81, 94, 98–104, 140, 141
   age 46–48
Cullaidh Shale Formation 25, 29, 30
Culnamean Member 62
cumulate 101, 115, 119
curling stones 138, 173
cyclic sedimentation 31, 34
cyclothems 22

dacite 144
Dactylioceras [ammonite] 28
Dalradian Supergroup 5, 11–12
Danian 6
debris flow 75, 121
depositional environment 3–6, 65, 68
   fluvial 62, 69, 72
   lacustrine 49, 62, 67, 72, 170
   Mesozoic 21–22, 29, 31–33
Derenenach Mudstone Formation 17, 19
Derrynaculen 126
Devensian glaciation 153–163
Devonian 5, 6, 15–16
Diabaig Formation 9, 10
diatomite 170, 176
Dibidil 26, 60, 110, 112, 114
Dimlington Stadial 153, 155–166
dinosaur tracks 27, 29, 31, 33
Dippin Head 80
Dippin sill 85, 89–90
doggers *see* carbonate concretions
dolerite 97, 99, 102, 103, 122
   plug 84
   sill 88, 91, 173
dolostone 12, 14, 15, 108, 109, 174
domed uplift 94, 150
Dougarie 167
dropstone 117
Druim Buidhe 84
Druim Hain 102
Druim na Fhuarain Sandstone Member 29
Druim nan Ramh 102
Drumadoon 19, 90
Drumadoon sill 78, 91

drumlin 160
drusy 132, 134, 135
Drynoch 83
Duart Castle 166
Duirinish 32, 66
Dùn Ard an t-Sabhail 62
Dùn Caan 38, 67
Dun Caan Shale Member 29
dunite 101, 102
Duntulm 30
Duntulm Castle 87, 88
Duntulm Formation 31–34, 86, 175
Durness Group 12, 14, 15, 174
dyke 6, 15, 18, 44, 47, 108
dyke swarm 2, 6, 23, 43, 59, 78–83
   and cone sheets 93, 94
   magma 143, 144

Eas Mor 105
Eastern Layered Intrusion 115, 116, 143
Eastern Red Hills Centre 47, 104, 106, 108–109
Echioceras [ammonite] 28
Eigg 1, 6, 44, 77, 84, 149
   glaciation 157–158
   landslip 168
   minerals and aggregate 169, 170, 175, 176
   pre-Mesozoic geology 30, 33, 34, 36, 38, 42
   sea level change 164, 167
Eigg Lava Formation 45, 47, 49, 51, 57, 59–60, 76
Eilean Chaluim Chille 170
Eilean Dubh Formation 12, 14
Eilean Flodigarry 87
electrum 171
Elgol Sandstone Formation 29, 30, 31
elpidite 138
Eriboll Sandstone Formation 12, 14
erosion 93, 110, 135, 155
erratics 155–158, 160, 161, 166
Euagassiceras [ammonite] 27
eucrites 114
Euestheria [arthropod] 22
Exogyra [bivalve] 38, 40
Eynort Mudstone Formation 62–63

Fallen Rocks, Arran 169
False Bedded Quartzite Member 12, 14
faults 62, 148, 150, 151
   see also thrust
felsite 47, 105, 108, 109, 122, 128
fergusonite 134

fiamme 110, 113, 134
Fingal's Cave 56, 73
finger structures 103, 116
Fionchra 55, 69
fish fragments 21, 22, 30, 31
Fiskavaig Formation 63–64
fissures 6, 44, 49, 71, 120
fissure-type eruption 43
Fiurnean 52
flint 71, 75
Flodigarry 168, 176
flood basalt lavas 43
flow structure, in lavas 55–56
    in intrusions 115
fluorite 134
fluvial environment 62, 69, 72
folds 95, 108, 126, 130, 150
foraminifera 38
Forse River 67
fractional crystallisation 92, 125, 139–141, 143–145, 146
Fucoid Beds Member 12, 14
fulgurite 17

gabbro 47, 97, 108, 134, 137, 143
    Ardnamurchan 123–126
    gravity anomaly 95
    Mull 127, 129–131, 133
    Rum 114, 115, 120
    Skye 99–102, 105–107
gadolinite 134
Garantiana Shale Member 29
Garbh-choire 102
Gars-bheinn 99
Gars-bheinn sill 85, 91
gas brecciation 134
gas cavities see drusy
gas streaming 105
gastropod 12, 31, 30, 32, 34
gemstones 89, 122, 172
geophysics anomalies 95–96, 134, 148
geos 166
Ghrudaidh Formation 12, 14
glacial erratics 138, 155–157, 158, 160, 161, 166
glacial striae 156, 157, 160, 161, 163, 166
glaciation 153–163
glaciofluvial deposits 7, 158, 174
glacio-isostatic depression 163
Glamaig 22, 26, 105
Glamaig Granite 107, 147
Glas Bheinn 126

Glas Bheinn Mhór 106, 168
Glas Bheinn Mhor Granite 108
Glas Eilean vent 124
glass 89, 97, 124, 174
glass-sand 40
glauconitic sandstone 36, 37, 38, 39, 40
Gleann Oraid Formation 62–64, 65, 66
Gleann Torra-mhichaig 161
Glen Bay 97
Glen Brittle Member 63
Glen Caladale Formation 63
Glen Cannel 164
Glen Cannel Granite 132
Glen Cloy 160
Glen Drynoch 67
Glen Dubh Sandstone Formation 17, 19
Glen Forsa 164, 167
Glen Iorsa 160
Glen More 155
Glen More Centre 126–130
Glen More Ring-dyke 131, 147
Glen Osdale 66, 67
Glen Rosa 162
Glen Sannox 15, 167, 172
Glen Shurig 15
Glen Sligachan 102, 105
Glendrian caves 166
Glenfinnan Group 11
Glenmore 84
gneiss 1, 8, 146
Goat Fell 135, 162
Gometra 69
Gondwana 5
Grampian Terrane 5
granite 2, 5–7, 97, 98, 114, 138, 143, 145,
    age 46, 47, 48
    Arran 43, 82, 135–137
    gravity anomaly 95
    Mull 126, 132–134
    Skye 100, 104–108, 147
granite quarries 173
graphite 175
gravity anomalies 95–97, 134
gravity survey 6
Great Estuarine Group 23, 28, 29–34, 175
Great Eucrite 47, 126, 127
Great Glen Fault 36, 41, 43, 148
Greensand Formation 37–39
Gribun 19, 21, 39
Gribun Chalk Formation 37
Gribun Conglomerate Formation 36, 39
Gribun Mudstone Member 71

groundwater 176
Grudaidh Formation 12, 14
Gryphaea [bivalve] 23, 26, 28
Gualainn na Sgurra 84
Guirdil 54
Guirdil Member 70

Hallaig Sandstone Member 27
Hallival 115, 116
halo, alteration 93
Hamra River 66, 67
Harker's Gully 107
Harpoceras [ammonite] 28
harrisite 115, 120
harrisitic olivine 119
hawaiite 53, 54, 55, 83
Healabhal Bheag 156
Healabhal Mhòr 66
Hebridean Igneous Province 1, 19, 85
　regional setting 43–49
　stratigraphy 49–77
Hebridean Terrane 41
Hedbergella [foraminifera] 38
Helen's Reef 138
Highland Border Complex 5, 15, 41
Highland Boundary Fault 41, 42, 43
Hildoceras [ammonite] 28
Hippopodium [bivalve] 28
Holm Island 27
Holocene 153–154
Holy Island sill 85, 89–90
honeycomb weathering 30, 31
hornfels 23, 99, 122, 134
Hugh Miller's reptile bed 34
hummocky moraine 7, 155, 161
Hutton's unconformity 16
hyaloclastite 49, 52, 62, 65, 68, 110
Hybodus [shark's tooth] 34
hydrocarbon maturation 23
hydrothermal alteration 67, 73, 76, 93, 127
hydrothermal mineral zones 72, 73
hypersthene-normative magmas 142

Iapetus Ocean 5
ice dome 157, 156, 158
ice movement 158
Icelandite 53
igneous activity, timing of 1, 44–48
Imachar 18
Inch Kenneth 19
Inner Bytownite Troctolites 102
Inner Gabbros 102

Inner Hebrides 1, 6
Inninmore 17, 22
Inninmore Fault 149
inninmorite 121
Inoceramus [bivalve] 38, 40
intrusions, Carboniferous and Permian 18
Inverarish Burn 28
Inverclyde Group 16
iron-ore 171
ironstone 24, 26, 28, 29
Isastrea [coral] 26
Islay 15, 84
Islay dyke swarm 83, 84
isostatic movement 163–164
isotope data 141, 142, 143, 145, 146

jet 27, 175
joints 56–57, 135
　see also columnar jointing
Jura 15
Jura–Kintyre dyke swarm 83
Jurassic 6, 20, 22–36

kame terrace 160
karst 49
kettlehole 158
Kilchoan 26
Kilchrist 109, 110
Kilchrist Member 14
Kildonnan 80, 83
Kildonnan Member 30, 34
Kilmaluag 30, 35
Kilmaluag Formation 32, 34, 86, 175
Kilmory Water 160
Kilt Rock 88
Kinloch 84, 153
Kinloch Formation 12
Kinlochspelve 160, 161
Kishorn Thrust 5, 9, 10, 12, 13
Knock Granite 132
Knoydartian Orogeny 5

lacustrine facies 49, 62, 67, 72, 170
Lag a'Bheith Formation 17, 19
Laggan Bay 75
Laig Gorge 34, 37, 38
Laig Gorge Fault 42
Laig Gorge Sandstone Member 38
Lamlash Bay 90
Lamlash Sandstone Formation 17, 19
lamprophye 15, 18, 172
landslip 22, 156, 167, 168, 176

Largybeg Point 19
Late-glacial Clyde Beds 160
lateritic weathering 36, 60, 71
Laurentia/Laurussia 5
lava 16, 21, 47, 93
  age of 44
  field characteristics 52–57
  stratigraphy 51, 57–77
  thickness 54
  types 53
lava field 2, 6, 43, 49–77, 149
Laxfordian Event 8
layered intrusion 91, 114, 115, 117, 119, 122, 130
  gabbro 97–99
  peridotite 101, 102
  sills 86, 87
leaf impressions 67, 68
Lealt 31
Lealt Shale Formation 30, 31, 34, 86
Levencorroch Mudstone Formation 17, 19
Lewis 172
Lewisian gneiss 1, 8, 146
Lias Group 23, 24, 26
lignite 36, 38, 40, 175
limestone 174
Limestone Coal Formation 17
Liostrea [bivalve] 27
Lithothamnion [algae] 169
Little Minch Sill-complex 85–87, 145
lobster 28
Loch Ainort 105
Loch Ainort Granite 107
Loch Aline 24, 37, 160, 174
Loch Aline White Sandstone Formation 36, 37, 39
Loch an Sgùrr Mhóir Member 63
Loch an t-Suidhe 160
Loch Arienas 160
Loch Assapol Fault 149
Loch Bà Centre 126, 132, 133
Loch Bà Felsite 47
Loch Bà Ring-dyke 133–134, 146
Loch Bay 31
Loch Bracadale 67
Loch Brittle 62
Loch Cuithir 176
Loch Don 161
Loch Don Sand Moraine 159, 161
Loch Dubh 55
Loch Dubh Formation 63, 64
Loch Eishort 27

Loch Harport 66
Loch Kilchrist 12
Loch Lomond Stadial 7, 153, 157, 158, 160–164, 166
Loch Mòr 32
Loch Mudle Fault 150
Loch na Dal 8
Loch na Keal 81, 159
Loch Roag 83, 144
Loch Scavaig 171
Loch Screapadal Fault 148
Loch Scridain 87, 89
Loch Scridain Sill-complex 85, 87–89, 144
Loch Slapin 26, 27
Loch Sligachan 67
Loch Spelve 160
Loch Teacuis 24
Loch Uisg Granite-Gabbro 130
Loch, Fiskavaig 55
Lochaline mine 40
Lochalsh Syncline 12
Lochan an Aodainn 122
Lochan nan Dunan 55
Lonachan Member 14
Lonfern Member 30, 34
Long Loch Fault 42, 119, 150
Lorn Plateau Volcanic Formation 16
Lower Fionchra Member 70
Lynn of Lorn Fault 148

MacCulloch's Tree 73–74
Machrie Breccias 15
Machrie Sandstone Formation 17
Maclean's Nose 52
magma 93, 114, 139–147
  contamination 141
  generation 142
  mixing 130, 146–147
  parental 87, 142, 143
magma types 143–145
magnesium 102, 171, 172
magnetic age *see under* age
magnetic anomaly 78–79, 95, 134, 148
magnetic survey 6
magnetisation, reversed 110
magnetite 84, 102, 108, 134
Main Late Devensian glaciation 153, 155–158, 160, 164
Main Postglacial Shoreline 167
Main Ring Fault 110, 112, 114
Main Rock Platform 164–166
Malcolm's Point 71, 72

mammallian remains 32, 34
mantle 18, 92, 142
mantle plume 7, 43, 48
Maol Buidhe 34
Maol na Gainmhich 105
marble 108, 175
marine-cut benches 153
Mariopteris [plant] 17
marl 19, 71
Marsco Hybrids 105, 107, 146, 147
Maternity Hollow 68
McFarlane's Rock Member 55, 63
Meacnaish Member 62, 63
Meall a'Mhaoil 105
Meall Breac 110, 112
Meall Buidhe 105
Meall Dearg granite 105
Meall Tuath 87
meltwater channel 160
Mercia Mudstone Group 17, 19
Merkland Burn 17
Mesozoic basins 19, 20, 22, 43, 44, 148
Mesozoic megablocks 137
metamorphism, Archaean 8
metasomatism 75
microgranite 85, 90, 132, 134, 137
Midland Valley Terrane 6
Mid-Ocean Ridge Basalt 63
millet seed grains 36
Minch Magnetic Anomaly 78–79
mine 40
Mingary 22
Minginish Conglomerate Formation 62–64
miospores 19, 21
mixing of magmas 97, 98, 114, 137
  Ardnamurchan 122–126
  Mull 132, 134
  Skye 107–109
Modiolus [bivalve] 19, 27
Moidart 166
Moine Supergroup 5, 10–11, 21, 41
Moine Thrust 3, 5, 8, 9, 12, 13, 41
molluscan fauna 160
Monadh Dubh Sandstone Formation 22
monchiquite 83, 144
monzogranite 145
monzonite 126
moraine 158, 160, 161, 162, 174
Morar Group 11
Morvern 1, 6, 11, 17, 18, 164, 169, 174
  lava field 71, 72, 73, 149
  Mesozoic 21, 24, 36–39, 40

Muck 1, 6, 33, 44, 47
  glaciation 157, 158, 164, 169
Muck Dyke Swarm 59, 78, 80, 81, 83, 84,
mugearite 53, 54, 83
  type locality 66
Mugeary 66
Mull 1, 2, 5, 6, 7, 126, 139
  glaciation 156, 158, 161–170
  magma-types 139, 141–142, 144–145
  Mesozoic 19, 23, 34, 36, 39
  minerals and aggregate 172, 173, 174, 175
  pre-Mesozoic geology 11–12, 15, 16,
Mull Central Complex 87, 92, 93, 95, 126–134,
  149, 150
  regional setting 43–48
Mull Central Lava Formation 70, 71, 75–77
Mull Dyke Swarm 71, 81, 82, 83, 144
Mull Lava Field 44, 47, 49, 51, 69, 71–75
Mull Lava Group 45, 48, 57, 69–77
Mull Plateau Lava Formation 70, 71, 73–75
  magma type 139, 141, 142, 145
Mullach Ard 110, 111, 150
Mullach Mór 11
Mullach Sgar Intrusion-complex 97, 98, 147
mullite 84, 89

Neomiodon [bivalve] 31, 32
Neoproterozoic 5
nepheline 90, 142
nephelinite 18
neptunian dyke 65
Neuropteris [plant] 17
New Red Sandstone 19, 22, 160, 173
North Arran Granite Pluton 43, 48, 82, 134,
  137, 151
North Atlantic Igneous Superprovince 1, 43
North Glen Sannox 15, 42
North Uist 91
North West Highlands 5
Northern Highlands Terrane 3
Northern Porphyritic Felsite 105
nunataks 155, 156, 158, 161

Ob Breakish 26
Ob Lusa 26
ocean floor spreading 43
Ockle Point 22
offshore basins 1, 17, 19
  Carboniferous 17
  Neogene 7
Oigh-sgeir 77
oil and gas 1

oil shale 175
Old Red Sandstone 5, 6, 15
Oligocene folding 149
olivine 53
olivine basalt 53, 54, 56, 83, 85, 87, 139
olivine-dolerite 84, 87, 145
olivine-gabbro 97, 131
opencast workings 171
Ophimottled Basalt Member 67
Ord 13, 14
Ord Window 12
Ordovician 5, 12, 14
Ornsay 8
orogeny, Caledonian 1–5, 15, 41
Orval 161
Orval Member 68, 70
Osdale Formation 66, 67
ostracod 30, 31, 32, 34
Outer Bytownite Gabbros 102, 106
Outer Bytownite Troctolites 99
Outer Gabbros 99
Outer Granite 108
Outer Hebrides 1, 3
Outer Hebrides Fault-zone 5
Oxyrhina [shark's tooth] 38
oyster 31, 34

Pabay Shale Formation 23, 24, 25, 26, 27
pahoehoe 56, 60, 75
palaeogeography 68
palaeokarst 26
palaeomagnetism 44, 46, 97, 110, 160
palaeosols 43, 49, 54–55
palaeovalley 44, 54, 62, 67, 68, 73, 77
Paleocene 6
palynological age 34, 44, 45
partial melting 145
peat 169, 175
Pecten [bivlave]? 23, 38
pedogenesis 16
pegmatite 87
Penarth Group 19, 21
peridotite 91, 95, 101, 102, 103, 115, 130
    plug 84
periglacial deposits 163
Permian 6, 16, 17–18, 20
Permian–Triassic boundary 19
petrography 52, 53
p-form channel 159, 166
photomicrograph 29, 116
phreatic explosion 77
picrite 67, 83, 86, 87, 143, 145

Picrite Basalt Member 67
picrobasalt 53, 83
picrodolerite 86, 87, 88, 145
pillow lavas 15, 49, 65, 68, 76, 77
Pipe Rock Member 12, 14
pitchstone 53, 76, 83, 121
    Sgurr of Eigg Pitchstone Formation 45, 47, 49, 57, 59, 77
placer deposits 170, 171
plant (fossil) 47, 62, 65, 66, 67, 68, 76
    Mesozoic 22, 30,
    Palaeozoic 15, 17
    tree 73–74, 75, 175
Plateau Group 71
platinum 171
Platypleuroceras [ammonite] 27
Pleistocene 153–167
Plesiosaurus [reptile] 34
Pleuroceras [ammonite] 28
plugs 6, 18, 44
pluton 107, 114, 135
ponding 54, 56
Port an t-Seilich 54
Port Haunn 54
Port Mór 56, 60
Port nam Marbh 23, 34
Portree Hyaloclastite Formation 52, 65, 66
Portree Shale Formation 23, 25, 27, 28
Praeexogyra [oyster] 31, 34
Praemytilus [bivalve] 34
Precambrian 4
Preshal Beg 56, 64, 65
Preshal Beg Conglomerate Formation 46, 47, 63–64
Preshal More 56, 64, 65
Preshal More lava 63, 140
Preshal More magma-type 140, 141, 142, 143, 144, 146
Prestwick–Mauchline Sill-complex 91
Productus [brachiopod] 16
Promicroceras [ammonite] 27
Protocardia [bivalve] 19, 21
Pseudoglyphea [lobster] 28
Pseudograpta [arthropod] 34
Pseudopecten [bivalve] 28
Psilophyton [plant] 15

quarry 15, 164, 173, 174, 175
quartz porphyry 173
quartz sand 37, 40
quartz-dolerite 132
quartzite 15

Quaternary   7, 153–170
Quinish   75
Quiraing   66, 168

Raasay   1, 8, 9, 85, 175, 176
   Mesozoic   22, 23, 27, 28, 38, 42
   Quaternary   156, 168
Raasay Ironstone Formation   23, 25, 27, 28, 29, 171
radiometric age   44, 45, 46, 59, 60, 126, 153, 155
raft, country rock   62, 87
raised beach   153, 160, 163, 164, 167, 174
Ramasaig Formation   66
rare-earth-element   145
Rb-Sr age   15
Red Burn   67
Red Hills   106, 150, 156, 157, 161, 169
Red Hills Granite Centre   104–108
regolith   156
remanié deposit   23, 37, 38, 40
reptile fauna   32, 34
reverse magnetisation   110
Rhaetavicula [bivalve]   19, 21
rheomorphism   116
Rhynchonella [brachiopod]   40
rhyodacite   49, 53, 110–114
rhyolite   85, 109, 110, 127, 134, 144
riebeckite   89, 137, 138, 173
ring-complex   *see* Arran Central Ring Complex
ring-dyke   92–95, 107, 147
   Ardnamurchan   122–126
   Mull Central Complex   127, 130–134
   Skye Central Complex   98, 102, 105, 109
roadstone   173
roches moutonnées   160, 161
rock platform   163–166
Rockall   1, 138
rockallite   138
rockfall   168, 176
Rona, Archaean   8
Ross of Mull   5
Ross of Mull Pluton   11, 15
Ruadh Stac Granite   105
Ruaival Drift   154, 160
Rubh' an Dùnain Formation   62–63
Rubh' an Eireannaich   109
Rubha na'Leac   22
Rubha nan Clach Member   63
Rudh' a'Chromain   87
Rudh' a'Chromain sill   89
Rum   1, 2, 8–9, 22, 26, 42, 68, 176
   dykes and plugs   60, 80–84
   glaciation   7, 153, 157, 158, 161
   landslip and scree   168, 169
   minerals   170, 171, 175
   sea level change   163, 164, 166, 167
Rum Central Complex   43, 49, 92, 110–120, 150
   emplacement and unroofing   45–48, 59, 62, 68, 81

'S Airde Beínn   84
Sailmhor Formation   12, 14
St Kilda   1, 2, 155, 160, 163
St Kilda Central Complex   43, 48, 97–98
Salterella [gastropod]   12
Salterella Grit Member   12, 14
Samaropsis [plant]   17
sand and gravel   174
sand dunes   169
Sanday   68, 83, 157, 158
Sandstone Member   34
Sangomore Formation   14
sanidinite-facies   67, 78
sapphire (corundum)   89, 122, 172
Scalpay   9, 26, 27, 35, 38, 42, 156
Scalpay Sandstone Formation   23, 25, 28
Scandian Event   5
scoria   49, 110
Scotch Topaz (quartz)   134
Scourian Event   8
Scourie Dyke Swarm   8
Screapadal Fault   42
screen, country rock   11, 21, 34, 93, 105
sea level change   153, 163–167
Sea of the Hebrides   69
sea stack   164, 166
seatearth   16
serpentine   93, 174
Sgurr an Duine   62
Sgurr Buidhe Member   63
Sgurr Dearg   127, 161
Sgurr na Stri   102
Sgurr nam Meann Ring-dyke   122, 125
Sgurr of Eigg   49, 76, 84
Sgurr of Eigg Pitchstone Formation   45, 47, 49, 57, 59, 77
shark's tooth   34, 38
sheet complex   23
shelly fauna   160
   see also bivalve and brachiopod
Sherwood Sandstone Group   17, 19

Shiant 1, 28, 85, 86
shield volcano 6
shoreline 169
    Jurassic 26
        postglacial 166–167
        raised 160, 167
silica sand 174
silica-cementation 36
sill 21, 49, 85, 90, 91
sill-complex 2, 6, 44, 85–91, 143, 144
Silurian 5
Singing Sands 169
Sithean Sluaigh 84
skarn 108, 171
Skerryvore Fault 42, 43, 148
Skridan Member 62, 63
Skudiburgh 30
Skudiburgh Formation 33, 86
Skye 1, 2, 5, 8–9, 10, 11, 42
    glaciation 7, 153, 156, 162
    landslip and scree 168, 169
    Mesozoic 22, 23, 26, 28, 35, 38
    minerals and aggregate 170–171, 173–175
    sea level change 163, 164, 166, 167
    sills 85, 91
Skye Central Complex 43, 45, 47, 49, 92, 95, 98–110
Skye Dyke Swarm 62, 83, 140, 144
Skye Lava Field 44, 49, 51, 84, 141, 149
Skye Lava Group 45, 46, 47, 57, 60–70
Skye Main Lava Series 139, 141, 142, 143, 145
Skye Marble 15
Sleadale Member 62
Sleadale Tuff 46
Sleat Group 9, 10
Sliddery Water 17, 160
slumping 117
Small Isles 1, 47, 68, 164, 167, 173
    see also Canna, Eigg, Muck and Rum
Soay 9, 37, 38, 62, 156
solution-weathering 38
South Arran Dyke 80
Southern Highland Group 12
spinel 84, 122
spinel lherzolite 142
sponges 12
Srath na Creitheach 47, 109
Srath na Creitheach Centre 104, 105
Sròn an t-Saighdeir 114, 161
Sròn Bheag 26, 34
Stac a'Mheadais Member 63
Staffa 58, 69, 72

Staffa Lava Formation 70–73, 140, 141
Staffin Bay 32, 33, 36
Staffin Bay Formation 34, 35
Staffin Shale Formation 35, 36
Stoer Group 9
stone chutes 169
stone stripes 161
Stornoway Formation 22, 25
Storr, The 66
Strath Suardal Formation 14, 15
Strathaird 31, 33, 60, 78, 156
    Mesozoic 27, 29, 32, 35, 37, 38
Strathaird Lava Formation 67
Strathaird Limestone Formation 37–38
Strathclyde Group 17
stress field 7
striae, glacial 156, 157, 160–161, 163, 166
Strollamus 35, 37, 38, 164
Stromatolite [algae] 30
structure 41–42, 148–151
subvolcanic emplacement 105
Suishnish Pier 28
Suishnish Sandstone Member 27
Swordle 22, 26
syenite 145

Talisker Bay 55, 56
Talisker Lava Formation 46, 47, 48, 63, 65
talus (scree) 64, 156, 163, 169
Tancredia [bivalve] 34
Tarskavaig 8, 22
Tarskavaig Group 5, 11
Tarskavaig Thrust 8
Taxodioxylon [conifer] 74
Tea Green Marl 19
terrane, structural 1–2, 41
Tertiary Volcanic (or Igneous) Province 1
teschenite 90
Thecosmilia [coral] 26
Theriosynoecum [ostracod] 32
thermal metamorphism 23, 24, 67, 78, 84
tholeiitic basalt 53, 83, 85, 87, 102, 103, 140–141
thorium 172
thrust 10, 41
    Moine 3, 5, 8, 9, 12, 13
Tighvein intrusion-complex 90
till 7, 154, 158, 160, 161
tin 172
Tobermory 71
tombolo 166, 167
topaz 134

topography, pre-Palaeogene 49, 55
Tormore 78
Torness felsite 128
Torosay 36, 37
Torosay Sandstone Member 23
Torr nam Fitheach 54
Torridon Group 9–10
Torridonian 3, 5, 9–10, 12, 146
trace-element signature 146
trace-fossil 12
trachyte 53, 54, 56, 83–85
trap topography 57, 60, 69
trees see under plants
Treshnish 69, 163
Triassic 6, 19–22, 24
trimline 155, 156, 161
troctolite 99, 101, 102, 109, 115, 116, 130
Trotternish 31, 156, 176
　lava field 55, 65, 66
　　Mesozoic 25, 27, 28, 30, 31, 32, 35
　　sills 86, 87, 88
tuff 35, 46, 56, 62, 93, 109, 110, 111, 137
　welded 77
Tungadal 66
tungsten 172
Tusdale Member 63

Uig 35
ultrabasic rocks 91, 97, 110, 114, 143, 171
　sills and dykes 83, 85
Ulva 69
unconformity 16, 24, 28, 42
Unio [bivalve] 32, 34
unroofing 95, 110
　Rum Central Complex 69
uplift 8, 110, 128
Upper Fionchra Member 68, 70
Upper Lava Member 67
Upper Ostrea Member 34
uranium 172
valley fill 65
Valtos Sandstone Formation 30–32, 34, 86, 168

vein deposits 171
vent 11, 75, 110, 124
vertebrate remains 30, 31, 32, 34, 38
Village Bay Till 160
Vininodiceras [ammonite] 28
Viviparus [gastropod] 32, 34
volcanic plug 78, 83–84, 143, 144
volcanic rocks, Palaeozoic 16, 17
volcaniclastic rocks 105
　breccia 75, 104, 109–110, 121, 127, 137
　pipe 104
volcanism 49, 92–93

water supply 176
water power 176
Waternish 26, 32, 37, 38, 66
Waterstein 31
Waterstein Head
wave-cut platform 163
weathering 30, 31, 54, 57, 73, 137, 152
　lateritic 36, 60, 71
Welshman's Rock 110, 111, 150
Westbury Formation 17, 19
Western Gabbro 97
Western Granite 46, 114
Western Layered Intrusion 120
Western Red Hills Centre 47, 95, 104, 105–108
whitewash 176
Wilderness, The 72
Windermere Interstadial 160

xenocrysts 18, 83, 90, 108, 109, 146
xenolith 104, 134
　cognate 89, 90, 101, 102, 130, 145, 146
　crust and mantle 18, 83, 87, 89, 144
　gneiss 8, 11, 15, 107
Xipheroceras [ammonite] 27

zeolite 54, 90
　zones 71, 72, 95
zircon 9, 98, 134, 172

INDEX

INDEX